The Methods, Marvels, and Madness

of

Magic Squares

A Look at a Multitude of Methods that can be used to Build a Million-Billion Magic Squares

by Ronald J. Wendel

Library of Congress Control Number: 2022912570

HARDBACK: 978-1-957575-87-2
PAPERBACK: 978-1-95757578-0
EBOOK: 978-1-95757579-7

Ordering Information:

For orders and inquiries, please contact:
1-888-404-1388
www.goldtouchpress.com
book.orders@goldtouchpress.com

Printed in the United States of America

Table of Contents

Introduction

The magic square isn't governed by chance. If you wish to haphazardly place numbers into an array and in that way hope to build a magic square, then maybe you should try your hand at winning the lottery – for that is how lotteries are won, but not how magic squares are built. Magic squares are built by meticulous methods and they contain precise patterns.

The magic square has been around for at least a thousand years (and possibly much longer). There is strong evidence to support the claim that the Chinese first discovered the magic square; however, the magic square has appeared in European as well as Chinese art, literature and mathematics over the past millennia. Great mathematic and scientific minds have played with them including Benjamin Franklin who found an 8th order and a 16th order semi-magic square that carries his name. (He also confessed to have "wasted" a year playing with these numbers, although it actually appears that he might have spent much more than a year with these arrays of numbers, and it doesn't really appear that he always thought it was a waste of time.)

Magic squares have a lot of diversity. In spite of this diversity, there are various patterns and symmetries within most (if not all) magic squares. Higher order magic squares have similar patterns that can be found within lower order magic squares. And there are magic squares where lower order magic squares can be found inside of higher order magic squares.

Many who start to play with these squares often wonder how common these magic squares are. And this is a good question; so, just how many magic squares are there?

If we were to look at all of the possible number arrangements in a 3 x 3 array of numbers, we would find 9! possible combinations (a nine with an exclamation mark is read as nine factorial), which is equal to 362,880 combinations. Of course out of all of those possibilities there will be instances that are rotations and mirrorings of other instances. By removing these duplications, the possibilities reduce to 45,360. And out of all those possibilities, only one of those is a 3rd order magic square.

There are 16! possible number arrangements in a 4 x 4 array of numbers (which is equal to 20,922,789,888,000 combinations). Again, there are duplicate instances if we consider that rotations and mirrorings represent the same arrangement. Eliminating these duplications gives us only 2,615,348,736,000 unique number arrangements, and out of those possibilities only 880 are 4th order magic squares.

The numbers only get exponentially larger as the order increases. For a 5 x 5 array of numbers, there are 25! possible number arrangements (which is equal to 15,511,210,043,330,985,984,000,000 combinations). Again, after taking into consideration the rotations and mirrorings, this number reduces to 1,938,901,255,416,373,248,000,000. And out of these it has been estimated that only 275,305,224 are 5th order magic squares.

And for a 6 x 6 array of numbers, there are 36! possible number arrangements (which is about $3.71993326 \times 10^{41}$) and of those it is estimated that a mere 1.775×10^{19} are 6th order magic squares.

It is hard to tell by just looking, but if these numbers are even close to correct, then a few quick calculations show that the ratio of the number of magic squares to the number of possible number arrangements keeps getting increasingly smaller as the order ever increases. In other words, even though there appears to be an over-abundance of magic squares, the actual percentage of magic squares compared to the amount of possible number arrangements is relatively small, and keeps getting smaller as the order increases.

In spite of this fact, once we understand a few magic square principles, methods and techniques, it becomes entirely possible to build a million-billion unique magic squares (if you have that kind of time). However, the purpose of this book isn't to teach you, the reader, a few parlor tricks that will make you the star of some late night mathematical party (although you actually could do that with the things we will discuss within these pages). Instead, this book endeavors to look at some of the patterns, relationships and symmetries that make up the methods, marvels, and madness we call magic squares.

And now, in the spirit of rare Math books everywhere, we turn our attention to the ongoing tale of Antonio and his shipmates:

> "It was a calm and peaceful night. The sea was smooth and a gentle breeze caused the sails to flap mildly in the wind. The crew had been gathered, and spent the early evening listening to Rubio's romantic rescues, and Seymour's swash-buckling sagas. But now, the crew was ready for some real mind-manipulating mathematics, and they hoped Antonio would take center stage (so-to-speak).
>
> The Captain had been reflecting on the strange crew of pirates that sailed under his flag. Most pirates that he had known spent their time drinking and fighting, but not this crew. For some reason, these men had a strange affinity to mathematics and science. And fate seemed to favor this formidable group of free-thinking sailors. Why it was just a few weeks ago that they had run across an abandoned ship filed with calculators, computers and white boards. He had never seen anything like it; and the pirates themselves took to these commodities like a rat to a pound of cheese.
>
> The captain felt the anticipation in the air, and with all eyes upon him, he made a resolve to set bygones behind him, and forgive Antonio of the past. After all, it has now been several weeks since he asked Antonio to tell them a story, and then Antonio

went off on some "dark and stormy night" monolog, and ended up by telling the crew about various patterns, relationships, and symmetries in mathematics.

Again, the captain realized that the eyes of the crew were resting heavily upon him. And aware that he might be losing his mind with all this talk of mathematics; and yet understanding that as a good captain he needed to help his crew to better themselves, as well as provide the kind of entertainment that they really enjoy, the captain turned to Antonio and then cleared his throat as he cleared his mind, and said, "I've decided to give you another chance. Antonio, tell us about Magic Squares."

And Antonio began . . . "

Chapter 1
The Magic Square

A magic square is a square array of numbers where the rows, columns and diagonals all add up to the same number. The smallest magic square is a 1 x 1 array consisting of a single number. Technically, this may be considered a magic square, but it is also a trivial magic square. After all, with only a single row and a single column, it doesn't take much for a number to add up to itself. There are no 2 x 2 arrays of numbers that will create a magic square; therefore, the smallest magic square without a trivial solution is a 3 x 3 array of numbers.

8	1	6
3	5	7
4	9	2

Figure 1.1   The basic 3rd order magic square.

Using the consecutive numbers 1 through 9, and not counting rotations or mirrorings, there is only one 3 x 3 magic square. However, as we consider larger and larger arrays, we find that there are literally trillions and trillions of magic squares. We will not look at even a fraction of these; however, with the methods that we will consider in this book, it is possible to create millions (and even billions) of these magic squares (if we care to take the time).

Definitions
Before a person begins any endeavor, it is always a good idea to make sure that those involved speak the same language. This book is no different. In the world of magic squares there are certain words that have definitions. The terms that will be covered include such words as the order of the magic square, the magic square constant, the main diagonals and the minor diagonals (or broken diagonals). Then toward the end of this chapter, I will briefly explain the several different types of magic squares.

First on our list of words is the *order* of the magic square. Magic squares are square arrays of numbers where the number of rows and columns are the same. The order of the magic square is the number of rows and columns in that square. For example, a 4th order magic square consists of an array of numbers that has 4 rows and 4 columns. Similarly, a 7th order magic square has 7 rows and 7 columns.

Each number position within the magic square is called a ***cell***. Usually, the cell is comprised of only one number. Therefore, if we have a 5th order magic square, we have an array of 5 rows and 5 columns, and there are 25 cells within this array. In later chapters we will also include the meaning of the cell such that when we talk about larger magic squares being built from smaller magic squares that the smaller magic square is referred to as a cell of the larger magic square.

Cells can be referenced by the number within that cell, or by the cell's location within the array. When describing a cell's location, the same notation is used as that of matrix notation. A cell on row 2 and column 4 is written as the cell located at (2, 4). The row is the first number, and the column is the second number.

Sometimes the number within the cell is referred to as the ***element***. The element always refers to just a single number and never a group of numbers or array of numbers.

The magic square ***constant*** is the value that the rows, columns and diagonals add up to within the array. Of course, this value is dependent on several things, like whether you start with zero, one, or some other number as the lowest number; and whether you use consecutive numbers, evenly spaced numbers, or some other sequence of numbering.

Except for a few special cases, the magic squares in this book will start with the number one and then use consecutive numbers to fill the square. With those conditions being set, the following listing shows the magic square order, and its corresponding constant.

Order	Constant	Middle #	First/Last #'s
3rd	15	5	1 . . . 9
4th	34	8.5	1 . . . 16
5th	65	13	1 . . . 25
6th	111	18.5	1 . . . 36
7th	175	25	1 . . . 49
8th	260	32.5	1 . . . 64
9th	369	41	1 . . . 81
10th	505	50.5	1 . . . 100
11th	671	61	1 . . . 121
12th	870	72.5	1 . . . 144
13th	1105	85	1 . . . 169
14th	1379	98.5	1 . . . 196
15th	1695	113	1 . . . 225
16th	2056	128.5	1 . . . 256
17th	2465	145	1 . . . 289
18th	2925	162.5	1 . . . 324
19th	3439	181	1 . . . 361
20th	4010	200.5	1 . . . 400

Table 1.1   Magic square orders and constants.

Notice that the middle number and the first and last numbers of the sequence are also listed in the table above. The magic square constant can be easily calculated by multiplying the order of the magic square by its middle number. For both even and odd-ordered magic squares, the middle number is the sum of the first and last number divided by two. In the table above, the even-ordered magic squares have a middle number half way between two integers.

$$\text{Middle number} \; = \; (\text{first number} \; + \; \text{last number}) \, / \, 2$$

$$\text{Magic square constant} \; = \; \text{order} * \text{middle number}$$

This method of determining the magic square constant will work as long as the sets of numbers used to build the magic square are regularly spaced. For example, some magic square enthusiasts like to start with the number 0 instead of 1. Other magic squares are built using consecutive even numbers, or numbers that are multiples of five. In each case, if the sets of numbers used to build the magic square are regularly spaced, the constant for that magic square can be calculated by taking the first and last number of the sequence; adding them together and dividing by two (to get the middle number); and then finally multiplying the middle number by the order of the square. For example, in chapter 3 there is a third order magic square using the numbers:

$$5, 10, 15, \quad 30, 35, 40, \quad 55, 60, 65$$

Using the formulas above, the middle number is equal to the first number plus the last number divided by two, or

$$\text{Middle number} \; = \; (5 + 65) \, / \, 2 \; = \; 35$$

And the magic square constant is then equal to the order multiplied to the middle number, or

$$\text{Magic square constant} \; = \; 3 * 35 \; = \; 105$$

However, if the magic square is just a regular magic square that starts at 1 and only uses consecutive numbers, then another way to find the constant of the magic square is to square the order, add one, divide that by two, and then multiply your result by the order again. In other words, the constant is equal to:

$$\text{magic square constant} \; = \; \text{order} * (\text{order}^2 + 1) \, / \, 2$$

It is interesting to note that when all the even-ordered magic squares are considered, half of them have a constant that is an even number while the other half have a constant that is an odd number. All even-ordered magic squares that have an order that is a multiple of 4 are called ***doubly even*** magic squares (like 4, 8, 12, etc.) and they have a constant that is an even number. The other even-ordered magic squares (like 6, 10, 14, etc.) are ***singly even***, and have a constant that is an odd number. We will see in later chapters that there is a difference in the way these two even-order magic squares are constructed.

Each magic square has two **main diagonals**. One diagonal runs from the upper left corner of the square to the lower right corner, and the second diagonal runs from the lower left corner to the upper right.

A	B	C	D
E	F	G	H
I	J	K	L
M	N	O	P

Figure 1.2   A 4 x 4 array of letters representing a magic square.

All the other diagonals in the array are **minor diagonals** (or **broken diagonals**). The minor diagonals also have the same number of cells as the main diagonals; however these diagonals are continued as if the magic square were wrapped upon itself. Consider the following example.

In the array above, the cells containing the letters [A, F, K, P] make up one of the main diagonals and the cells containing the letters [M, J, G, D] make up the other main diagonal. There are also six broken diagonals, with three running parallel to one main diagonal and the other three running parallel to the other main diagonal. The first three broken diagonals are [E, J, O, D], [I, N, C, H] and [M, B, G, L]. The other three are parallel to the other main diagonal and are [A, H, K, N], [B, E, L, O] and [C, F, I, P].

A	B	C	D
E	F	G	H
I	J	K	L
M	N	O	P

A	B	C	D
E	F	G	H
I	J	K	L
M	N	O	P

A	B	C	D
E	F	G	H
I	J	K	L
M	N	O	P

Figure 1.3   Examples of broken diagonals running parallel to the [A,F,K,P] diagonal.

As we talk about methods of building magic squares, we sometimes talk about swapping numbers, and sometimes we talk about swapping cell positions. When we swap cell positions, we swap the values between two positions, regardless of what numbers are in those cells. And if we are swapping numbers, we swap the actual two values regardless of where they are located.

In order to distinguish between these two concepts, whenever we want to swap two numbers we will write it as $2 \leftrightarrow 15$ which means that the number 2 is exchanged with the number 15 regardless of where these two numbers are located within the array. When we want to describe cell exchanges we will write it as $(1, 2) \leftrightarrow (4,3)$ which means that the number in cell location $(1, 2)$ is exchanged with the number in cell location $(4, 3)$ regardless of what the actual numbers are within those cells.

Finally, when discussing magic squares (mostly odd-ordered magic squares), we need to define the difference between **sets** and **sequences**, as used in this book. The sequence consists of all the numbers from 1 to n^2 for an nth order magic square. A set consists of n numbers in the nth order magic square. For example, for a 5th order magic square, the sequence of numbers are the numbers from 1 to 25, while a set within that sequence would be five numbers within the sequence; where specifically, the numbers 1 thru 5 make up one set, the numbers 6 thru 10 make up another set, etc. We need to make this distinction because in later chapters we will be rearranging the order of the numbers within the sets, as well as rearranging the order of the numbers within the sequence, and they are not the same thing.

Types of Magic Squares
The beauty of the magic square can be found in the patterns and diversity that they have within themselves as well as with each other. Many similar patterns can be grouped together into categories. Some of these categories even have names like the nested magic squares or the diabolic magic squares.

The very basic magic square consists of every row and every column adding up to the magic square constant. In some circles this is considered a magic square; however, most magic square societies require that both main diagonals also add up to the constant. In this book, magic squares where one or both diagonals fail to add up to the constant are called *semi-magic squares*; whereas, an array of numbers will be considered to be a magic square only when the rows, columns, and main diagonals add up to the constant.

Benjamin Franklin developed some magic squares that were only semi-magic; but these semi-magic squares had other properties that made them even more amazing. A slight twist to those semi-magic squares would have made them regular magic squares, but it appears that Benjamin Franklin might have known that, yet wanted to keep them semi-magic so as to better see the other unique properties of those squares. We will look at those magic squares in a later chapter.

Bi-magic magic squares are regular magic squares that are also magic squares when each number in the array is squared. The following magic square is an 8th order bi-magic square. The square on the left is the magic square using the consecutive numbers from 1 to 64 and has a constant equal to 260. The square on the right is the resulting square when each of the elements from the left-hand square is squared. The right-hand square has a constant equal to 11,180.

2	58	13	53	19	43	32	40
7	63	12	52	22	46	25	33
48	24	35	27	61	5	50	10
41	17	38	30	60	4	55	15
29	37	18	42	16	56	3	59
28	36	23	47	9	49	6	62
51	11	64	8	34	26	45	21
54	14	57	1	39	31	44	20

4	3364	169	2809	361	1849	1024	1600
49	3969	144	2704	484	2116	625	1089
2304	576	1225	729	3721	25	2500	100
1681	289	1444	900	3600	16	3025	225
841	1369	324	1764	256	3136	9	3481
784	1296	529	2209	81	2401	36	3844
2601	121	4096	64	1156	676	2025	441
2916	196	3249	1	1521	961	1936	400

Figure 1.4   An example of a bi-magic square.

A *nested* magic square is a magic square that has one or more magic squares nested inside of it. There are several ways to create a larger magic square from smaller nested magic squares. We will consider several types of these nested magic squares in the chapters ahead.

Border, or ***concentric*** magic squares are a type of nested magic square where if the outer ring of cells of the magic square are removed, the remaining array is another magic square. It is possible to create several concentric magic squares where several rings can be removed, and each leaves the remaining array as a magic square. Within the pages ahead, we will devote a chapter to this type of magic square.

A magic square is considered ***pan-diagonal*** if all of the minor (or broken) diagonals also add up to the constant. This would include all of the minor diagonals running in both directions (all of the minor diagonals running parallel to both main diagonals).

A ***most-perfect pan-diagonal*** magic square is a pan-diagonal magic square where the several sub-squares also add up to the magic square constant. The following is a 4th order most-perfect pan-diagonal magic square. Notice that every 2 x 2 cell also adds up to the 4th order constant.

4	14	7	9
15	1	12	6
10	8	13	3
5	11	2	16

Figure 1.5   An example of a 4th order most-perfect pan-diagonal magic square.

In some circles, a ***diabolic*** magic square is nothing more than another name for a pan-diagonal magic square. However, some pan-diagonal magic squares are truly diabolical because it is possible to find other groupings of numbers within the magic square that also add up to the constant. Later on we will look at an example of a 4th order and a 5th order diabolic magic square.

A ***self-similar*** magic square is where a magic square goes through a transformation of some kind, and the resulting magic square is the same as the original except that it has been mirrored or rotated in some way from its original orientation. There are several types of transforms that will result in another magic square; however, self-similar magic squares are usually associated with compliment transforms. We will look at this transform and a couple of other transforms in the next chapter.

There are also several types of ***symmetries*** associated with these magic squares. They include (but are not limited to) center-symmetries, mirror symmetries and rotational symmetries. A center-symmetry is where a number is symmetrical with itself (in a second magic square) or where it is symmetrical with its compliment, by passing through the center of the array. A mirror symmetry is a reflection of the array such that the first row (or column) is now the last, the second row (or column) is next to the last, etc. And a rotational symmetry is where the array of numbers is identical with itself, only the array has been rotated 90 or 180 degrees either clockwise or counter-clockwise. Sometimes an array of numbers can have more than one type of symmetry in relation to a second array of numbers.

A pan-diagonal magic square that is also symmetric is called an ***ultra-magic*** magic square. The following magic square is pan-diagonal. It is also symmetrical such that if you take any number in the array and add it to its corresponding number through the center of the array, the sum of these two numbers is 50. Because of this symmetry, the following magic square is also ultra-magic:

1	27	46	16	42	12	31
39	9	35	5	24	43	20
28	47	17	36	13	32	2
10	29	6	25	44	21	40
48	18	37	14	33	3	22
30	7	26	45	15	41	11
19	38	8	34	4	23	49

Figure 1.6   A 7th order ultra-magic magic square.

Finally, there appears to be two general types of even-ordered magic squares, and two general types of odd-ordered magic squares. The even-ordered magic squares are divided into those that are singly even and those that are doubly even. There are a few significant differences in the construction methods of these two types of even-ordered magic squares. And earlier in this chapter we also saw the differences between the singly even and doubly constants of these even-ordered magic squares.

Similarly, there appear to be significant differences between the construction methods of prime number odd-ordered magic squares, and composite number odd-ordered magic squares.

	Even-Ordered		Odd-Ordered	
Order	**Singly-Even**	**Doubly-Even**	**Prime-Odd**	**Composite-Odd**
3rd			X	
4th		X		
5th			X	
6th	X			
7th			X	
8th		X		
9th				X
10th	X			
11th			X	
12th		X		
13th			X	
14th	X			
15th				X
16th		X		
17th			X	
18th	X			
19th			X	
20th		X		

Table 1.2   Magic square types by their orders.

There are also methods of constructing a magic square that can be used on composite-number-ordered magic squares regardless of whether the order is odd or even. We will examine several singly-even, doubly-even, prime, and composite order methods in later chapters in this book.

And as we look at magic squares in the following chapters, we will also see that there are two parts to constructing any magic square. The first is the initial state, and the second is the method. The initial state can be thought of as the set-up for the magic square (the set-up is different for even-order verses odd-order squares). The method is the algorithm that is used on the initial state to then create the magic square itself.

Finally, once we have a magic square, there are certain operations that can be performed on that magic square that will result in another magic square, which will make it unique from the first. We'll look at some of these operations next.

Chapter 2
Magic Square Operations

Once you have a magic square, there are several operations that can be performed on the magic square that will result in another magic square. Some of these operations will result in a magic square that is similar to the original magic square. Other times these operations yield a completely different and unique magic square, and sometimes these operations merely result in a rotation or mirroring of the original magic square.

A magic square is not considered unique if it is a rotation or mirroring of its original square. Therefore, if we perform an operation on a magic square and the result is not a rotation or mirroring from the one we started with, then we have a unique magic square.

Row and Column Swapping

Once you have a magic square, you can swap corresponding rows or columns, and the magic square will still be magic. In a magic square, corresponding rows refers to the same number of rows from the top as from the bottom. Likewise, corresponding columns refers to the same number of columns from the left as from the right.

Therefore, if you have an 8th order magic square, the second row and the seventh rows are corresponding rows because each is the second row from the top and the bottom respectively. Swapping these two rows of an 8th order magic square will result in another magic square that is unique, yet similar to the original magic square. Let's look at the following example.

1	2	62	60	61	59	7	8
9	55	11	53	52	14	50	16
48	18	46	20	21	43	23	41
32	39	27	37	36	30	34	25
40	31	35	29	28	38	26	33
24	42	22	44	45	19	47	17
49	15	51	13	12	54	10	56
57	58	6	4	5	3	63	64

Figure 2.1   A typical 8th order magic square.

The square above is an 8th order magic square. Now, if we swap the second row with the seventh row, we have another 8th order magic square that is unique from the original square, yet very similar to it.

1	2	62	60	61	59	7	8
49	15	51	13	12	54	10	56
48	18	46	20	21	43	23	41
32	39	27	37	36	30	34	25
40	31	35	29	28	38	26	33
24	42	22	44	45	19	47	17
9	55	11	53	52	14	50	16
57	58	6	4	5	3	63	64

Figure 2.2   A unique magic square created by swapping symmetrical rows equal distances from the center; which in this case are rows 2 and 7.

The swap we performed on the magic square above was a symmetrical swap where the row swapped on the top of the magic square was symmetrical (through the center of the square) with the row on the bottom of the magic square. There are also several types of semi-symmetrical row and column swaps that will still give us a magic square. One example is to swap row one with row six, and then swap row eight with row three. In this swap, we are trading the first row on the top with the third to the last on the bottom, and then we complete the swap by trading the last row with the third row from the top. The following is magic square from figure 2.2 after these swaps have been performed.

24	42	22	44	45	19	47	17
49	15	51	13	12	54	10	56
57	58	6	4	5	3	63	64
32	39	27	37	36	30	34	25
40	31	35	29	28	38	26	33
1	2	62	60	61	59	7	8
9	55	11	53	52	14	50	16
48	18	46	20	21	43	23	41

Figure 2.3   A unique magic square created by performing an asymmetrical swap between rows 1 and 6 (highlighted), and then rows 3 and 8.

Another asymmetrical swap we can perform is to swap two rows or columns on one side of the magic square with each other, and then swap the corresponding two rows or columns on the other side of the magic square with each other. Again, we'll start with the magic square in figure 2.2 and swap columns one and two with each other, and then swap columns seven and eight with each other. This is our resulting magic square.

2	1	62	60	61	59	8	7
15	49	51	13	12	54	56	10
18	48	46	20	21	43	41	23
39	32	27	37	36	30	25	34
31	40	35	29	28	38	33	26
42	24	22	44	45	19	17	47
55	9	11	53	52	14	16	50
58	57	6	4	5	3	64	63

Figure 2.4   A unique magic square where columns 1 and 2 were swapped with each other followed by columns 7 and 8 swapping with each other.

Quadrant Swapping

Quadrant swapping is a little different from just swapping the corresponding rows and columns. When quadrant swapping, we keep the numbers within the quadrant in their same relative positions, and then swap that quadrant with the quadrant diagonally opposite through the middle of the magic square. At the same time, we also diagonally swap the other set of quadrants with each other.

Quadrant swapping a 4 x 4 magic square is like swapping the first and third columns with each other, swapping the second and fourth columns with each other, then swapping the first and third rows with each other and then finally swapping the second and fourth rows with each other. For example, let's start with the following 4 x 4 magic square:

1	15	14	4
12	6	7	9
8	10	11	5
13	3	2	16

Figure 2.5   An initial 4th order magic square.

As we swap the quadrants, we will keep the numbers 1, 15, 12 and 6 in the same position relative to each other, and replace them with the numbers 11, 5, 2 and 16 (which we will also keep in their same relative positions). We then do the same thing with the numbers 14, 4, 7 and 9 while we swap them with 8, 10, 13 and 3. After the swap, we get this magic square:

11	5	8	10
2	16	13	3
14	4	1	15
7	9	12	6

Figure 2.6   The preceding magic square, after the quadrants were swapped.

Abnormal Row and Column Swapping

The quadrant swapping described above doesn't always work on even-order magic squares; however, there is an abnormal type of column and row swapping that does appear to always work on odd-ordered magic squares. For example, let's take a 5th order magic square. This abnormal swap is equivalent to swapping the first column with the fourth, and the second with the fifth. Then we swap the first row with the fourth, and finally the second row with the fifth. The magic square on the left below is our initial square, and the square on the right is the resulting square after the swaps were performed.

1	20	9	23	12
19	8	22	11	5
7	21	15	4	18
25	14	3	17	6
13	2	16	10	24

17	6	3	25	14
10	24	16	13	2
4	18	15	7	21
23	12	9	1	20
11	5	22	19	8

Figure 2.7   The initial 5th order magic square is on the left while the magic square on the right is the result after the abnormal quadrant swap.

This abnormal type of row and column swapping has the appearance of a quadrant swap, but it does have a noticeable difference. We can see how the 2 x 2 cells in each corner swapped places with the corresponding cell through the center of the square. We are also swapping the top two cells of the third column with the bottom two cells of the same column. Likewise, we swap the left two cells of the third row with the right two cell from the same row. While this isn't exactly quadrant swapping, it does have a very similar appearance to the even-order quadrant swapping described earlier.

There are also other types of odd-ordered quadrant swapping variations for larger magic squares. For example, with a 7th order magic square, we can divide the array so that it has four 2 x 2 quadrants at each corner, and four 2 x 3 center sections along the sides, top and bottom that trade places with each other, leaving a center 3 x 3 section that doesn't move. Or we could divide it so that there are four 3 x 3 quadrants at each corner, and four 1 x 3 center sections along the sides, top and bottom that swap, and one number in the middle that stays right where it is. The next magic square shows these two methods:

First method (four 2 x 2 corners, four 2 x 3 sides, center 3 x 3):

1	34	11	37	21	47	24
33	10	36	20	46	23	7
9	42	19	45	22	6	32
41	18	44	28	5	31	8
17	43	27	4	30	14	40
49	26	3	29	13	39	16
25	2	35	12	38	15	48

Second method (four 3 x 3 corners, four 1 x 3 sides, one center number):

1	34	11	37	21	47	24
33	10	36	20	46	23	7
9	42	19	45	22	6	32
41	18	44	28	5	31	8
17	43	27	4	30	14	40
49	26	3	29	13	39	16
25	2	35	12	38	15	48

Figure 2.8   Two alternate methods of creating quadrants within the same 7th order magic square.

Here are these two magic squares after their respective quadrant swaps have been performed (spaces have been removed):

39	16	3	29	13	49	26
15	48	35	12	38	25	2
6	32	19	45	22	9	42
31	8	44	28	5	41	18
14	40	27	4	30	17	43
47	24	11	37	21	1	34
23	7	36	20	46	33	10

30	14	40	4	17	43	27
13	39	16	29	49	26	3
38	15	48	12	25	2	35
5	31	8	28	41	18	44
21	47	24	37	1	34	11
46	23	7	20	33	10	36
22	6	32	45	9	42	19

Figure 2.9   The resulting magic squares after performing their respective quadrant swaps (from the figure above).

Again, when performing quadrant swaps like these on a square that is already magic, (odd or even-ordered), the resulting magic square becomes a completely new and distinct magic square. It isn't a rotation or mirroring of the magic square from which it originated.

Single-Pair Quadrant Swaps
Finally, sometimes it is possible to get away with only swapping one set of quadrants without swapping the other; however, there are not as many magic squares that will let you do this. Here is one that will.

1	15	14	4
12	6	7	9
8	10	11	5
13	3	2	16

1	15	8	10
12	6	13	3
14	4	11	5
7	9	2	16

Figure 2.10   A single pair quadrant swap on a 4th order magic square.

In this example, we only swapped the quadrant containing the numbers 14, 4, 7 and 9 with the quadrant containing the numbers 8, 10, 13 and 3; yet we still created another magic square.

Compliment Transforms of the Magic Square
There are also various types of number transforms possible within a magic square. Here, we'll take the order of the magic square, square it and add one; and then subtract the elements in the magic square from this number to obtain another magic square. Let's look at an example.

We'll start with a 4 x 4 magic square. The order of the square is 4, so we square this number and add 1 to give us 17. Now we subtract every member in the original magic square from 17 to give us a new magic square.

9	16	5	4
6	3	10	15
12	13	8	1
7	2	11	14

becomes

17-9	17-16	17-5	17-4
17-6	17-3	17-10	17-15
17-12	17-13	17-8	17-1
17-7	17-2	17-11	17-14

which is

8	1	12	13
11	14	7	2
5	4	9	16
10	15	6	3

Figure 2.11   The compliment transform process on a 4th order magic square.

Sometimes this method will give us a rotation or mirroring of the original square. This time it didn't. Our resulting magic square is unique from its originator.

With regard to the number 17, the numbers 1 and 16 are compliments of each other, as are 2 and 15; and 3 and 14; and so on. All the numbers in the new magic square have traded places with their respective compliments. There is another way to look at this type of transform. In the example above, the 1 and the 16 traded places, as did the 2 and the 15, and so on. Therefore, we could also say that the first and last numbers swapped places, as did the second and next-to-last, and so on. In other words, a compliment transform could also be called a first-to-last swap.

This method of compliment swapping appears to work on any magic square and give a result that is also a magic square. However, due to the symmetry of a magic square, often the resulting magic square will be self-similar (it will be some type of mirror or rotation of itself).

Multiplicative Transforms

To create a magic square from a magic square using this method, we take a magic square, and multiply each value in the square by the square's order. We then divide each value by the order squared plus one, and then replace the value with the remainder. After doing this for each value in the array, we have our new magic square. Let's try this with a 4th order magic square.

Multiply each value in the left array by 4 to
generate the values in the array on the right:

1	15	14	4
12	6	7	9
8	10	11	5
13	3	2	16

4	60	56	16
48	24	28	36
32	40	44	20
52	12	8	64

Figure 2.12   An example of an order product transform on a 4th order magic square.

Now we divide each value in the right array by 17 (which is $4^2 + 1$) and replace value in that cell with the remainder of that division.

4	9	5	16
14	7	11	2
15	6	10	3
1	12	8	13

Figure 2.13   The resulting array after an order product transform on the previous 4th order magic square. (Notice that this magic square is a rotation of the original magic square.)

The resulting magic square is a 90° counter-clockwise rotation of the original magic square. In this case, this transformation created a magic square that is self-similar.

We'll look at one more example of this before we move on. This time we'll perform our order transform on a 5th order magic square:

17	24	1	8	15
23	5	7	14	16
4	6	13	20	22
10	12	19	21	3
11	18	25	2	9

multiply each value
by 5 to get:

85	120	5	40	75
115	25	35	70	80
20	30	65	100	110
50	60	95	105	15
55	90	125	10	45

Figure 2.14   An example of an order product transform on a 5th order magic square.

Now divide each value by 26 (or $5^2 + 1$) and replace the element in the cell with the remainder:

7	16	5	14	23
11	25	9	18	2
20	4	13	22	6
24	8	17	1	15
3	12	21	10	19

Figure 2.15   The resulting array after an n-order product transform has been completed on the previous 5th order magic square. (Notice that this magic square is unique from the original magic square.)

In this example, the resulting magic square is not a rotation or mirroring of the original magic square. Instead, it is unique from the original. And there are still other transforms that can be performed on a magic square. Some work all of the time, and some work only some of the time. In a later chapter we will look at a doubling transform that creates a closed set of pan-diagonal magic squares.

Odd/Even Number Swap for Even-Order Magic Squares
Even-ordered magic squares have a symmetry within themselves that allows us to swap the odd numbers with the even numbers, and still maintain the magic square. In fact, it appears (from testing that the author has done) that if the magic square is pan-diagonal, and odd and even numbers are swapped, that the magic square will still be pan-diagonal, and if the magic square is just a standard magic square, that it will still be a standard magic square after the swap transform.

The way that the swapping works is like this: the 1 is swapped with the 2, the 3 with the 4, the 5 with the 6, etc. In other words, each odd number is exchanged with the even number just above it. However, an even easier way to describe this exchange is to individually consider each value in the square, and if the number is odd, then add one to it, and if the number is even, then subtract one from it. The result will be the same either way. The following figure shows two 4th order magic squares. The square on the left is our original square, and the square on the right is the one we get after performing the odd/even number swap:

5	10	11	8
16	3	2	13
4	15	14	1
9	6	7	12

6	9	12	7
15	4	1	14
3	16	13	2
10	5	8	11

Figure 2.16   Related magic squares where the right square is the result after performing an odd/even number swap on the left square.

Both of the squares shown above are standard magic squares and are unique to each other. It should also be noted that this type of transform always appears to work on doubly-even-order magic squares (whose orders are 4, 8, 12, etc.), but may not always work if the magic square order is singly-even (orders of 6, 10, 14, etc.). The following is an example of a 6th order magic square where the odd/even number swap does work:

1	12	13	24	30	31
35	29	14	20	11	2
33	10	22	16	27	3
4	9	21	15	28	34
32	26	23	17	8	5
6	25	18	19	7	36

2	11	14	23	29	32
36	30	13	19	12	1
34	9	21	15	28	4
3	10	22	16	27	33
31	25	24	18	7	6
5	26	17	20	8	35

Figure 2.17 The magic square on the right was created form the one on the left by performing an odd/even number swap as explained above.

Again, both of the squares shown above are standard 6th order magic squares and are unique to each other.

Shifting the Rows and Columns of a Magic Square
Probably the easiest way to transform a magic square is to simply shift the rows and/or columns of the magic square. Columns are shifted left and right, and rows are shifted up or down. For example, when shifting the columns of a magic square right one column, each column moves to the right one column, and the right-most column moves to become the left-most column. Similarly, if shifting the rows down one row, each row moves down, and the bottom row now becomes the top row.

Shifting the rows and/or columns of a magic square can be a very useful tool. Often if we have a semi-magic square, it can become a regular magic square simply by shifting the rows and columns. For example, the 5 x 5 array on the left and below is a semi-magic square; yet when we shift the columns to the right by one column, it becomes a regular magic square (as shown on the right below):

18	6	24	12	5
9	22	15	3	16
25	13	1	19	7
11	4	17	10	23
2	20	8	21	14

5	18	6	24	12
16	9	22	15	3
7	25	13	1	19
23	11	4	17	10
14	2	20	8	21

Figure 2.18 The same array of numbers with the array on the left being a semi-magic square while the array on the right is shifted right one column making it now a regular magic square.

Usually if we have an odd-ordered semi-magic square, it will become a regular magic square when we shift it so that the middle number is in the middle cell of the array. Such is the case in the example above, 13 is the middle number of a fifth order magic square that starts with 1 and goes to 25.

There are also many regular magic squares that remain regular magic squares when shifted diagonally. By shifting diagonally, I mean that the shift includes moving one row and one column to give a result of one diagonal shift. For example, the following magic square on the left uses the same array of numbers as the array on the right, with the difference being that the array on the right has been shifted diagonally from the array on the left. Specifically, the array on the right has been shifted right one column, and up one row.

<table>
<tr><td>1</td><td>19</td><td>7</td><td>25</td><td>13</td></tr>
<tr><td>17</td><td>10</td><td>23</td><td>11</td><td>4</td></tr>
<tr><td>8</td><td>21</td><td>14</td><td>2</td><td>20</td></tr>
<tr><td>24</td><td>12</td><td>5</td><td>18</td><td>6</td></tr>
<tr><td>15</td><td>3</td><td>16</td><td>9</td><td>22</td></tr>
</table>

<table>
<tr><td>4</td><td>17</td><td>10</td><td>23</td><td>11</td></tr>
<tr><td>20</td><td>8</td><td>21</td><td>14</td><td>2</td></tr>
<tr><td>6</td><td>24</td><td>12</td><td>5</td><td>18</td></tr>
<tr><td>22</td><td>15</td><td>3</td><td>16</td><td>9</td></tr>
<tr><td>13</td><td>1</td><td>19</td><td>7</td><td>25</td></tr>
</table>

Figure 2.19 Both of the arrays above are regular 5th order magic squares. The array on the right has been shifted right one row, and up one column from the array on the left.

It is interesting to note that in the world of magic squares, the two magic squares shown above are considered unique to each other since neither is a mirroring or rotation of the other. However, they were still made from the same array of numbers which were basically in the same order, except for the diagonal shift (shifted right one row and up one column). We will look at this property of shifting rows and columns in more detail in a later chapter when we discuss a way of mapping similar magic squares.

A Note on Addition
The only way to check a magic square is to do the addition. The rows, columns and diagonals must add up to the same number to be a magic square. Unless you have a spreadsheet or some other software program to do the checking for you, it is easy to become bogged down by the addition. However, if you must find the totals by hand, there are some tricks that can make the addition easier and much quicker.

Let's take the 4 x 4 magic square as an example. It has a constant equal to 34, which means that all four numbers in each row, column and diagonal must add to 34. But that also means each group of four numbers must have pairs that add to 17 + 17, or 16 + 18, or 15 + 19, and so on. If we can simply find these matching pairs in a row, column or diagonal, then we know that the total is equal to 34. Let's look at the following magic square (again).

<table>
<tr><td>1</td><td>15</td><td>14</td><td>4</td></tr>
<tr><td>12</td><td>6</td><td>7</td><td>9</td></tr>
<tr><td>8</td><td>10</td><td>11</td><td>5</td></tr>
<tr><td>13</td><td>3</td><td>2</td><td>16</td></tr>
</table>

Figure 2.20 A basic 4th order magic square.

Notice that within each of the rows, we can quickly find two numbers that add to 16, while the other two numbers in that row add to 18. I know that 16 + 18 equals 34, so I have just checked, and proved to myself that all the rows add up to the constant. As for the columns, the first column has pairs of numbers that add to 20 + 14, the second and third columns have pairs that add to 16 + 18, and the fourth column has pairs of that add to 20 + 14. With that, I have now quickly checked and found that all the columns add to

34. Finally, I look at the diagonals, and I can see that they each have pairs of numbers that add to $17 + 17$.

In this example, I have proved to myself that it is a magic square, and I have done it much faster than I could have done it with a calculator. This is a simple but powerful trick, and with practice, it can be used on larger magic squares.

Chapter 3
The 3rd Order and Other Unique Magic Squares

The 3rd Order Magic Square

If we take the numbers 1 through 9 and arrange them so that we get a magic square, there is only one combination of numbers that will give us a 3rd order magic square. It looks like this:

8	1	6
3	5	7
4	9	2

Figure 3.1   The basic 3rd order magic square.

Any other 3rd order magic square (that uses the numbers 1 thru 9) is either a rotation and/or mirroring of this original magic square. In fact many magic square enthusiasts consider all magic squares that are rotations and/or mirrorings of an original magic square to be the same magic square. We will adhere to this same restriction throughout this book; although as we look at various constructions and patterns of squares we will see combinations of magic squares that will blur this distinction.

There are many variations to this 3rd order magic square. In fact, the rules that govern the variations of a third order magic square can also be applied to any order magic square. Let's look at some of these principles.

Any nine consecutive numbers can be used to create a magic square. Also, any nine evenly spaced numbers will work, including consecutive even or consecutive odd numbers (which are evenly spaced numbers with a difference of 2 between each number). The next magic square is made from the consecutive odd numbers ranging from 1 to 17. The magic square constant is 27.

15	1	11
5	9	13
7	17	3

Figure 3.2   A 3rd order magic square made of consecutive odd numbers.

An "n x n" magic square can also be made from n sets of n numbers where the numbers within each set are evenly spaced, and where the difference between each set is also evenly spaced. The numbers used in such a magic square are still considered to be regularly spaced; therefore, the method used to calculate the constant (as explained in chapter one) will also calculate the constant for these magic squares.

As an example, the numbers (5, 10, 15), (30, 35, 40) and (55, 60, 65) are three sets of three numbers that are evenly spaced within each set, and then also have an even spacing between each set. Within the three sets each number has a difference of 5; and the difference between the last number of the preceding set and the first number of the following set is 15. Arranged as a magic square, we have:

60	5	40
15	35	55
30	65	10

Figure 3.3   A 3rd order magic square made from three sets of evenly spaced multiples of five.

Using this pattern, a variety of magic squares can be created. The following magic square is the prime magic square with the lowest constant (which is equal to 111). The numbers used are (1, 7, 13), (31, 37, 43) and (61, 67, 73). There is a difference of 6 between each number within the three sets, and the difference between the last number of a preceding set and the first number of the following set is 18:

67	1	43
13	37	61
31	73	7

Figure 3.4   A 3rd order magic square made from three sets of evenly spaced prime numbers.

Of course, in the magic square above we are making the assumption that the number one is a prime number. In most mathematical circles the number one is not considered a prime number. However, we won't even touch that argument here; and therefore, for the sake of the magic square listed above, we'll simply say it is prime, and move on.

Alternate Perspectives on Magic Squares

Let's return to our original 3rd order magic square, and create a variation where 5 is subtracted from each element. Our resulting magic square looks like this:

3	-4	1
-2	0	2
-1	4	-3

Figure 3.5   A zero-centered 3rd order magic square.

In this variation, we have centered our sequence of numbers around the number zero. Normally, we use the numbers one through nine to create our 3rd order magic square. In this case the magic square shown above uses the numbers -4 through 4. With this change it is much easier to see the balance in the magic square. Every row, column and diagonal adds up to the number zero. Of course, we expect every row, column and diagonal to add up to the same number, but by centering our sequence around the number zero, we see a unique symmetry that isn't as apparent with the usual numbers.

Another way to look at our original magic square is to assign a letter pair for each number in the square. The way that these letter pairs are assigned works like this. The first three numbers all start with an uppercase A (the numbers 1, 2 and 3). The next three numbers start with an uppercase B (the numbers 4, 5 and 6), and the final three numbers start with an uppercase C (the numbers 7, 8 and 9).

For the second letter in each letter pair, we will assign a lowercase a to the first number of each triplet (the numbers 1, 4 and 7), a lowercase b to the second number in each triplet (numbers 2, 5 and 8), and a lowercase c to the third number in each triplet (numbers 3, 6 and 9). Using this notation, the numbers one through nine, written in sequence, looks like this:

Aa, Ab, Ac, Ba, Bb, Bc, Ca, Cb, Cc

And arranging these letter pairs as a magic square, gives us the following:

Cb	Aa	Bc
Ac	Bb	Ca
Ba	Cc	Ab

Figure 3.6   The 3rd order magic square using a letter pair of capital and small letters to represent the numbers 1 through 9.

Notice that each row and column has exactly one capital A, B and C and also exactly one small a, b and c. The diagonals have a pattern of their own. One diagonal has exactly one capital A, B and C while the small letters are all b's; and the other diagonal has exactly one small a, b and c while the capital letters are all B's.

Other patterns can be seen when we highlight just the odd numbers or just the even numbers in a magic square. The figure below shows the 3rd order magic square with the odd numbers highlighted.

8	1	6
3	5	7
4	9	2

Figure 3.7   The 3rd order magic square highlighting the odd numbers within the square.

We will look at these and other patterns like these in more detail in chapter 14. And there are still many other patterns that can be found within these magic squares.

In the next figure there are five graphic patterns. The pattern on the left is what we get when we draw lines from a numbers original position to its current position (the dot represents a number that did not move from its original position).

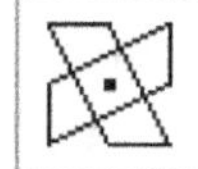  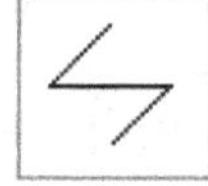  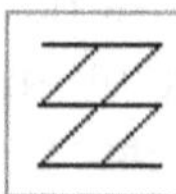

Figure 3.8   The 3rd order magic square showing various methods of connecting sequences of numbers within the squares with lines.

Moving right, we have the pattern that is created when the numbers are connected sequentially (from 1 to 2 to 3, etc.). The center pattern connects the odd numbers sequentially, and the pattern to the right of that connects the even numbers sequentially. Finally, the pattern on the right is what we get when the previous two patterns are overlain on top of each other. We will look at these types of patterns in more detail in chapter 15.

One-of-a-Kind Magic Squares
The 3rd order magic square is a one-of-a-kind. There are also several other magic squares that are unique. We'll end this chapter by looking at a few of them. The first magic square we'll consider is one that was created by Royal V. Heath. It is unique because it is still a magic square even if turned upside down. Turning it upside down causes some numbers to change while others remain the same; for example, the number 98 becomes 86 when turned upside down, while the number 88 is the same in both orientations. Here is the magic square:

96	11	89	68
88	69	91	16
61	86	18	99
19	98	66	81

Figure 3.9   A 4th order magic square using numbers that have an upside-down symmetry. This 4 x 4 array of numbers is a magic square in both orientations. This magic square has a constant equal to 264.

And here is the same magic square after being turned upside down:

18	99	86	61
66	81	98	19
91	16	69	88
89	68	11	96

Figure 3.10   This is the same 4 x 4 array of numbers as in the previous figure; however the numbers in this figure are upside-down when compared with the previous figure such that numbers like 18 becomes 81, and 98 becomes 86. In both orientations the array is still a 4th order magic square. This magic square also has a constant equal to 264.

The following 12th order magic square is the smallest possible magic square made of consecutive odd primes where the rows, columns and diagonals add to 4514 (this again assumes that the number one is a prime number).

1	823	821	809	811	797	19	29	313	31	23	37
89	83	211	79	641	631	619	709	617	53	43	739
97	227	103	107	193	557	719	727	607	139	757	281
223	653	499	197	109	113	563	479	173	761	587	157
367	379	521	383	241	467	257	263	269	167	601	599
349	359	353	647	389	331	317	311	409	307	293	449
503	523	233	337	547	397	421	17	401	271	431	433
229	491	373	487	461	251	443	463	137	439	457	283
509	199	73	541	347	191	181	569	577	571	163	593
661	101	643	239	691	701	127	131	179	613	277	151
659	673	677	683	71	67	61	47	59	743	733	41
827	3	7	5	13	11	787	769	773	419	149	751

Figure 3.11   The smallest possible magic square made of consecutive odd prime numbers.

The following magic square is made from the first 18 digits of the repeating decimal representation of the fractions 1/19 through 18/19. It is the only perfect magic square (where the rows, columns, and diagonals all add to the same number) that is generated by a cyclic number where the fractions are listed in ascending order. Notice that within each row (and column) the numbers 1 through 8 are each used twice, while the 0 and the 9 are only used once. Thus the constant of this magic square is 81.

	0	5	2	6	3	1	5	7	8	9	4	7	3	6	8	4	2	1
1/19 =	0	5	2	6	3	1	5	7	8	9	4	7	3	6	8	4	2	1
2/19 =	1	0	5	2	6	3	1	5	7	8	9	4	7	3	6	8	4	2
3/19 =	1	5	7	8	9	4	7	3	6	8	4	2	1	0	5	2	6	3
4/19 =	2	1	0	5	2	6	3	1	5	7	8	9	4	7	3	6	8	4
5/19 =	2	6	3	1	5	7	8	9	4	7	3	6	8	4	2	1	0	5
6/19 =	3	1	5	7	8	9	4	7	3	6	8	4	2	1	0	5	2	6
7/19 =	3	6	8	4	2	1	0	5	2	6	3	1	5	7	8	9	4	7
8/19 =	4	2	1	0	5	2	6	3	1	5	7	8	9	4	7	3	6	8
9/19 =	4	7	3	6	8	4	2	1	0	5	2	6	3	1	5	7	8	9
10/19 =	5	2	6	3	1	5	7	8	9	4	7	3	6	8	4	2	1	0
11/19 =	5	7	8	9	4	7	3	6	8	4	2	1	0	5	2	6	3	1
12/19 =	6	3	1	5	7	8	9	4	7	3	6	8	4	2	1	0	5	2
13/19 =	6	8	4	2	1	0	5	2	6	3	1	5	7	8	9	4	7	3
14/19 =	7	3	6	8	4	2	1	0	5	2	6	3	1	5	7	8	9	4
15/19 =	7	8	9	4	7	3	6	8	4	2	1	0	5	2	6	3	1	5
16/19 =	8	4	2	1	0	5	2	6	3	1	5	7	8	9	4	7	3	6
17/19 =	8	9	4	7	3	6	8	4	2	1	0	5	2	6	3	1	5	7
18/19 =	9	4	7	3	6	8	4	2	1	0	5	2	6	3	1	5	7	8

Figure 3.12   An 18th order magic square created from the fractions 1/19 through 18/19 where the fractions are listed in ascending order.

Other sets of fractions will create semi-magic squares where the rows and columns will add to the same constant, but the diagonals won't. For example, here is a square created form the fractions 1/7 through 6/7. Each row and column adds up to 27, but the diagonals don't. And even rearranging the rows (so that they are not in any particular order) will not yield anything but a semi-magic square.

1/7 =	1	4	2	8	5	7
2/7 =	2	8	5	7	1	4
3/7 =	4	2	8	5	7	1
4/7 =	5	7	1	4	2	8
5/7 =	7	1	4	2	8	5
6/7 =	8	5	7	1	4	2

Figure 3.13   A 6th order semi-magic square created from the fractions 1/7 through 6/7 where the fractions are listed in ascending order. In this array, the diagonals do not add up to the magic square constant.

With the completion of this chapter we've now completed our initial look at magic squares. These first three chapters have laid a foundation of defining words, understanding operations, and looking at patterns. Now we will use this foundation as we consider several methods that can be used to build a million-billion magic squares.

The 4th Order Magic Square
In this chapter we will introduce two concepts about building magic squares. The first is
the initial state (or initial square); and the second is the method used to create the magic
square. The initial square (or initial state) is our starting point; it is the array of numbers
that we consider in their original positions. The method is the process applied to the
numbers in their original position that will create the magic square. For now, we'll say
that the initial square is simply the array of numbers where we start with the number one
in the upper left-hand corner and we list the numbers sequentially from left to right and
from top to bottom so that the largest number in the array is in the lower right-hand
corner. The following is an example of an initial square for a 4 x 4 array of numbers.

1	2	3	4
5	6	7	8
9	10	11	12
13	14	15	16

Figure 4.1   An initial 4 x 4 array of numbers where the numbers are listed
sequentially from left to right and from top to bottom.

Notice that this array isn't a magic square; however, if we simply swap four pairs of
numbers, we have a magic square. (The shaded numbers in the following square are the
numbers that didn't move):

1	15	14	4
12	6	7	9
8	10	11	5
13	3	2	16

Figure 4.2   A 4th order magic square (shaded numbers are those that are still in
their initial positions).

The numbers that were swapped with each other are:

$$2 \leftrightarrow 15$$
$$3 \leftrightarrow 14$$
$$5 \leftrightarrow 12$$
$$8 \leftrightarrow 9$$

This swapping of numbers is the method that we used to go from our initial square to a magic square. However, there are in fact several ways to swap pairs of numbers to create a 4th order magic square from an initial 4 x 4 array of numbers. Here are two other magic squares that were created from our initial square:

1	14	15	4
8	11	10	5
12	7	6	9
13	2	3	16

1	14	15	4
12	7	6	9
8	11	10	5
13	2	3	16

Figure 4.3   Two other 4th order magic squares that are unique to each other.

Each of these magic squares started with the initial square in Figure 4.1. The two columns below correspond with the two magic squares above, the column on the left shows the numbers that traded places in the magic square on the left, and the column on the right shows the numbers that traded places in the magic square on the right. The swapped numbers are:

$$2 \leftrightarrow 14 \qquad\qquad 2 \leftrightarrow 14$$
$$3 \leftrightarrow 15 \qquad\qquad 3 \leftrightarrow 15$$
$$5 \leftrightarrow 8 \qquad\qquad 5 \leftrightarrow 12$$
$$6 \leftrightarrow 11 \qquad\qquad 6 \leftrightarrow 7$$
$$7 \leftrightarrow 10 \qquad\qquad 8 \leftrightarrow 9$$
$$9 \leftrightarrow 12 \qquad\qquad 10 \leftrightarrow 11$$

With these two magic squares, and the one before it, we now have three distinct magic squares all created from the same initial square. We could go on, but for now, we'll stop and consider our next concept.

As mentioned before, these magic squares can be thought of as a 4 x 4 array of cells. Each cell is identified by its corresponding row and column. In our initial square, the number 1 is in the first row and the first column, or cell (1, 1). The number 2 is in the first row and second column, or cell (1, 2); and so on. We could then say that our first magic square was created by swapping the contents of the following cells with each other:

$$(1, 2) \leftrightarrow (4, 3)$$
$$(1, 3) \leftrightarrow (4, 2)$$
$$(2, 1) \leftrightarrow (3, 4)$$
$$(3, 1) \leftrightarrow (2, 4)$$

The next two magic squares could be described as each starting with the same initial square and having the contents of the following cells swap with each other:

$$(1, 2) \leftrightarrow (4, 2) \qquad\qquad (1, 2) \leftrightarrow (4, 2)$$
$$(1, 3) \leftrightarrow (4, 3) \qquad\qquad (1, 3) \leftrightarrow (4, 3)$$
$$(2, 1) \leftrightarrow (2, 4) \qquad\qquad (2, 1) \leftrightarrow (3, 4)$$
$$(2, 2) \leftrightarrow (3, 3) \qquad\qquad (2, 2) \leftrightarrow (2, 3)$$
$$(2, 3) \leftrightarrow (3, 2) \qquad\qquad (2, 4) \leftrightarrow (3, 1)$$
$$(3, 1) \leftrightarrow (3, 4) \qquad\qquad (3, 2) \leftrightarrow (3, 3)$$

What we have done is define two different ways of creating a magic square. The first is a swap-by-number method, and the second is a swap-by-position method. By describing our construction this way, we now have two methods that we can use to build similar magic squares even though several numbers might be in a different position within the array. To illustrate, let's take our initial listing of the numbers 1 thru 16 and change the way we generate that list within the array, so that instead of starting with the initial square that we had at the beginning of this chapter, we now start with:

1	2	5	6
3	4	7	8
9	10	13	14
11	12	15	16

Figure 4.4   Another initial arrangement of numbers within a 4 x 4 array.

This new initial square was created by listing the numbers 1 - 4 in the upper left corner of the array. The numbers 5 - 8 are in the upper right corner. And following the same pattern, the numbers 9 - 12 are in the lower left corner and the numbers 13 - 16 in the lower right corner. Now, we'll use the first swap-by-number algorithm (from the beginning of the chapter) to create a new magic square (the shaded numbers are those that didn't move from their position within the initial square).

1	2	5	6
3	4	7	8
9	10	13	14
11	12	15	16

2	↔	15
3	↔	14
5	↔	12
8	↔	9

1	15	12	6
14	4	7	9
8	10	13	3
11	5	2	16

Figure 4.5   An alternate initial 4 x 4 array using a swap-by-number scheme to create a unique 4th order magic square.

Now, we'll try using our second algorithm of swapping the contents of the cells (the swap-by-position method) to create yet another magic square from our alternate initial square:

1	2	5	6
3	4	7	8
9	10	13	14
11	12	15	16

(1, 2)	↔	(4, 3)
(1, 3)	↔	(4, 2)
(2, 1)	↔	(3, 4)
(3, 1)	↔	(2, 4)

1	15	12	6
14	4	7	9
8	10	13	3
11	5	2	16

Figure 4.6   The initial 4 x 4 array from above using a swap-by-position scheme to create the same 4th order magic square.

In the example above, the swap-by-number method and the swap-by-position method actually created the same magic square from the same initial square. This won't always be the case as we will see in our next example.

The next two magic squares will use the swap-by-number method and the swap-by-position method from the second magic square (figure 4.3, on the left); and apply it to the initial square from figure 4.4. In this case, we will create two unique magic squares from the two distinct methods. First, here is our swap-by-number method used to create the magic square from the alternate initial square:

<table>
<tr><td>1</td><td>2</td><td>5</td><td>6</td></tr>
<tr><td>3</td><td>4</td><td>7</td><td>8</td></tr>
<tr><td>9</td><td>10</td><td>13</td><td>14</td></tr>
<tr><td>11</td><td>12</td><td>15</td><td>16</td></tr>
</table>

2	↔	14
3	↔	15
5	↔	8
6	↔	11
7	↔	10
9	↔	12

<table>
<tr><td>1</td><td>14</td><td>8</td><td>11</td></tr>
<tr><td>15</td><td>4</td><td>10</td><td>5</td></tr>
<tr><td>12</td><td>7</td><td>13</td><td>2</td></tr>
<tr><td>6</td><td>9</td><td>3</td><td>16</td></tr>
</table>

Figure 4.7   The alternate initial 4 x 4 array using a different swap-by-number scheme to create a unique 4th order magic square.

And now, our swap-by-position method used to create another magic square:

<table>
<tr><td>1</td><td>2</td><td>5</td><td>6</td></tr>
<tr><td>3</td><td>4</td><td>7</td><td>8</td></tr>
<tr><td>9</td><td>10</td><td>13</td><td>14</td></tr>
<tr><td>11</td><td>12</td><td>15</td><td>16</td></tr>
</table>

(1, 2)	↔	(4, 2)
(1, 3)	↔	(4, 3)
(2, 1)	↔	(2, 4)
(2, 2)	↔	(3, 3)
(2, 3)	↔	(3, 2)
(3, 1)	↔	(3, 4)

<table>
<tr><td>1</td><td>12</td><td>15</td><td>6</td></tr>
<tr><td>8</td><td>13</td><td>10</td><td>3</td></tr>
<tr><td>14</td><td>7</td><td>4</td><td>9</td></tr>
<tr><td>11</td><td>2</td><td>5</td><td>16</td></tr>
</table>

Figure 4.8   The alternate initial 4 x 4 array using a different swap-by-position scheme to create still another unique 4th order magic square.

What we just did is we started with an initial square and we used two different (but related) methods of moving the values around within the array, and we thus generated two unique magic squares. And we can just as easily do the same thing with the third magic square that we saw earlier in this chapter. The key point is that we can take alternate initial squares, and apply various methods to create unique magic squares. In the examples above, we defined a swap-by-number method and a swap-by-position method and used them to create magic squares from alternate initial squares.

Other 4th Order Initial Squares

There are in fact several other ways to create initial squares. One is to list the first four odd numbers on the first row, followed by the first four even numbers on the second row. The third row has the remaining odd numbers, and the fourth row has the remaining even numbers. Another initial square can be created by dividing the 4 x 4 array into 2 x 2 quadrants. The upper left quadrant has the numbers from the array that have a remainder of 1 when divided by 4. The upper right quadrant has the numbers that have a remainder of 2 when divided by 4, and so on. These two initial squares are shown below:

<table>
<tr><td>1</td><td>3</td><td>5</td><td>7</td></tr>
<tr><td>2</td><td>4</td><td>6</td><td>8</td></tr>
<tr><td>9</td><td>11</td><td>13</td><td>15</td></tr>
<tr><td>10</td><td>12</td><td>14</td><td>16</td></tr>
</table>

<table>
<tr><td>1</td><td>5</td><td>2</td><td>6</td></tr>
<tr><td>9</td><td>13</td><td>10</td><td>14</td></tr>
<tr><td>3</td><td>7</td><td>4</td><td>8</td></tr>
<tr><td>11</td><td>15</td><td>12</td><td>16</td></tr>
</table>

Figure 4.9   Two additional initial arrays that can create 4th order magic squares.

Still another way to create an initial square is to take the initial square from figure 4.1, and perform row and/or column swaps on the initial square itself, and in this way create a new initial square. As mentioned earlier, if we have a magic square and then swap corresponding rows and/or columns, we still have a magic square. In this case, we are swapping the corresponding rows and/or columns of the initial square before we apply the swap algorithm to create the magic square. For example, we could start with:

1	2	3	4
5	6	7	8
9	10	11	12
13	14	15	16

and swap the first and fourth rows

13	14	15	16
5	6	7	8
9	10	11	12
1	2	3	4

Figure 4.10   Creating a new initial square from a previous initial square.

In the example above, we have swapped the top row with the bottom row to create an initial square that is yet again different from any previous initial squares we have created so far. Now, using this new initial square along with our swap-by-position method, we can create the following magic square.

13	14	15	16
5	6	7	8
9	10	11	12
1	2	3	4

(1, 2)	↔	(4, 3)
(1, 3)	↔	(4, 2)
(2, 1)	↔	(3, 4)
(3, 1)	↔	(2, 4)

13	3	2	16
12	6	7	9
8	10	11	5
1	15	14	4

Figure 4.11   Using a previous swap-by-position scheme on the new initial square to create a 4th order magic square.

Notice that we can create a unique magic square by using a unique swap algorithm on an initial square; and we can also create unique magic squares by having a unique initial square, and then applying one of our swap algorithms to that initial square. In this next section we'll continue by looking at an interesting 4th order magic square.

The 4-Way Swap
Sometimes we can build magic squares where the numbers don't swap places with just a single number. Instead, the first number moves to the place of a second number, the second number moves to the place of a third number, the third number moves to the place of a fourth number, and finally the fourth number moves to the place of the first number.

1	14	15	4
8	11	10	5
12	7	6	9
13	2	3	16

14	1	4	15
11	8	5	10
7	12	9	6
2	13	16	3

Figure 4.12   The magic square on the right is create from the one on the left by swapping columns 1 and 2 with each other, and then swapping columns 3 and 4 with each other.

In the figure above, we started with the magic square on the left, and created the magic square on the right by swapping columns 1 with 2, and columns 3 with 4. We used a magic square to generate another magic square. Now, we can describe the magic square on the right by comparing both the numbers and their positions with their original location from the initial square in figure 4.1. The left side shows the swap-by-number method while the right side shows the swap-by-position method:

$$1 \rightarrow 2 \rightarrow 13 \rightarrow 14 \rightarrow 1 \qquad\qquad (1, 1) \rightarrow (1, 2) \rightarrow (4, 1) \rightarrow (4, 2) \rightarrow (1, 1)$$
$$3 \rightarrow 16 \rightarrow 15 \rightarrow 4 \rightarrow 3 \qquad\qquad (1, 3) \rightarrow (4, 4) \rightarrow (4, 3) \rightarrow (1, 4) \rightarrow (1, 3)$$
$$5 \rightarrow 7 \rightarrow 9 \rightarrow 11 \rightarrow 5 \qquad\qquad (2, 1) \rightarrow (2, 3) \rightarrow (3, 1) \rightarrow (3, 3) \rightarrow (2, 1)$$
$$6 \rightarrow 12 \rightarrow 10 \rightarrow 8 \rightarrow 6 \qquad\qquad (2, 2) \rightarrow (3, 4) \rightarrow (3, 2) \rightarrow (2, 4) \rightarrow (2, 2)$$

Figure 4.13   A 4-way swap-by-number method on the left, with a corresponding 4-way swap-by-position method on the right created from observing the change of position of the numbers in the two magic squares in the previous figure.

Again, the left is a swap-by-number algorithm. This means that regardless of where the numbers are positioned within the initial square, those are the numbers that will be swapped with each other. The algorithm on the right is a swap-by-position algorithm which means that regardless of what number is in that position in the array, those are the positions that will swap with each other.

In other words, we can take the initial square on the left below and use either swap method to create the magic square on the right below.

1	2	3	4
5	6	7	8
9	10	11	12
13	14	15	16

14	1	4	15
11	8	5	10
7	12	9	6
2	13	16	3

Figure 4.14   Applying either 4-way swap on the initial square on the left will create the 4th order magic square on the right.

However, if we start with a different initial square, and perform either of these swap methods on this new initial square, we'll get two different magic squares. In the following example, we'll first use the swap-by-number method on our initial square to create one magic square, and then we'll use the swap-by-position method to create a second, unique-from-the-first, magic square. Here is the process using the swap-by-number method:

1	2	5	6
3	4	7	8
9	10	13	14
11	12	15	16

$$1 \rightarrow 2 \rightarrow 13 \rightarrow 14 \rightarrow 1$$
$$3 \rightarrow 16 \rightarrow 15 \rightarrow 4 \rightarrow 3$$
$$5 \rightarrow 7 \rightarrow 9 \rightarrow 11 \rightarrow 5$$
$$6 \rightarrow 12 \rightarrow 10 \rightarrow 8 \rightarrow 6$$

14	1	11	8
4	15	5	10
7	12	2	13
9	6	16	3

Figure 4.15   The 4-way swap-by-number method is applied to the initial square on the left which creates the unique magic square on the right.

The magic square on the right was created from the initial square on the left by using the four-number swap method where the numbers themselves traded places. For the next example, we'll start with the same initial square, but this time we'll swap-by-position, and thereby create another unique magic square:

1	2	5	6
3	4	7	8
9	10	13	14
11	12	15	16

$(1, 1) \rightarrow (1, 2) \rightarrow (4, 1) \rightarrow (4, 2) \rightarrow (1, 1)$
$(1, 3) \rightarrow (4, 4) \rightarrow (4, 3) \rightarrow (1, 4) \rightarrow (1, 3)$
$(2, 1) \rightarrow (2, 3) \rightarrow (3, 1) \rightarrow (3, 3) \rightarrow (2, 1)$
$(2, 2) \rightarrow (3, 4) \rightarrow (3, 2) \rightarrow (2, 4) \rightarrow (2, 2)$

12	1	6	15
13	8	3	10
7	14	9	4
2	11	16	5

Figure 4.16 The 4-way swap-by-position method is applied to the same initial square on the left which creates another unique 4th order magic square on the right.

In order to illustrate the concept one more time, we'll start with yet another initial square, and then use these two swap methods to create two other unique magic squares. Again, the initial square is on the left, the swap method is shown in the middle, and the resulting magic square is on the right.

1	3	5	7
2	4	6	8
9	11	13	15
10	12	14	16

$1 \rightarrow 2 \rightarrow 13 \rightarrow 14 \rightarrow 1$
$3 \rightarrow 16 \rightarrow 15 \rightarrow 4 \rightarrow 3$
$5 \rightarrow 7 \rightarrow 9 \rightarrow 11 \rightarrow 5$
$6 \rightarrow 12 \rightarrow 10 \rightarrow 8 \rightarrow 6$

14	4	11	5
1	15	8	10
7	9	2	16
12	6	13	3

1	3	5	7
2	4	6	8
9	11	13	15
10	12	14	16

$(1, 1) \rightarrow (1, 2) \rightarrow (4, 1) \rightarrow (4, 2) \rightarrow (1, 1)$
$(1, 3) \rightarrow (4, 4) \rightarrow (4, 3) \rightarrow (1, 4) \rightarrow (1, 3)$
$(2, 1) \rightarrow (2, 3) \rightarrow (3, 1) \rightarrow (3, 3) \rightarrow (2, 1)$
$(2, 2) \rightarrow (3, 4) \rightarrow (3, 2) \rightarrow (2, 4) \rightarrow (2, 2)$

12	1	7	14
13	8	2	11
6	15	9	4
3	10	16	5

Figure 4.17 The top arrangement applies the 4-way swap-by-number method to yet another initial square on the left and thereby creating a unique 4th order magic square on the right. The bottom arrangement applies the 4-way swap-by-position method to the same initial square on the left and creates yet another unique 4th order magic square on the right.

In the example above, the resulting magic squares originated from the same initial square. The top sequence shows the swap-by-number method, and the bottom sequence shows the swap-by-position method. The interesting point is that even though both methods were created from some other magic square, when these methods were applied to yet another initial square, these methods still created two unique magic squares.

A 4th Order Diabolic Magic Square

The following magic square contains some very unique properties. It is called a "diabolic" magic square because not only do the rows, columns, major diagonals and broken diagonals all add up to 34, but so does any group of 2 x 2 cells within the array.

14	11	5	4
1	8	10	15
12	13	3	6
7	2	16	9

Figure 4.18   A "diabolic" 4th order magic square.

You can also take the numbers from the four corners of the square, and they also add up to 34. Not only that, but it is also possible to take the four corners of any 3 x 3 array within this magic square, and they also add up to 34.

Because of the properties that this magic square has, several other magic square variations can be made by simply shifting the rows or columns as well as the traditional swapping of rows and columns. For example, you can take a column off the right side and place it on the left, or take a row off the top and place it on the bottom. The resulting 4 x 4 squares are still magic and still retain their "diabolic" properties such that any 2 x 2 cell within the square as well as the four corners will continue to add to 34.

14	11	5	4
1	8	10	15
12	13	3	6
7	2	16	9

4	14	11	5
15	1	8	10
6	12	13	3
9	7	2	16

Figure 4.19   A diabolic magic square is pan-diagonal so that even when the right-most column is placed on the left side (or the rows or columns are shifted in any other way), the array is still a magic square.

In the example above, we took the original diabolic magic square, and removed the far right column and placed it on the left side, thus shifting the other columns to the right. This variation is not a rotation or mirroring of the first magic square, thus making it a unique magic square from the original.

Most other magic squares don't allow you to shift rows and columns like this. Doing so throws off the diagonal totals so that even though the rows and columns keep their totals equal to each other, the diagonals don't.

Another way to consider this property is to place several of these diabolic magic squares next to each other.   In the figure below, we've copied this diabolic magic square and placed the copies next to each other so that they create a 12 x 12 array of numbers. Now, we can select any 4 x 4 square within this array, and it also will be a diabolic magic square.

The figure below is called a magic carpet. By simply selecting any number and making it the upper left corner of a 4 x 4 array of numbers gives us a 4th order magic square. In fact, this arrangement will give us sixteen unique magic squares.

14	11	5	4	14	11	5	4	14	11	5	4
1	8	10	15	1	8	10	15	1	8	10	15
12	13	3	6	12	13	3	6	12	13	3	6
7	2	16	9	7	2	16	9	7	2	16	9
14	11	5	4	14	11	5	4	14	11	5	4
1	8	10	15	1	8	10	15	1	8	10	15
12	13	3	6	12	13	3	6	12	13	3	6
7	2	16	9	7	2	16	9	7	2	16	9
14	11	5	4	14	11	5	4	14	11	5	4
1	8	10	15	1	8	10	15	1	8	10	15
12	13	3	6	12	13	3	6	12	13	3	6
7	2	16	9	7	2	16	9	7	2	16	9

Figure 4.20   The shaded areas show various 4th order diabolic magic squares within the magic carpet.

In the figure above, I've taken three 4 x 4 sets of numbers and shaded them. Each of these 4 x 4 arrays is a pan-diagonal magic square.

This property of the magic carpet will work with any pan-diagonal magic square of any order. Later on we will see more examples of these fascinating magic squares, and we'll even look at some methods of creating pan-diagonal magic squares from certain types of regular magic squares. But for now, let's move on to the 5th order magic square.

Chapter 5
The 5th Order Magic Square

The 5th Order Magic Square
In this chapter we are going to reintroduce the two concepts that we looked at in the previous chapter. I'm actually using the term "reintroduce" loosely because even though the basic ideas will be the same, we are now dealing with odd-ordered magic squares, and therefore, the way we apply these concepts will be different. Again, the first concept is the initial state from which the magic square will be built; and the second is the method that is used to create the magic square.

The initial state for an odd-ordered magic square is the initial sequence of numbers. For now, the initial sequence of a 5th order magic square are the numbers 1 thru 25 listed in ascending order. As we will see later in this chapter, the initial sequence can change; yet, by using a pre-determined method, we can still create a magic square. With that in mind, let's begin by looking at a couple of methods. We'll start by looking at the following 5th order magic square and use it to explain our method.

1	20	9	23	12
19	8	22	11	5
7	21	15	4	18
25	14	3	17	6
13	2	16	10	24

Figure 5.1   A 5th order magic square.

In the magic square above, we used the numbers 1 thru 25 in ascending order to create a magic square. This was our initial sequence. Our method uses the following process. We take five numbers at a time, and start by placing the number 1 in the upper left corner of the square. Each succeeding number is placed to the right one column, and up one row from the preceding number. When a number is on the top row, then the next number is continued on the bottom row. Likewise, when a number is on the right-most column, the next number is continued on the left-most column.

The first number of the next set of five numbers is placed in relation to the last number of the preceding set of five numbers. In this example, the number 5 is the last number of the first set, and the number 6 is the first number of the next set. The 6 is placed two rows below the 5 and then the remaining numbers of that set (numbers 7 thru 10) are placed as described above for the numbers 2 thru 5. The number 10 is the last number of the second group of five and the number 11 is the first number of the next group of five numbers. Therefore, the 11 is placed in relation to the 10, being placed two rows below the 10. This process continues until all twenty-five numbers have been placed in the square. This method of placing numbers gives us a magic square.

Now let's create a second magic square that uses the same initial sequence and a similar method; however, the method will be different enough that our resulting magic square will be unique from the first.

17	24	1	8	15
23	5	7	14	16
4	6	13	20	22
10	12	19	21	3
11	18	25	2	9

Figure 5.2   A second 5th order magic square.

As seen in the magic square above, we again take five numbers at a time, but this time we start by placing the number 1 in the third column of the first row. Each succeeding number is placed to the right one column, and up one row from the preceding number. In this example, the placement of the 6 in relation to the 5 is different. This time the 6 is placed directly below the 5, or more accurately, we could say, the 6 is placed one row below the 5. If we continue this process until all twenty-five numbers have been placed in the square, we have another magic square. This second magic square is unique from the first magic square in this chapter.

Changing the Sequence
We've now seen two methods that will create a 5th order magic square. For those magic squares we used the numbers 1 thru 25 in ascending order as the initial sequence. In this section we will change the initial sequence while keeping the method the same to create two new magic squares. We'll start by taking a look at the magic square from figure 5.1 again.

1	20	9	23	12
19	8	22	11	5
7	21	15	4	18
25	14	3	17	6
13	2	16	10	24

Figure 5.3   Our original 5th order magic square again.

By simply exchanging five pairs of numbers, we'll use this magic square to build another magic square. Keep in mind that we're not swapping number positions; instead, we will be exchanging the actual numbers in the sequence. Specifically, we will swap the 4 with the 5, the 9 with the 10, the 14 with the 15, the 19 with the 20 and the 24 with the 25. The resulting magic square looks like this:

1	19	10	23	12
20	8	22	11	4
7	21	14	5	18
24	15	3	17	6
13	2	16	9	25

Figure 5.4   A variation of the initial 5th order magic square where five pairs of numbers were swapped.

What we have now is the same magic square that we would get if we had used the first method, along with the following sequence as our initial sequence of numbers:

1 2 3 5 4   6 7 8 10 9   11 12 13 15 14   16 17 18 20 19   21 22 23 25 24

The sequence above is mostly in ascending order. Notice how the numbers are grouped into five sets of five, and the last two numbers of each set are swapped, i.e., the 4 and 5 are swapped, the 9 and 10 are swapped, etc. This magic square is then built just as the first one from this chapter was built: the 1 is placed in the upper left corner and the following numbers are placed to the right one and up one from the preceding number. The last number of the first set of five numbers is a 4. The 6 is still the first number of the next set of five; and therefore, the 6 is placed two rows below the 4. We then continue following the same pattern of placing the remaining numbers in the sequence. As we continue, we create a new magic square using a previous method and a new sequence of numbers.

If we use this new sequence of numbers, and follow the second method (where the one is placed in the third cell of the top row), we get this unique magic square:

17	25	1	8	14
23	4	7	15	16
5	6	13	19	22
9	12	20	21	3
11	18	24	2	10

Figure 5.5   A variation of the second 5th order magic square.

Yet another possible sequence lists the first three odd numbers, followed by the two even numbers from the first set of five; then lists the three even numbers followed by the two odd numbers from the second set of five. The third set of five numbers repeats the pattern set by the first group, etc. The entire listing looks like this:

1 3 5 2 4   6 8 10 7 9   11 13 15 12 14   16 18 20 17 19   21 23 25 22 24

We can use this new sequence along with the two methods mentioned previously to create yet two other unique magic squares. The magic square on the following page and on the left uses our latest sequence along with the first construction method, while the magic square on the right also uses this same sequence along with the second construction method.

1	19	7	25	13
17	10	23	11	4
8	21	14	2	20
24	12	5	18	6
15	3	16	9	22

18	22	1	10	14
25	4	8	12	16
2	6	15	19	23
9	13	17	21	5
11	20	24	3	7

Figure 5.6   Two new magic squares using the previous construction
methods and the new sequence.

A Concise Description of the Method and Sequence

In the previous sections, we looked at two methods and three sequences that can be used
to create similar, yet unique magic squares. So, just how many methods and sequences
are there? Let's take a look at the sequences first.

There are at least 24 sequences for a fifth order square. To describe these 24 sequences,
we can use the first five numbers as a template. The order of the first five numbers set
the pattern that will be used for the next five numbers, and then the next, and so on
throughout all 5 sets of 5 numbers. Therefore, if we describe a pattern as 1 2 4 5 3, it
means that the entire pattern is:

1 2 4 5 3   6 7 9 10 8   11 12 14 15 13   16 17 19 20 18   21 22 24 25 23

Each group of five numbers lists the first number, the second, the fourth, the fifth and
finally the third number of that set. With that understanding, here are the first five
numbers of 24 magic square sequences:

1 2 3 4 5	1 4 2 3 5
1 2 3 5 4	1 4 2 5 3
1 2 4 3 5	1 4 3 2 5
1 2 4 5 3	1 4 3 5 2
1 2 5 3 4	1 4 5 2 3
1 2 5 4 3	1 4 5 3 2
1 3 2 4 5	1 5 2 3 4
1 3 2 5 4	1 5 2 4 3
1 3 4 2 5	1 5 3 2 4
1 3 4 5 2	1 5 3 4 2
1 3 5 2 4	1 5 4 2 3
1 3 5 4 2	1 5 4 3 2

You'll notice that we're always using the number one as our initial number. (In other
words we're not considering the other 96 sequences that we could get when we start with
the numbers two, three, four or five). There is a reason for this that should become
apparent a little later when we discuss how to identify that the various methods and
sequences are unique.

Also keep in mind that there are still other sequences that will also have their own
variations of sequences. One such sequence is:

1 6 11 16 21   2 7 12 17 22   3 8 13 18 23   4 9 14 19 24   5 10 15 20 25

From this sequence we could generate 23 more sequences using the patterns from the previous 24 sequences above as a template. These sequences could then be applied with various other methods to create a multitude of fifth order magic squares.

Describing the Magic Square

Now let's take a look at our methods. In the first method mentioned in this chapter, we started by placing the number 1 in the upper left corner of the square. Each succeeding number was placed to the right one and up one from the preceding number. Again, if the preceding number was on the top row, then the next number continued on the bottom row; and similarly, if the preceding number was on the right-most column, then the next number continued on the left-most column.

This took care of the placement of the first five numbers. The number 6 was then placed in relation to the number 5 (the last number in the first set); and in this first method, the number 6 was placed two rows below the number 5. The rest of the numbers in this group (numbers 7 thru 10) were again placed right one and up one from the preceding number; and so on.

The second method placed the one in the center cell of the top row. The remaining numbers in the first group of five were also placed right one and up one. The six again was placed in relation to the five, and this time it was placed right below the 5, and so on.

You've probably already noticed that there is a pattern to these methods. In fact, we could summarize the methods in this way:

> Determine the actual sequence of the first five numbers.
> Begin with the placement of the first number
> Next, place the second number (in relation to the first number)
> Then show the placement of the sixth number (in relation to the fifth)

Using these steps, we can create an abbreviated notation to describe our first method as well as any similar method. (However, instead of showing the sequence of numbers first, we will show it last in our abbreviated notation for the description.)

P1 (1, 1)	The placement of the first number (its location in the array)
P21 (rt 1, up 1)	The placement of the second number in relation to the first
P65 (dn 2)	The placement of the sixth number in relation to the fifth
Seq 1 2 3 4 5	The sequence of the first five numbers.

When describing the position of the number 1, we'll use regular matrix notation of listing the row first followed by the column. Therefore, position (1, 1) means the number is listed on row 1, column 1. Our second magic square had the one located on row 1, column 3, or in other words, position (1, 3). Here is the second magic square of this chapter along with its abbreviated description:

17	24	1	8	15
23	5	7	14	16
4	6	13	20	22
10	12	19	21	3
11	18	25	2	9

P1 (1, 3)
P21 (rt 1, up 1)
P65 (dn 1)
Seq 1 2 3 4 5

Figure 5.7   A previous 5th order magic square on the left along with the description of how it was created on the right.

This notation is actually very powerful in that it lets a person quickly see patterns between different types of magic squares. For example, the next magic square has a different starting location for the number one, but other than that, the position of the two in relation to the one, the position of the six in relation to the five and the sequence are the same as the magic square above.

18	25	2	9	11
24	1	8	15	17
5	7	14	16	23
6	13	20	22	4
12	19	21	3	10

P1 (2, 2)
P21 (rt 1, up 1)
P65 (dn 1)
Seq 1 2 3 4 5

Figure 5.8   A similar 5th order magic square with its description.

In fact, there are three other magic squares that share this similar characteristic. By also moving the position of the number one to locations (3, 1), (4, 5) and (5, 4) and keeping the other properties that same, we have three more unique magic squares. The following is a concise way to describe these five magic squares:

P1 (1, 3) (2, 2) (3, 1) (4, 5) (5, 4)
P21 (rt 1, up 1)
P65 (dn 1)
Seq 1 2 3 4 5

This method of describing magic squares becomes even more powerful when we run across groupings of magic squares that share several characteristics in common. For example, the following two descriptions will create ten unique magic squares.

P1 (1, 4) (2, 3) (3, 2) (4, 1) (5, 5)   P1 (1, 4) (2, 3) (3, 2) (4, 1) (5, 5)
P21 (rt 1, up 1)   P21 (rt 1, up 1)
P65 (rt 2, dn 1)   P65 (lf 2)
Seq 1 2 3 4 5   Seq 1 2 3 4 5

The descriptions above consist of two sets of five magic squares. Both sets share the same starting locations for the number one. They also both use the same method of placing the 2 in relation to the 1 (right one cell, and up one cell). However, the left-side description places the six in relation to the five in a location that is right 2 columns and down 1 row; while the right-side description places the six (in relation to the five) left 2 columns. With these two descriptions side-by-side we can easily see what these ten magic squares have in common, and where they are different.

Revisiting the Number Sequence and Identifying Magic Squares
With this method of describing squares, let's reexamine the previous magic square from
figure 5.8. In that magic square, we show the position of the first number at (2, 2) and
the sequence as 1 2 3 4 5. However, we also could have said that this magic square has
the first number starting at position (1, 3) with the sequence being 2 3 4 5 1. The number
1 is now the last number of the first five numbers. This also makes the number 7 the first
number of the second group of five numbers (with the sequence of the second group of
five numbers being 7 8 9 10 6).

18	25	2	9	11
24	1	8	15	17
5	7	14	16	23
6	13	20	22	4
12	19	21	3	10

P1 (2, 2)
P21 (rt 1, up 1)
P65 (dn 1)
Seq 1 2 3 4 5

Figure 5.9   The same magic square that we saw in the previous figure.

Earlier in this chapter we mentioned that we would only consider sequences that started
with the number one, and this is an example of the reason why. Simply by changing the
start number/location within a sequence, we can effectively describe the same magic
square in two (or more) different ways. Therefore, by having the restriction that all
sequences will start with the number one, we reduce the chances of this notation
accidentally describing the same magic square in more than one way.

Of course, there are millions of 5th order magic squares (no one yet knows for sure
exactly how many unique 5th order magic squares there are). But from our notation, it is
apparent that many magic squares can be grouped together according to characteristics
that they share. These magic squares can also be placed into larger groupings. For
example, all the 5th order magic squares that we've looked at and described up to this
point can be placed in the larger grouping of magic squares that have the numbers 11, 12,
13, 14 and 15 (not necessarily in that order) in the main diagonal running from the lower
left to the upper right corner.

The following is a description of five magic squares that would not be a part of this larger
grouping because they have different numbers in the main diagonals. Notice that the
position of the two with respect to the one is not next to the one in any way. To get to the
two (from the one) we move as a knight would in chess. Because of this we end up with
the numbers 1, 7, 13, 19, and 25 along one main diagonal and the numbers 3, 8, 13, 18,
and 23 along the other main diagonal instead of the numbers 11, 12, 13, 14, and 15.

3	24	20	11	7
12	8	4	25	16
21	17	13	9	5
10	1	22	18	14
19	15	6	2	23

P1 (1, 4) (2, 5) (3, 1) (4, 2) (5, 3)
P21 (rt 2, dn 1)
P65 (lf 2, dn 2)
Seq 1 2 3 4 5

Figure 5.10   A 5th order magic square that is unique from the other 5th order
magic squares that we've considered so far in this chapter.

A Note about the Notation

The description of the method as presented in this chapter will be referred to in later chapters as the construction method, or just the magic square description. As we saw earlier in this chapter, the description has four parts to it with the first three parts labeled P1, P21, and P65. Again, the P1 describes the position of the first number of the sequence (which will always be the number one). The P21 describes the position of the second number in relation to the first number (note that the second number won't always be the number 2). It is also describing the position of the third number in relation to the second number, the fourth number in relation to the third, etc. This also applies to the five numbers in each of the subsequent sets of five numbers.

The P65 describes the position of the first number of the next set of five in relation to the last number from the previous set of five. The P65 relation is specific for 5th order magic squares. The 7th order magic square will have a P87 relation. This describes the position of the eighth number in relation to the seventh number. Similarly, the 9th order magic squares have a P10 9 relation which describes the position of the tenth number in relation to the ninth number in the sequence. Finally, the sequence (Seq) only shows the first five numbers of the actual sequence for 5th order magic squares; it shows the first seven numbers for 7th order magic squares, etc. In almost all cases, this should be enough information to represent the entire sequence. Therefore, if we have the following description of a magic square:

P1 (1, 3)
P21 (rt 1, up 1)
P65 (dn 1)
Seq 1 3 5 2 4

The actual sequence for all 25 numbers of this magic square is:

1 3 5 2 4   6 8 10 7 9   11 13 15 12 14   16 18 20 17 19   21 23 25 22 24

For this sequence, the P1 label describes the location in the array of the first number of the entire sequence (which is still 1). The P21 label describes the position of the second number in relation to the first, which in this case describes the position of the 3 in relation to the 1. It also describes the position of the 5 in relation to the 3, the 2 in relation to the 5, and so on. The P65 label describes the position of the first number of the next set in relation to the last number of the previous set (which in this example is the position of the 6 in relation to the 4; as well as the position of the 11 in relation to the 9, and so on). This notation gives us enough information to build the magic square; and it also gives us enough information to compare similar magic squares.

A 5th Order Diabolic Magic Square

Before we leave this chapter, let's look at a 5th order diabolic magic square. This magic square is diabolical because not only is it a regular magic square where the rows, columns and main diagonals all add to 65, but so do the broken diagonals. And, in addition to that, there are still other patterns of five numbers that also add up to the magic square constant. Here is a 5th order diabolic magic square:

1	15	24	8	17
23	7	16	5	14
20	4	13	22	6
12	21	10	19	3
9	18	2	11	25

Figure 5.11   A "diabolic" 5th order magic square.

Using the notation we developed earlier in this chapter, we can describe the magic square shown above like this:

P1 (1, 1)
P21 (rt 2, up 1)
P65 (rt 1, dn 1)
Seq : 1 2 3 4 5

However, because it is diabolic, it is also pan-diagonal. This means that the one could actually be placed anywhere within the array. Therefore, we can show the P1 parameter of a 5th order pan-diagonal magic square like this:

P1 (1 .. 5, 1 .. 5)

The above notation would describe all 25 positions for the number 1, thus creating 25 unique magic squares. Keep in mind that these 25 unique magic squares are unique in that they are not rotations or mirrorings of each other even though, as our notation shows, they are definitely all related to each other.

The following is a magic carpet created from this diabolic magic square. Again, we can select any 5 x 5 cell within this magic carpet, and it will be a pan-diagonal magic square (as can be seen with the shaded areas within this large array):

1	15	24	8	17	1	15	24	8	17	1	15	24	8	17
23	7	16	5	14	23	7	16	5	14	23	7	16	5	14
20	4	13	22	6	20	4	13	22	6	20	4	13	22	6
12	21	10	19	3	12	21	10	19	3	12	21	10	19	3
9	18	2	11	25	9	18	2	11	25	9	18	2	11	25
1	15	24	8	17	1	15	24	8	17	1	15	24	8	17
23	7	16	5	14	23	7	16	5	14	23	7	16	5	14
20	4	13	22	6	20	4	13	22	6	20	4	13	22	6
12	21	10	19	3	12	21	10	19	3	12	21	10	19	3
9	18	2	11	25	9	18	2	11	25	9	18	2	11	25
1	15	24	8	17	1	15	24	8	17	1	15	24	8	17
23	7	16	5	14	23	7	16	5	14	23	7	16	5	14
20	4	13	22	6	20	4	13	22	6	20	4	13	22	6
12	21	10	19	3	12	21	10	19	3	12	21	10	19	3
9	18	2	11	25	9	18	2	11	25	9	18	2	11	25

Figure 5.12   A 15 x 15 magic carpet created by placing 5th order diabolic magic squares next to each other.

And finally let's look at some of the other patterns within this magic square that make it diabolical. As we can see below, there are many patterns where we have 5 numbers that add to 65. Many of these groups are arranged as an 'x' or as a '+'. For example, the numbers 1, 7, 13, 20 and 24 add to 65; as do the numbers 15, 7, 4, 23 and 16. We also find that the corner numbers with the center number (1, 17, 13, 9 and 25), as well as the middle side numbers with the center number (24, 20, 13, 6 and 2) also add to 65. Keep in mind that these groups of numbers are in addition to the rows, columns, main diagonals and broken diagonals that also all add to 65.

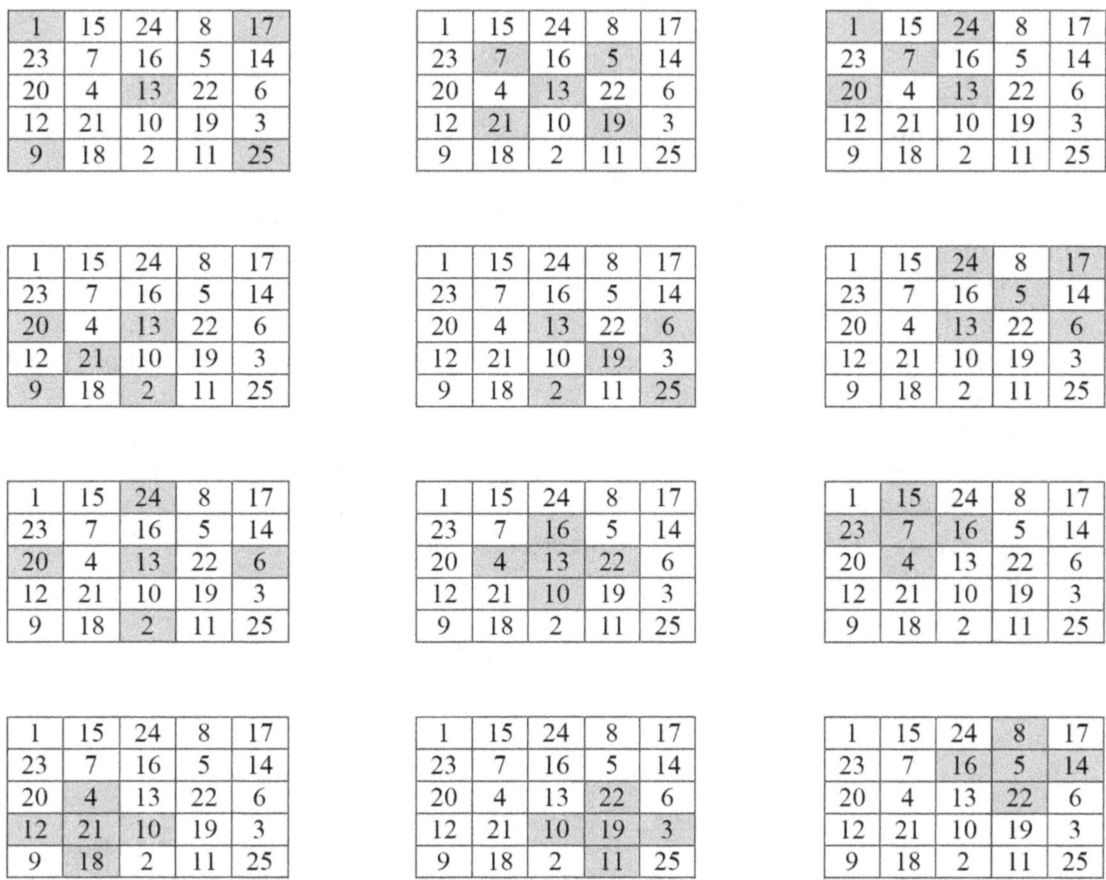

Figure 5.13   The shading in the copies of the magic square above show the various ways that this diabolic magic square adds up to 65 (which are in addition to the rows, columns, main diagonals and broken diagonals which also all add to 65).

Chapter 6
The 6th Order and Singly Even-Ordered Magic Squares

The 6th Order Magic Square

Life would be much easier if all even-order magic squares followed the same rules, but they don't. As mentioned earlier, even-order magic squares appear to be divided into two groups. The first are called the "double even" magic squares. These are magic squares that have an order that is a multiple of four (they are doubly even). The other group is singly even. They include the 6th, 10th, 14th and so on orders of magic squares.

The construction rules for the singly even magic squares are different than those for the doubly even magic squares. With the singly even it is possible to swap numbers diagonally through the middle or straight across through the middle (as we can with the doubly even magic squares), however, the singly even also require an additional "cross swap" to complete the magic square. This cross swap is not a symmetrical move within the magic square. This can be explained better by looking at an example. We'll start with our standard initial square:

1	2	3	4	5	6
7	8	9	10	11	12
13	14	15	16	17	18
19	20	21	22	23	24
25	26	27	28	29	30
31	32	33	34	35	36

Figure 6.1   An initial 6 x 6 array of numbers.

To create our first magic square in this chapter, most of our swaps will be "straight through" swaps. We'll swap a number with its corresponding number straight through the magic square. For example, the 2 is swapped with the 32, the 3 with the 33 and so on. (The shaded numbers are those that did not move and are still in their original position.)

1	35	33	4	32	6
30	8	28	27	11	7
13	23	22	21	14	18
24	17	16	15	20	19
12	26	9	10	29	25
31	2	3	34	5	36

$2 \leftrightarrow 32, \quad 5 \leftrightarrow 35, \quad 3 \leftrightarrow 33$, then $32 \leftrightarrow 35$
$7 \leftrightarrow 12, \quad 25 \leftrightarrow 30, \quad 19 \leftrightarrow 24$, then $30 \leftrightarrow 12$
$9 \leftrightarrow 27, \quad 10 \leftrightarrow 28$, then $27 \leftrightarrow 28$
$14 \leftrightarrow 17, \quad 20 \leftrightarrow 23$, then $17 \leftrightarrow 23$
$15 \leftrightarrow 22, \quad 16 \leftrightarrow 21$

Figure 6.2   A 6th order magic square along with its construction method.

Altogether there are twelve swaps involving 24 numbers; and within those 24 numbers, eight of them swap again among themselves (performing a cross swap). The construction method above (listed to the right of the magic square) has a "then" swap that is an additional cross swap that is performed after the previous swaps (in that row or column). The swaps follow a pattern, and while the pattern isn't perfectly symmetrical, it does have some symmetry.

Similarly, instead of swapping straight thru the array, we can also take our initial square and swap numbers diagonally thru the center of the square. Again, in this example we have twelve swaps involving 24 numbers followed by four more cross swaps involving eight of the 24-already-swapped numbers.

<table>
<tr><td>1</td><td>32</td><td>33</td><td>4</td><td>35</td><td>6</td></tr>
<tr><td>12</td><td>8</td><td>27</td><td>28</td><td>11</td><td>25</td></tr>
<tr><td>13</td><td>17</td><td>22</td><td>21</td><td>20</td><td>18</td></tr>
<tr><td>24</td><td>23</td><td>16</td><td>15</td><td>14</td><td>19</td></tr>
<tr><td>30</td><td>26</td><td>10</td><td>9</td><td>29</td><td>7</td></tr>
<tr><td>31</td><td>5</td><td>3</td><td>34</td><td>2</td><td>36</td></tr>
</table>

$2 \leftrightarrow 35, \quad 5 \leftrightarrow 32, \quad 3 \leftrightarrow 33,$ then $32 \leftrightarrow 35$
$7 \leftrightarrow 30, \quad 25 \leftrightarrow 12, \quad 19 \leftrightarrow 24,$ then $30 \leftrightarrow 12$
$9 \leftrightarrow 28, \quad 10 \leftrightarrow 27, \quad$ then $27 \leftrightarrow 28$
$14 \leftrightarrow 23, \quad 20 \leftrightarrow 17, \quad$ then $17 \leftrightarrow 23$
$15 \leftrightarrow 22, \quad 16 \leftrightarrow 21$

Figure 6.3   Another 6th order magic square along with its construction method.

The swaps after the "then" are the same for both magic squares shown above. I did that on purpose to help show the differences as well as the similarities between the two magic squares. In reality, each of the "then" swaps actually has two possibilities. Therefore, for the straight-through swaps, we can define our method as:

$2 \leftrightarrow 32, \quad 5 \leftrightarrow 35, \quad 3 \leftrightarrow 33,$ then $2 \leftrightarrow 5$ or $32 \leftrightarrow 35$
$7 \leftrightarrow 12, \quad 25 \leftrightarrow 30, \quad 19 \leftrightarrow 24,$ then $7 \leftrightarrow 25$ or $30 \leftrightarrow 12$
$9 \leftrightarrow 27, \quad 10 \leftrightarrow 28,$ then $9 \leftrightarrow 10$ or $27 \leftrightarrow 28$
$14 \leftrightarrow 17, \quad 20 \leftrightarrow 23,$ then $14 \leftrightarrow 20$ or $17 \leftrightarrow 23$
$15 \leftrightarrow 22, \quad 16 \leftrightarrow 21$

This method above lends itself nicely to be arranged in the following table.

2↔32	5↔35	3↔33	then	2↔5, or 32↔35
7↔12	25↔30	19↔24	then	7↔25, or 30↔12
9↔27	10↔28		then	9↔10, or 27↔28
14↔17	20↔23		then	14↔20, or 17↔23
15↔22	16↔21			

Table 6.1   A 6th order construction methods showing the number swaps.

The table makes it easier to look at multiple methods. The order of operation is such that we start from the left and go to the right across each row. Each row is independent of the others, so that we can complete any row in any order. The only constraint is that if the row has a "then" on it, the operations to the left of the "then" must be completed before the operations to the right of it.

The choice of one swap or the other to the right of the "then" doesn't affect the rest of the magic square; therefore, the method above will develop up to 12 unique magic squares. And because of the way the swaps are done, a pair of diagonal swaps with one "then" swap is equivalent to a pair of straight-through swaps with the other "then" swap. This means that the 12 possible diagonal combinations are equivalent to the 12 straight-through combinations.

In both examples above, we swapped 3 with 33, and 19 with 24. In each case, we could have left these numbers in their original places, and instead swapped 4 with 34, and swapped 13 with 18. (Or we could have created two more variations by swapping 3 with 33, and 13 with 18; or we could have swapped 4 with 34, and 19 with 24.) Here now is the straight-through swap algorithm with all of the alternate swap possibilities.

2↔32	5↔35	3↔33 or 4↔34	then	2↔5 or 32↔35
7↔12	25↔30	13↔18 or 19↔24	then	7↔25 or 30↔12
9↔27	10↔28		then	9↔10 or 27↔28
14↔17	20↔23		then	14↔20 or 17↔23
15↔22	16↔21			

Table 6.2   A more refined definition of several 6th order swap combinations.

As we can see in the example above, by writing our swap algorithms with the "then" and "or" expressions, we can more concisely list many more magic squares that are all related to each other within just a few lines.

Defining a Swap-by-Position Method
We'll take the straight-through swap method mentioned above, and instead of defining it by the numbers that are swapped, we will defined it by the positions that are swapped (as we did in chapter 4 with the 4th order magic squares). Then by using any one of several types of initial squares, we can create other unique magic squares. Any one of the "or" swaps can be performed without affecting the outcome of the magic square (other than the fact that doing so creates another unique magic square).

$(1, 2) \leftrightarrow (6, 2)$	$(1, 5) \leftrightarrow (6, 5)$	$(1, 3) \leftrightarrow (6, 3)$ or $(1, 4) \leftrightarrow (6, 4)$	then	$(1, 2) \leftrightarrow (1, 5)$ or $(6, 2) \leftrightarrow (6, 5)$
$(2, 1) \leftrightarrow (2, 6)$	$(5, 1) \leftrightarrow (5, 6)$	$(3, 1) \leftrightarrow (3, 6)$ or $(4, 1) \leftrightarrow (4, 6)$	then	$(2, 1) \leftrightarrow (5, 1)$ or $(2, 6) \leftrightarrow (5, 6)$
$(2, 3) \leftrightarrow (5, 3)$	$(2, 4) \leftrightarrow (5, 4)$		then	$(2, 3) \leftrightarrow (2, 4)$ or $(5, 3) \leftrightarrow (5, 4)$
$(3, 2) \leftrightarrow (3, 5)$	$(4, 2) \leftrightarrow (4, 5)$		then	$(3, 2) \leftrightarrow (4, 2)$ or $(3, 5) \leftrightarrow (4, 5)$
$(3, 3) \leftrightarrow (4, 4)$	$(3, 4) \leftrightarrow (4, 3)$			

Table 6.3   A swap-by-position definition of multiple 6th order swap methods.

By writing the algorithm in a table, we can more easily see a type of symmetry within the swap algorithm. The first two rows of the table swap the numbers on the top and bottom rows as well as the first and last columns of the magic square. There is a symmetry that can be seen between the row swaps and the column swaps. The same type of symmetry can be seen between the third and fourth rows of the table.

As we saw in chapter 4, this algorithm works with our initial square where the 1 is in the upper left corner, and the numbers are written sequentially in ascending order from left to right and from top to bottom. However, we can also apply this swap-by-position method with other initial squares to create additional unique magic squares.

Alternate Initial Squares
In a 6th order square we have more numbers to work with than we did with the 4 x 4 array; therefore, we can also generate more initial squares to work with. Here are two initial squares that can be used to create unique magic squares:

1	2	5	6	9	10
3	4	7	8	11	12
13	14	17	18	21	22
15	16	19	20	23	24
25	26	29	30	33	34
27	28	31	32	35	36

1	2	3	10	11	12
4	5	6	13	14	15
7	8	9	16	17	18
19	20	21	28	29	30
22	23	24	31	32	33
25	26	27	34	35	36

Figure 6.4   Two initial 6 x 6 arrays that can be used to create 6th order magic squares.

We'll use each initial square along with the method from above to create two unique magic squares. Of course the method actually has several variations, so we'll select one swap from each of the "or" options. This is the method that we will use:

(1, 2) ↔ (6, 2)	(1, 5) ↔ (6, 5)	(1, 3) ↔ (6, 3)	then	(1, 2) ↔ (1, 5)
(2, 1) ↔ (2, 6)	(5, 1) ↔ (5, 6)	(4, 1) ↔ (4, 6)	then	(2, 6) ↔ (5, 6)
(2, 3) ↔ (5, 3)	(2, 4) ↔ (5, 4)		then	(2, 3) ↔ (2, 4)
(3, 2) ↔ (3, 5)	(4, 2) ↔ (4, 5)		then	(3, 5) ↔ (4, 5)
(3, 3) ↔ (4, 4)	(3, 4) ↔ (4, 3)			

Table 6.4   A single swap-by-position definition to a 6th order magic square.

Here are the magic squares. The magic square below and on the left was created from the initial square above and on the left. Likewise the magic square below and on the right was created from the initial square above and on the right. Both magic squares were created using the swap-by-position method listed in the table above.

1	35	31	6	28	10
12	4	30	29	11	25
13	21	20	19	16	22
24	23	18	17	14	15
34	26	7	8	33	3
27	2	5	32	9	36

1	35	27	10	26	12
15	5	31	24	14	22
7	17	28	21	20	18
30	29	16	9	8	19
33	23	6	13	32	4
25	2	3	34	11	36

Figure 6.5   Two magic squares created form the initial squares above using the swap-by-position method listed above.

Of course other variations of our algorithm along with alternate initial squares can produce quite a few unique magic squares.

Defining a Swap-by-Number Method
We could have also taken the first magic square from this chapter and defined our swap method without the "then" statements. Instead, we would end up with four pairs of numbers that swap with each other, along with four other sets of four number swaps.

1	35	33	4	32	6
30	8	28	27	11	7
13	23	22	21	14	18
24	17	16	15	20	19
12	26	9	10	29	25
31	2	3	34	5	36

3 ↔ 33, 19 ↔ 24

2 → 32 → 5 → 35 → 2

7 → 12 → 25 → 30 → 7

9 → 27 → 10 → 28 → 9

14 → 17 → 20 → 23 → 14

15 ↔ 22, 16 ↔ 21

Figure 6.6   A previous 6th order magic square along with its construction description which includes four 4-way number swaps.

Note that the two-way arrows in the figure above show two numbers that trade places with each other (such as the 3 and the 33); while the one-way arrows show numbers that go to the location of the next number (like the 2 which goes to the position of the 32, which goes to the position of the 5, which goes to the position of the 35, which finally goes to the position of the 2 – thus creating a 4-way number swap).

The second magic square in this chapter was very similar to the first. When we looked at it we also used a couple of "then" statements to describe how the numbers moved from the initial square to their magic square location. Here is the second magic square (from this chapter) with its corresponding swap-by-number algorithm:

1	32	33	4	35	6
12	8	27	28	11	25
13	17	22	21	20	18
24	23	16	15	14	19
30	26	10	9	29	7
31	5	3	34	2	36

$3 \leftrightarrow 33, 19 \leftrightarrow 24$

$2 \rightarrow 35 \rightarrow 5 \rightarrow 32 \rightarrow 2$

$7 \rightarrow 30 \rightarrow 25 \rightarrow 12 \rightarrow 7$

$9 \rightarrow 28 \rightarrow 10 \rightarrow 27 \rightarrow 9$

$14 \rightarrow 23 \rightarrow 20 \rightarrow 17 \rightarrow 14$

$15 \leftrightarrow 22, 16 \leftrightarrow 21$

Figure 6.7   Another previous 6th order magic square along with its construction description which includes four 4-way number swaps.

While this swap-by-number method appears to be more-straight forward, it doesn't work reliably on other types of initial squares; and it isn't as concise when describing multiple magic squares, and how they relate to each other.

The 10th Order Magic Square

Writing the swap-by-position algorithm helped us see patterns within the 6th order magic square. We can apply the same idea, and use this as a pattern to develop a swap method for 10th order magic squares.

You will notice in the table below that within this method there are quite a few more "or" possibilities. Again, this means that we can build two identical magic squares except that in one of the magic squares we use one of the "or" options, and in the second we use the other "or" option; and then with that being the only difference, we then will have two magic squares that are almost identical and yet are very unique to each other.

I've had to alter our table to get the width of it to fit within the width of a single page. Each row of the table is still independent of the other rows; and the first column lists multiple swaps that must be performed first. Most of the remaining columns have "or" swaps which means that only one of the swaps needs to be performed, but any combination of "or" can be used within a single row of the table. And as before, the swaps to the left of the "then" must be done before the swaps to the right.

Here is a 10th order swap-by-position algorithm:

$(1, 2) \leftrightarrow (10, 2)$ $(1, 3) \leftrightarrow (10, 3)$ $(1, 8) \leftrightarrow (10, 8)$ $(1, 9) \leftrightarrow (10, 9)$ $(2, 1) \leftrightarrow (2, 10)$ $(3, 1) \leftrightarrow (3, 10)$ $(8, 1) \leftrightarrow (8, 10)$ $(9, 1) \leftrightarrow (9, 10)$	$(1, 4) \leftrightarrow (10, 4)$ or $(1, 5) \leftrightarrow (10, 5)$ or $(1, 6) \leftrightarrow (10, 6)$ or $(1, 7) \leftrightarrow (10, 7)$	then	$(1, 2) \leftrightarrow (1, 9)$ or $(10, 2) \leftrightarrow (10, 9)$ $(2, 1) \leftrightarrow (9, 1)$ or $(2, 10) \leftrightarrow (9, 10)$	$(1, 3) \leftrightarrow (1, 8)$ or $(10, 3) \leftrightarrow (10, 8)$

(3, 2) ↔ (3, 9) (4, 2) ↔ (4, 9) (7, 2) ↔ (7, 9) (8, 2) ↔ (8, 9) (2, 3) ↔ (9, 3) (2, 4) ↔ (9, 4) (2, 7) ↔ (9, 7) (2, 8) ↔ (9, 8)	(4, 1) ↔ (4, 10) or (5, 1) ↔ (5, 10) or (6, 1) ↔ (6, 10) or (7, 1) ↔ (7, 10)	then	(2, 3) ↔ (2, 8) or (9, 3) ↔ (9, 8) (2, 4) ↔ (2, 7) or (9, 4) ↔ (9, 7)	(3, 1) ↔ (8, 1) or (3, 10) ↔ (8, 10)
(3, 4) ↔ (8, 4) (3, 7) ↔ (8, 7) (4, 3) ↔ (4, 8) (7, 3) ↔ (7, 8) (5, 4) ↔ (5, 7) (6, 4) ↔ (6, 7)	(5, 2) ↔ (6, 2) or (5, 9) ↔ (6, 9) (5, 3) ↔ (5, 8) or (6, 3) ↔ (6, 8)	then	(3, 2) ↔ (8, 2) or (3, 9) ↔ (8, 9) (3, 4) ↔ (3, 7) (4, 3) ↔ (7, 3)	(4, 2) ↔ (7, 2) or (4, 9) ↔ (7, 9) (8, 4) ↔ (8, 7) (4, 8) ↔ (7, 8)
(4, 5) ↔ (7, 5) (4, 6) ↔ (7, 6) (5, 5) ↔ (6, 6) (5, 6) ↔ (6, 5)	(3, 5) ↔ (8, 5) or (3, 6) ↔ (8, 6)	then	(4, 5) ↔ (4, 6) (5, 4) ↔ (6, 4)	(7, 5) ↔ (7, 6) (5, 7) ↔ (6, 7)
	(1, 5) ↔ (1, 6)* or (2, 5) ↔ (2, 6) or (9, 5) ↔ (9, 6) or (10, 5) ↔ (10, 6)			

* Select a pair that hasn't moved yet.

Table 6.5   A swap-by-position definition for 10th order swap methods.

Using the algorithm above, I have built the following two magic squares. These two magic squares are identical to each other except in three places where I used a different "or" option. See if you can find the difference (Hint: I shaded the numbers that are in their original positions – which should help identify numbers that haven't moved).

1	99	98	4	95	6	7	93	92	10
90	12	88	87	16	15	84	83	19	11
80	79	23	77	75	26	74	28	22	21
31	69	68	34	66	65	37	63	32	40
41	52	43	57	56	55	54	48	49	50
60	42	58	47	46	45	44	53	59	51
61	39	38	64	36	35	67	33	62	70
30	29	73	27	25	76	24	78	72	71
20	82	13	14	85	86	17	18	89	81
91	2	3	94	5	96	97	8	9	100

1	99	98	4	95	6	7	93	92	10
90	12	88	87	16	15	84	83	19	11
80	79	23	77	75	26	74	28	22	21
31	69	68	34	66	65	37	63	32	40
50	52	48	57	56	55	54	43	49	41
51	42	53	47	46	45	44	58	59	60
61	39	38	64	36	35	67	33	62	70
30	29	73	27	25	76	24	78	72	71
20	82	13	14	85	86	17	18	89	81
91	2	3	94	5	96	97	8	9	100

Figure 6.8   Two 10th order magic squares created from the preceding table.

Just as with the 6th order magic squares, we can also develop a swap-by-number algorithm to create yet other unique magic squares. The following swap-by-number method is based off of the preceding swap-by-position method. Note that due to the asymmetrical nature of the singly even magic squares, this method is not as reliable with various types of initial squares, and it can just as easily produce a semi-magic square, or even a non-magic square.

2 ↔ 92 3 ↔ 93 8 ↔ 98 9 ↔ 99 11 ↔ 20 21 ↔ 30 71 ↔ 80 81 ↔ 90	4 ↔ 94 or 5 ↔ 95 or 6 ↔ 96 or 7 ↔ 97	then	92 ↔ 99 or 2 ↔ 9 20 ↔ 90 or 11 ↔ 81	93 ↔ 98 or 3 ↔ 8
13 ↔ 83 14 ↔ 84 17 ↔ 87 18 ↔ 88 22 ↔ 29 32 ↔ 39 62 ↔ 69 72 ↔ 79	31 ↔ 40 or 41 ↔ 50 or 51 ↔ 60 or 61 ↔ 70	then	13 ↔ 18 or 83 ↔ 88 84 ↔ 87 or 14 ↔ 17	30 ↔ 80 or 21 ↔ 71
24 ↔ 74 27 ↔ 77 33 ↔ 38 63 ↔ 68 44 ↔ 47 54 ↔ 57	42 ↔ 52 or 49 ↔ 59 43 ↔ 48 or 53 ↔ 58	then	29 ↔ 79 or 22 ↔ 72 74 ↔ 77 47 ↔ 57	39 ↔ 69 or 32 ↔ 62 24 ↔ 27 35 ↔ 36
35 ↔ 65 36 ↔ 66 45 ↔ 56 46 ↔ 55	25 ↔ 75 or 26 ↔ 76	then	38 ↔ 68 65 ↔ 66	33 ↔ 63 44 ↔ 54
	5 ↔ 6* or 15 ↔ 16 or 85 ↔ 86 or 95 ↔ 96			

* Select a pair that hasn't moved yet.

Table 6.6   A swap-by-number definition for 10th order swap methods.

As an example, we will use the table above on the initial square below to create a magic square. (To create this initial square, pairs of rows have been swapped: the first with the second, the third with the fourth, etc.)

11	12	13	14	15	16	17	18	19	20
1	2	3	4	5	6	7	8	9	10
31	32	33	34	35	36	37	38	39	40
21	22	23	24	25	26	27	28	29	30
51	52	53	54	55	56	57	58	59	60
41	42	43	44	45	46	47	48	49	50
71	72	73	74	75	76	77	78	79	80
61	62	63	64	65	66	67	68	69	70
91	92	93	94	95	96	97	98	99	100
81	82	83	84	85	86	87	88	89	90

Figure 6.9   An alternate initial 10 x 10 array of numbers.

The swap-by-number method was used to create the following semi-magic square. Again, because of the many "or" options, this swap-by-number method is not as reliable as the swap-by-position method. Here is the semi-magic square (all the rows and columns add up to the magic square constant, but the main diagonals don't):

90	12	83	87	15	16	84	88	19	11
1	99	98	94	6	5	7	93	92	10
40	69	68	34	66	65	37	63	32	31
80	79	23	77	75	26	74	28	22	21
51	42	53	47	46	45	44	58	59	60
41	52	48	57	56	55	54	43	49	50
30	29	73	27	25	76	24	78	72	71
61	39	38	64	36	35	67	33	62	70
91	2	3	4	95	96	97	8	9	100
20	82	18	14	85	86	17	13	89	81

Figure 6.10 A semi-magic square created from the initial square above and the
swap-by-number table listed previously.

The swap algorithms developed for these 6th order and 10th order magic squares can be extended to the 14th order and higher single even-ordered magic squares. And of course, there are other swap algorithms that can be developed. And as you might well suppose, we have barely even scratched the surface of possibilities.

Singly Even Magic Squares Can Be Broken into Two Parts
Because the singly even magic squares are not symmetrical like the doubly even magic squares, we can create magic squares where some cells exchange places symmetrically through the middle of the square while other cells must be exchanged in some non-symmetrical way. In spite of this, we can still see a few general patterns with the magic squares that we've created. For example, the next two magic squares have the same numbers in the shaded areas of the array. The differences between the magic squares can be found in the non-shaded areas. In fact, there are dozens of similar magic squares that can be created simply by rearranging the values in the non-shaded area.

65	67	93	95	4	1	32	30	60	58
66	68	94	96	2	3	31	29	59	57
89	91	17	19	27	26	56	54	64	62
90	92	18	20	25	28	55	53	63	61
15	13	24	23	49	50	79	80	85	87
14	16	21	22	51	52	78	77	88	86
40	38	48	46	73	76	81	83	9	11
39	37	47	45	75	74	82	84	10	12
44	42	72	70	99	98	5	7	33	35
43	41	71	69	100	97	6	8	34	36

Figure 6.11 A magic square with the shaded areas showing the two different construction areas of a 10th order (singly even) magic square.

65	67	93	95	3	2	32	30	60	58
66	68	94	96	4	1	31	29	59	57
89	91	17	19	27	26	56	54	64	62
90	92	18	20	25	28	55	53	63	61
15	16	23	21	49	50	77	79	88	87
14	13	22	24	51	52	80	78	85	86
40	38	48	46	73	76	81	83	9	11
39	37	47	45	75	74	82	84	10	12
44	42	72	70	100	97	5	7	33	35
43	41	71	69	98	99	6	8	34	36

Figure 6.12 A magic square similar to the previous magic square. The shaded areas are identical between the two squares with the only differences between them in the white areas.

In addition to this, we can leave the un-shaded areas as they are, and rearrange the numbers within the shaded areas to create still more unique magic squares. We can then go even further by keeping the two groups distinct, but still rearrange the numbers in both the shaded and un-shaded areas to create even more magic squares that are unique from each other.

All the singly even magic squares share this characteristic. They can all be broken into two parts: the corner cells that swap one way, and the center two rows and two columns that swap another way. The following figure shows this division within the 6th and 10th order magic squares, and it can easily be extended to higher order singly even magic squares.

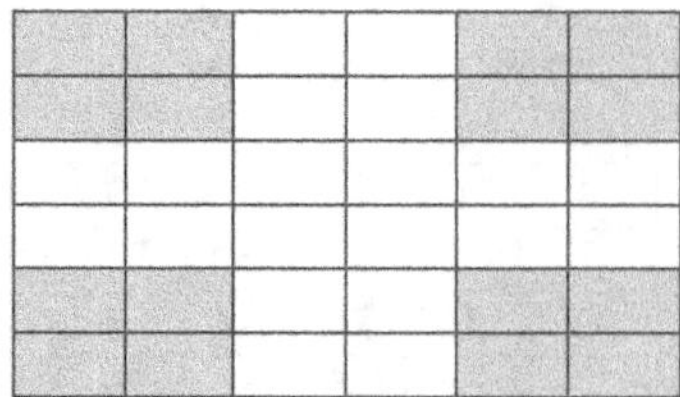

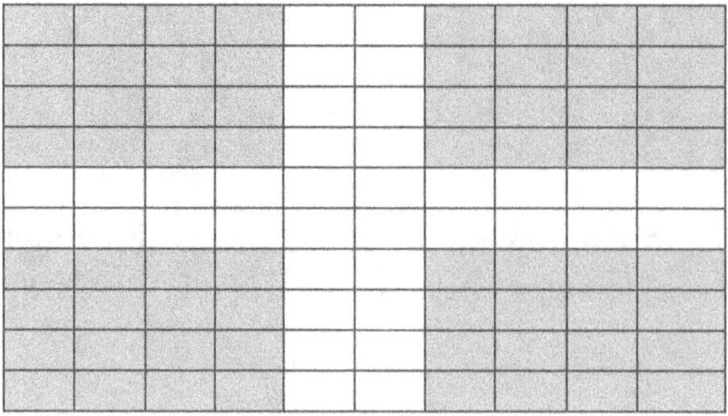

Figure 6.12 A 6th order and a 10th order example of the two parts of singly even magic squares.

Of course there are still other methods of creating singly even magic squares, and we will see one more in this chapter, and then a few more methods in later chapters.

An Alternate Method of Showing a Method

The following is known as the Devedec Method. It is a graphical method of showing the position swaps needed to create a series of even-ordered magic squares. The disadvantage of this method is that it only shows one method per order; however, the advantage is that it shows the similarity between multiple orders very clearly. The figure on the following page shows a method of creating magic squares for the fourth order through the 20th order. An added advantage to this method is that once a pattern is established like this, the pattern can easily be extended to orders much higher than those shown.

In the figure below, the positions marked with an O are those that do not swap (they stay in their original position). The diagonal line represents positions that swap diagonally through the center of the square. Vertical lines swap vertically through the square (from top to bottom, and from bottom to top); and horizontal lines swap horizontally through the square (from right to left, and from left to right).

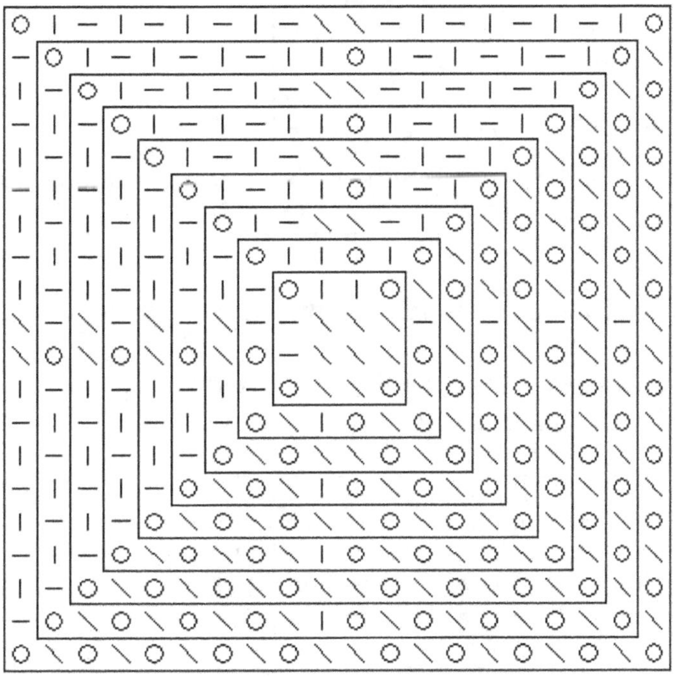

Figure 6.13   An alternate method of creating even-ordered magic squares.

Keep in mind that the figure above isn't a magic square, but it is an algorithm or method of creating magic squares and since this algorithm isn't symmetrical, we can also use rotations or mirrorings of this algorithm and in this way create other unique magic squares. To see exactly how this method works, we'll use the center portion of the figure above to create a 6th order magic square. First, we start with our initial square:

1	2	3	4	5	6
7	8	9	10	11	12
13	14	15	16	17	18
19	20	21	22	23	24
25	26	27	28	29	30
31	32	33	34	35	36

Figure 6.14   The same initial square we saw at the beginning of this chapter.

In the initial square above, we list the numbers in ascending order starting with 1 in the upper left corner, and move from left to right and from top to bottom, end with 36 in the lower right corner. Then we take the center 6 x 6 array from the Devedec method, and use it as our template on the initial square. Here is the center 6 x 6 array:

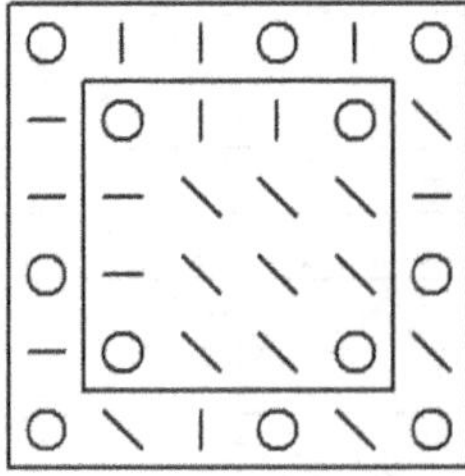

Figure 6.15   The center 6 x 6 array from the Devedec method.

Now we apply the position swaps from the figure above on to the initial square, and we create the following 6th order magic square:

1	35	33	4	32	6
30	8	28	27	11	7
18	23	22	21	14	13
19	17	16	15	20	24
12	26	9	10	29	25
31	2	3	34	5	36

Figure 6.16   The resulting magic square which was created by using the Devedec method on our initial square.

Even though the figure only shows one method of magic square per magic square order, we can still start with different initial squares, apply the Devedec method, and then create other unique magic squares. For example, here are two initial squares that we saw earlier in this chapter:

1	2	5	6	9	10
3	4	7	8	11	12
13	14	17	18	21	22
15	16	19	20	23	24
25	26	29	30	33	34
27	28	31	32	35	36

1	2	3	10	11	12
4	5	6	13	14	15
7	8	9	16	17	18
19	20	21	28	29	30
22	23	24	31	32	33
25	26	27	34	35	36

Figure 6.17   Two initial squares that we saw earlier in the chapter.

Then, after applying the Devedec method on each initial square, we create the following two magic squares. (Note that the initial square on the left above corresponds with the magic square on the left below; and the initial square on the right above corresponds with the magic square on the right below.)

1	35	31	6	28	10
34	4	30	29	11	3
22	23	20	19	14	13
15	21	18	17	16	24
12	26	7	8	33	25
27	2	5	32	9	36

1	35	27	10	26	12
33	5	31	24	14	4
18	29	28	21	8	7
19	17	16	9	20	30
15	23	6	13	32	22
25	2	3	34	11	36

Figure 6.18   The resulting magic squares after using the Devedec method.

Keep in mind that the Devedec method that we looked at isn't unique. We could use it as a pattern to create other similar diagrams that could then be used to create other magic squares. Of course, the possibilities start to become limitless.

Finally, the methods that we've seen in this chapter aren't the only ways to create and show singly even-order magic squares. In a later chapter where we discuss magic square templates, we will see that we can use these templates as yet another way to build singly even-order magic squares.

Chapter 7
The 7th Order and Odd-Ordered Constructions

The 7th Order Magic Square

In this chapter we're going to take a closer look at several construction patterns that work on odd-ordered magic squares. First, here is a 5th order magic square that we saw in an earlier chapter:

17	24	1	8	15
23	5	7	14	16
4	6	13	20	22
10	12	19	21	3
11	18	25	2	9

P1 (1, 3)
P21 (rt 1, up 1)
P65 (dn 1)
Seq 1 2 3 4 5

Figure 7.1   A 5th order magic square along with its construction method. Note that the one is in the middle of the top row.

A similar construction method can be used on a 7th order magic square.

30	39	48	1	10	19	28
38	47	7	9	18	27	29
46	6	8	17	26	35	37
5	14	16	25	34	36	45
13	15	24	33	42	44	4
21	23	32	41	43	3	12
22	31	40	49	2	11	20

P1 (1, 4)
P21 (rt 1, up 1)
P87 (dn 1)
Seq 1 2 3 4 5 6 7

Figure 7.2   A 7th order magic square along with its construction method. Again, the one is in the middle of the top row.

For this seventh order magic square, the one is placed in the middle of the top row. The position of the two in relation to the one is the same for this magic square as it was for the fifth order magic square. Likewise, the position of the eight in relation to the seven in this magic square is the same as the position of the six in relation to the five in the previous magic square. Finally, the sequence in each case is the same: they start with the lowest number and simply ascend in order to the highest number.

But why stop here? The same thing can be done with a 9th order magic square. The following is a 9th order magic square, with the one in the middle of the top row, and the rest of the magic square is created using the same construction technique as the previous two magic squares just shown above:

47	58	69	80	1	12	23	34	45
57	68	79	9	11	22	33	44	46
67	78	8	10	21	32	43	54	56
77	7	18	20	31	42	53	55	66
6	17	19	30	41	52	63	65	76
16	27	29	40	51	62	64	75	5
26	28	39	50	61	72	74	4	15
36	38	49	60	71	73	3	14	25
37	48	59	70	81	2	13	24	35

P1 (1, 5)
P21 (rt 1, up 1)
P10 9 (dn 1)
Seq 1 2 3 4 5 6 7 8 9

Figure 7.3   A 9th order magic square along with its construction method.

It appears that any odd-sided magic square can be constructed in a similar manner. In fact, we can even go back to the 3rd order magic square, and find that it follows this same pattern.

Now let's look at a variation of this construction for the 5th, 7th and 9th order magic squares. First, the 5th order magic square:

23	6	19	2	15
10	18	1	14	22
17	5	13	21	9
4	12	25	8	16
11	24	7	20	3

P1 (2, 3)
P21 (rt 1, up 1)
P65 (up 2)
Seq 1 2 3 4 5

Figure 7.4   A 5th order magic square along with its construction method. Note that the one is placed just one cell above the center number.

The construction technique used this time is a little different. In the figure above, the 1 is placed one cell above the center cell. The position of the two in relation to the one is the same as before. However, this time the six is placed two cells above the five. Now let's compare that with the 7th order square of numbers (as shown below).

46	15	40	9	34	3	28
21	39	8	33	2	27	45
38	14	32	1	26	44	20
13	31	7	25	43	19	37
30	6	24	49	18	36	12
5	23	48	17	42	11	29
22	47	16	41	10	35	4

P1 (3, 4)
P21 (rt 1, up 1)
P87 (up 2)
Seq 1 2 3 4 5 6 7

Figure 7.5   A 7th order magic square along with its construction method. Again, the one is placed just one cell above the center number.

Again, the 1 is placed one cell above the center cell. And the position of the two in relation to the one is the same as the previous 5th order square (above), as is the eight with respect to the seven. And, we can also see the same pattern in the 9th order magic square (below):

77	28	69	20	61	12	53	4	45
36	68	19	60	11	52	3	44	76
67	27	59	10	51	2	43	75	35
26	58	18	50	1	42	74	34	66
57	17	49	9	41	73	33	65	25
16	48	8	40	81	32	64	24	56
47	7	39	80	31	72	23	55	15
6	38	79	30	71	22	63	14	46
37	78	29	70	21	62	13	54	5

P1 (4, 5)
P21 (rt 1, up 1)
P10 9 (up 2)
Seq 1 2 3 4 5 6 7 8 9

Figure 7.6   A 9th order magic square along with its construction method. Once again, the one is placed just one cell above the center number.

This method of starting with the one just above the center cell also appears to work with any odd-ordered magic square. And in fact, if we go back and look at the 3rd order magic square, we again find that it also follows this same pattern.

Both methods mentioned above appear to work with any odd-ordered magic square. This means that we can use these two methods on 11th order, 13th order and even higher odd-ordered squares to create magic squares. And in addition to this, by changing the sequence of numbers used, (like we did in chapter 5 with the 5th order magic squares) we are able to create hundreds more magic squares using just these methods.

More Odd-Ordered Magic Square Patterns

There are several more similar methods that appear to transcend the order of the magic square, and work on any odd-sided square of numbers. And, there are some methods that only work on certain odd-orders of squares, but don't appear to work on other odd-orders of squares. Let's look at some of these. We'll begin again with a 5th order magic square.

1	20	9	23	12
19	8	22	11	5
7	21	15	4	18
25	14	3	17	6
13	2	16	10	24

P1 (1, 1)
P21 (rt 1, up 1)
P65 (dn 2)
Seq 1 2 3 4 5

Figure 7.7   A 5th order magic square along with its construction method. Note that the one is placed in the upper left corner cell.

We can use the same construction on a 7th order magic square.

1	34	11	37	21	47	24
33	10	36	20	46	23	7
9	42	19	45	22	6	32
41	18	44	28	5	31	8
17	43	27	4	30	14	40
49	26	3	29	13	39	16
25	2	35	12	38	15	48

P1 (1, 1)
P21 (rt 1, up 1)
P87 (dn 2)
Seq 1 2 3 4 5 6 7

Figure 7.8   A 7th order magic square along with its construction method. Again, the one is placed in the upper left corner cell.

With both the 5th order and the 7th order magic squares, we start with a 1 in the upper left corner cell. The position of the 2 with respect to the 1, as well as the position of the first number of the next sequence in relation to the last number of the previous sequence are the same for both magic squares.

Of course, if we look back at the 3rd order magic square, this method doesn't work (because there is only one possible 3rd order magic square, and it doesn't have the 1 in a corner); and interestingly enough, this method doesn't work on a 9th order or a 15th order magic square; but it does work on an 11th and a 13th order magic square.

Still, another similar construct can be used if the one is placed one column to the left of the center column on the top row. We'll compare the 5th order and the 7th order magic squares together.

7	1	25	19	13
5	24	18	12	6
23	17	11	10	4
16	15	9	3	22
14	8	2	21	20

17	9	1	49	41	33	25
8	7	48	40	32	24	16
6	47	39	31	23	15	14
46	38	30	22	21	13	5
37	29	28	20	12	4	45
35	27	19	11	3	44	36
26	18	10	2	43	42	34

Figure 7.9   A 5th order and a 7th order magic square. This is still another construction for most odd sided magic squares.

In this method, each sequence is completed as before; however, the next sequence begins one column to the left of the last number in the previous sequence. The next sequence is then completed, and again, the following sequence begins one column to the left of the last number of the previous sequence, and so on.

It is interesting to note that if this method is used on a 9th order square, the result is only a semi-magic square (the rows, columns and only one diagonal add up to 369). Then if this method is tried on an 11th or 13th order square, they are again regular magic squares. And then if this method tried on a 15th order square, it is again a semi-magic square (where the rows, columns and only one diagonal add up to 1695)

Seventh Order and Sequences to Higher Orders

With the fifth order magic square, we saw that there were several unique sequences that could be used with the various methods that were considered. The same is true with the higher odd-order magic squares; and as the order increases, so do the number of differing sequences that will work to produce a magic square.

However, as we saw in the previous section, there are several 9th order and 15th order magic squares that appear to take exception to many of these construction rules that work universally with 5th, 7th, 11th and 13th order magic squares.

Another example of this exception can be seen when we start with a one in the middle of the top row, and use a regular ascending sequence, except that we swap the last two numbers of each set within the sequence. Here is the seventh order magic square which does work.

30	39	49	1	10	19	27
38	47	6	9	18	28	29
46	7	8	17	26	34	37
5	13	16	25	35	36	45
14	15	24	33	41	44	4
20	23	32	42	43	3	12
22	31	40	48	2	11	21

P1 (1, 4)
P21 (rt 1, up 1)
P87 (dn 1)
Seq 1 2 3 4 5 7 6

Figure 7.10   A 7th order magic square along with its construction method.

The magic square above is identical to the magic square in figure 7.2, except that within the sequence, the last two numbers within each set of seven numbers have traded places. Now, if we do the same thing with a ninth order square, it creates a semi-magic square (where only one of the diagonals adds to 369).

47	58	69	81	1	12	23	34	44
57	68	79	8	11	22	33	45	46
67	78	9	10	21	32	43	53	56
77	7	17	20	31	42	54	55	66
6	18	19	30	41	52	62	65	76
16	26	29	40	51	63	64	75	5
27	28	39	50	61	71	74	4	15
35	38	49	60	72	73	3	14	25
37	48	59	70	80	2	13	24	36

P1 (1, 5)
P21 (rt 1, up 1)
P10 9 (dn 1)
Seq 1 2 3 4 5 6 7 9 8

Figure 7.11   A 9th order semi-magic square.

If we return to a seventh order square, and consider the example where the one was placed just above the center number and again swap the last two numbers of each sequence, we find that we have another 7th order magic square.

46	15	40	9	35	3	27
20	39	8	33	2	28	45
38	13	32	1	26	44	21
14	31	6	25	43	19	37
30	7	24	48	18	36	12
5	23	49	17	41	11	29
22	47	16	42	10	34	4

P1 (3, 4)
P21 (rt 1, up 1)
P87 (up 2)
Seq 1 2 3 4 5 7 6

Figure 7.12   A 7th order magic square along with its construction method.

The magic square above is identical to the magic square in figure 7.5, except within each sequence of seven, the last two numbers have traded places. Now, if we try the same thing on a 9th order square, we find that this time we also have a magic square.

77	28	69	20	61	12	54	4	44
35	68	19	60	11	52	3	45	76
67	26	59	10	51	2	43	75	36
27	58	17	50	1	42	74	34	66
57	18	49	8	41	73	33	65	25
16	48	9	40	80	32	64	24	56
47	7	39	81	31	71	23	55	15
6	38	79	30	72	22	62	14	46
37	78	29	70	21	63	13	53	5

P1 (4, 5)
P21 (rt 1, up 1)
P10 9 (up 2)
Seq 1 2 3 4 5 6 7 9 8

Figure 7.13   A 9th order magic square along with its construction method.

This time the magic square above is identical to the magic square in figure 7.6, except within the sequence, the last two numbers of each set of nine have traded places.

When trying various sequences, it appears that ninth order magic squares seem to be the most unforgiving. I have also found similar problems with 15th order magic squares, while the 5th, 7th, 11th and 13th orders appear to be the most workable. Why is that? It could be that there are two types of odd-ordered magic squares, just like there are two types of even-ordered magic squares. It appears that odd-ordered magic squares where the order is a prime number are more flexible than their composite-number-odd-ordered counterparts when altering the number sequence.

However, the shoe is on the other foot (so to speak) when looking at composite-number-odd-ordered magic squares. In chapter 10 we will show how to use templates to build larger magic squares from smaller magic squares. There we will see that it is easy to create composite-number-odd-ordered magic squares from various magic square templates (and these methods won't work on prime-number-ordered magic squares).

Other Sequences
As the order becomes larger, it is possible to try more variations of sequences. The following is a partial listing of 7th order and 9th order sequence variations. Seventh order squares appear to use these various sequences without discrimination; however, depending on the starting point and the method used for construction, only some of these sequence variations will work with 9th order squares to create unique magic squares.

7th Order Sequences	9th Order Sequences
1 2 3 4 5 6 7	1 2 3 4 5 6 7 8 9
1 2 3 4 5 7 6	1 2 3 4 5 6 7 9 8
1 2 3 4 6 5 7	1 2 3 4 5 6 8 7 9
1 2 3 4 6 7 5	1 2 3 4 5 6 8 9 7
1 2 3 4 7 5 6	1 2 3 4 5 6 9 7 8
1 2 3 4 7 6 5	1 2 3 4 5 6 9 8 7
1 2 3 5 4 6 7	1 2 3 4 5 7 6 8 9
1 2 3 5 4 7 6	1 2 3 4 5 7 6 9 8
1 2 3 5 6 4 7	1 2 3 4 5 7 8 6 9
1 2 3 5 6 7 4	1 2 3 4 5 7 8 9 6
1 2 3 5 7 4 6	1 2 3 4 5 7 9 6 8
1 2 3 5 7 6 4	1 2 3 4 5 7 9 8 6
. . ., etc.	. . ., etc.

Mapping Start Positions

Up until now we have used an abbreviated notation that is very concise to describe the construction method of odd-ordered magic squares. As mentioned earlier, this notation looks something like this:

P1 (1, 1)
P21 (rt 1, up 1)
P65 (dn 2)
Seq 1 2 3 4 5

The first item describes the position of the first number of the sequence (which is the number 1). We use regular matrix notation where the row is followed by the column to show the position of the number 1 in the square array. Therefore, P1 (1, 1) means that the position of the number 1 is located on row 1, column 1. The second item is the position of the second number in relation to the first number. In the example above, we describe this placement as P21 (rt 1, up 1) which means that the 2 is right one cell and up one cell from the 1, and the 3 is right one and up one from the 2, etc.

The third item is the placement of the first number of the next set of five numbers in relation to the last number from the previous set of five numbers (which in this case is the 6 in relation to the 5). The item P65 (dn 2) means that the 6 is down two cells from the 5. Finally, the last item in the notation above is the order of the sequence. And, as we have seen earlier, there are many combinations of sequences that we can try.

With this background, we can now create a mapping of several magic squares that is even more concise. Let's use the 5th order magic square as an example.

A	CH	B	DE	
CH	B	DEF		A
B	DE		AG	CH
DE		A	CH	B
	A	CH	B	DE

P21 (rt 1, up 1)
Seq 1 2 3 4 5

A: P1 (see chart)   **B**: P1 (see chart)   **C**: P1 (see chart)   **D**: P1 (see chart)
   P65 (dn 2)   P65 (dn 1)   P65 (lf 1)   P65 (rt 2, dn 1)

E: P1 (see chart)   **F**: P1 (see chart)   **G**: P1 (see chart)   **H**: P1 (see chart)
   P65 (lf 2)   P65 (up 2)   P65 (rt 2)   P65 (rt 2, dn 2)

Figure 7.14   A mapping of start positions for 5th order magic squares.

In the mapping above, we show 32 unique 5th order magic squares. All 32 of these magic squares use the same relation in describing the position of the 2 in relation to the 1 (the P21 value – which in this case is right 1, and up 1). Likewise, they all use the same sequence of 1 2 3 4 5. Therefore, we list these two items right under the array. The array itself shows the starting positions. (The starting position is where the 1 is placed.) For example, there are five A start positions. They are located at (1, 1), (5, 2), (4, 3), (3, 4) and (2, 5). Likewise, there are five starting positions for B, C, D, E and H. There is only one start position for the F and G magic squares. All the A magic squares are related to each other; as are the B's, C's, D's, E's and H's respectively. Yet none of these 32 magic squares are rotations or mirrorings of each other, thus making them unique.

Besides the starting position of the number 1, the only difference between each of these magic squares is the position of the 6 in relation to the 5. For the A magic squares, the P65 is described as down 2, which means that the 6 is placed two cells below the 5. Similarly, the P65 for the B magic squares is down one cell. All of the P65 positions are listed under their respective letters under the array.

This mapping becomes very interesting when we compare similar maps of differing orders. For example, the next grid shows a mapping of 7th order magic squares.

In the mapping below there are 44 unique 7th order magic squares. It is interesting to note that with the 5th order mapping, the C and H algorithms and the D and E algorithms shared the same cells, and now in the 7th order mapping they each have their own cells.

Again, all the magic squares that are represented by the same letter are related to each other; and even though they are related, still none are rotations or mirrorings of each other, thus making them unique.

A	H	C	B	D	E	
H	C	B	D	E		A
C	B	D	EF		A	H
B	D	E		AG	H	C
D	E		A	H	C	B
E		A	H	C	B	D
	A	H	C	B	D	E

P21 (rt 1, up 1)
Seq 1 2 3 4 5 6 7

A: P1 (see chart)   **B**: P1 (see chart)   **C**: P1 (see chart)   **D**: P1 (see chart)
    P87 (dn 2)   P87 (dn 1)   P87 (lf 1)   P87 (rt 2, dn 1)

E: P1 (see chart)   **F**: P1 (see chart)   **G**: P1 (see chart)   **H**: P1 (see chart)
    P87 (lf 2)   P87 (up 2)   P87 (rt 2)   P87 (rt 2, dn 2)

Figure 7.15   A mapping of start positions for 7th order magic squares.

Altogether, the mapping above represents 44 magic squares. As with the 5th order mapping, all of these 7th order magic squares share the same P21 value and the same number sequence with each other. Therefore these two items are listed right under the mapping grid. The start positions (the P1 positions) are represented by the letters, and the only difference between the various magic squares is the P87 parameter (the position of the 8 in relation to the 7).

I arranged the two mappings so that the P87 value for the A magic squares in this mapping are the same as the P65 value for the A magic squares in the 5th order mapping. Likewise, the P87 and the P65 value is the same for the B magic squares in both mappings (as well as all of the other letters shown here). It is interesting now to compare the C, D, E and H positions on the two maps (as well as the F and G positions).

And speaking of the F and G positions, both the 5th order mapping and the 7th order mapping only have one position for each (whereas these same mappings have multiple positions for A, B, C, etc.). In a later chapter we will see that it is possible to create pan-diagonal magic squares from magic squares created from the F and G positions.

Additional Start Position Mappings

Since the 7th order magic square has two more rows and columns than a 5th order magic square, there are several more mapping possibilities. This due to that fact that with seven rows and seven columns, it is possible to move up to three places to the left, right, up and/or down. Therefore, a more extended map of 7th order magic squares looks like this:

A	HKL	C	B	DJM	E	
HKL	C	B	DJM	E		A
C	B	DJM	EF		A	HKL
B	DJM	E		AG	HKL	C
DJM	E		A	HKL	C	B
E		A	HKL	C	B	DJM
	A	HKL	C	B	DJM	E

P21 (rt 1, up 1)
Seq 1 2 3 4 5 6 7

A: P87 (dn 2)	**B**: P87 (dn 1)	**C**: P87 (lf 1)	**D**: P87 (rt 2, dn 1)
E: P87 (lf 2)	**F**: P87 (up 2)	**G**: P87 (rt 2)	**H**: P87 (rt 2, dn 2)
J: P87 (dn 3)	**K**: P87 (lf 3)	**L**: P87 (up 3)	**M**: P87 (rt 3)

Figure 7.16   An extended mapping of start positions for 7th order magic squares.

The 7th order mapping above shows the parameters needed to create a total of 72 unique magic squares. The mapping also conveniently works for higher order magic squares. For example, the following mapping on the next page shows the parameters needed to create a total of 200 unique 11th order magic squares.

Notice also that the letters themselves show the P1 positions within the table; therefore, we don't need to list them again separately below the table.

A	HNR	TU	KL	C	B	DJM	SV	PQ	E	
HNR	TU	KL	C	B	DJM	SV	PQ	E		A
TU	KL	C	B	DJM	SV	PQ	E		A	HNR
KL	C	B	DJM	SV	PQ	E		A	HNR	TU
C	B	DJM	SV	PQ	EF		A	HNR	TU	KL
B	DJM	SV	PQ	E		AG	HNR	TU	KL	C
DJM	SV	PQ	E		A	HNR	TU	KL	C	B
SV	PQ	E		A	HNR	TU	KL	C	B	DJM
PQ	E		A	HNR	TU	KL	C	B	DJM	SV
E		A	HNR	TU	KL	C	B	DJM	SV	PQ
	A	HNR	TU	KL	C	B	DJM	SV	PQ	E

P21 (rt 1, up 1)
Seq 1 2 3 4 5 6 7 8 9 10 11

A: P12 11 (dn 2)	**B**: P12 11 (dn 1)	**C**: P12 11 (lf 1)	**D**: P12 11 (rt 2, dn 1)
E: P12 11 (lf 2)	**F**: P12 11 (up 2)	**G**: P12 11 (rt 2)	**H**: P12 11 (rt 2, dn 2)
J: P12 11 (dn 3)	**K**: P12 11 (lf 3)	**L**: P12 11 (up 3)	**M**: P12 11 (rt 3)
N: P12 11 (dn 4)	**P**: P12 11 (lf 4)	**Q**: P12 11 (up 4)	**R**: P12 11 (rt 4)
S: P12 11 (dn 5)	**T**: P12 11 (lf 5)	**U**: P12 11 (up 5)	**V**: P12 11 (rt 5)

Figure 7.17 An extended mapping of start positions for 11th order magic squares.

And this next mapping gives us the information needed to create a total of 288 unique 13th order magic squares.

A	HNR	WZ	TU	KL	C	B	DJM	SV	XY	PQ	E	
HNR	WZ	TU	KL	C	B	DJM	SV	XY	PQ	E		A
WZ	TU	KL	C	B	DJM	SV	XY	PQ	E		A	HNR
TU	KL	C	B	DJM	SV	XY	PQ	E		A	HNR	WZ
KL	C	B	DJM	SV	XY	PQ	E		A	HNR	WZ	TU
C	B	DJM	SV	XY	PQ	EF		A	HNR	WZ	TU	KL
B	DJM	SV	XY	PQ	E		AG	HNR	WZ	TU	KL	C
DJM	SV	XY	PQ	E		A	HNR	WZ	TU	KL	C	B
SV	XY	PQ	E		A	HNR	WZ	TU	KL	C	B	DJM
XY	PQ	E		A	HNR	WZ	TU	KL	C	B	DJM	SV
PQ	E		A	HNR	WZ	TU	KL	C	B	DJM	SV	XY
E		A	HNR	WZ	TU	KL	C	B	DJM	SV	XY	PQ
	A	HNR	WZ	TU	KL	C	B	DJM	SV	XY	PQ	E

P21 (rt 1, up 1)
Seq 1 2 3 4 5 6 7 8 9 10 11 12 13

A: P14 13 (dn 2)	**B**: P14 13 (dn 1)	**C**: P14 13 (lf 1)	**D**: P14 13 (rt 2, dn 1)
E: P14 13 (lf 2)	**F**: P14 13 (up 2)	**G**: P14 13 (rt 2)	**H**: P14 13 (rt 2, dn 2)
J: P14 13 (dn 3)	**K**: P14 13 (lf 3)	**L**: P14 13 (up 3)	**M**: P14 13 (rt 3)
N: P14 13 (dn 4)	**P**: P14 13 (lf 4)	**Q**: P14 13 (up 4)	**R**: P14 13 (rt 4)
S: P14 13 (dn 5)	**T**: P14 13 (lf 5)	**U**: P14 13 (up 5)	**V**: P14 13 (rt 5)
W: P14 13 (dn 6)	**X**: P14 13 (lf 6)	**Y**: P14 13 (up 6)	**Z**: P14 13 (rt 6)

Figure 7.18 An extended mapping of start positions for 13th order magic squares.

This mapping method can easily show how many construction methods are similar between several odd-orders of magic squares (as we can see with the 5th, 7th, 11th, and 13th order magic squares). It can also make deficiencies and differences become blaringly obvious. Upon examining the mappings of the 9th and 15th orders of magic squares we see that the same abundance does not exist for these orders as it does for the previous odd-orders that we have just considered.

Mapping Start Positions for 9th and 15th Order Magic Squares

For the 9th order magic square, I found that even though the A method was tested in all 81 cells, it didn't work in any of them. The B method was then tested in all 81 cells, and in this case it was found to work in only three of them. The same is true for the C, N, and P methods. The F and G methods only work once each (as they did with the previous odd-orders that we considered).

Where the algorithms are crossed out in the figures below, these methods don't work at all. In the case of the D, J, K, L, and M algorithms, as we build the pattern, we find that the pattern interferes with itself. (For example, when you go down three rows, you find that after a couple of iterations you are trying to write numbers back on top of other numbers while there are still other cells that remain blank.) With the A, E, H, Q, and R methods, we can build a complete 9 x 9 array (and fill all 81 cells); however, we find that there are rows and/or columns that don't add to the constant. Here is the mapping of methods that work for the 9th order magic squares:

	N			B				
		C			P			
	B			F			N	
		P			G			C
				N			B	
					C			P

P21 (rt 1, up 1)
Seq 1 2 3 4 5 6 7 8 9

A: ~~P10 9 (dn 2)~~   **B**: P10 9 (dn 1)   C: P10 9 (lf 1)   **D**: ~~P10 9 (rt 2, dn 1)~~
E: ~~P10 9 (lf 2)~~   F: P10 9 (up 2)   G: P10 9 (rt 2)   **H**: ~~P10 9 (rt 2, dn 2)~~
J: ~~P10 9 (dn 3)~~   **K**: ~~P10 9 (lf 3)~~   L: ~~P10 9 (up 3)~~   M: ~~P10 9 (rt 3)~~
N: P10 9 (dn 4)   **P**: P10 9 (lf 4)   Q: ~~P10 9 (up 4)~~   R: ~~P10 9 (rt 4)~~

Figure 7.19   A mapping of start positions for 9th order magic squares.

We find the same types of problems when we test and create a similar map of 15th order magic squares. In fact, after working with the 11th and 13th order magic squares, and seeing the many possible magic squares that can be made, it is surprising to see how few of them work with the 9th and 15th order magic squares.

							B			a				
		b			C									
				B			a							
		C												b
	B			a			F							
								G			b			C
	a												B	
								b			C			
										B			a	
					b			C						

P21 (rt 1, up 1)
Seq 1 2 3 4 5 6 7 8 9 10 11 12 13 14 15

~~**A:** P16 15 (dn 2)~~	**B:** P16 15 (dn 1)	**C:** P16 15 (lf 1)	~~**D:** P16 15 (rt 2, dn 1)~~
~~**E:** P16 15 (lf 2)~~	**F:** P16 15 (up 2)	**G:** P16 15 (rt 2)	~~**H:** P16 15 (rt 2, dn 2)~~
~~**J:** P16 15 (dn 3)~~	~~**K:** P16 15 (lf 3)~~	~~**L:** P16 15 (up 3)~~	~~**M:** P16 15 (rt 3)~~
~~**N:** P16 15 (dn 4)~~	~~**P:** P16 15 (lf 4)~~	~~**Q:** P16 15 (up 4)~~	~~**R:** P16 15 (rt 4)~~
~~**S:** P16 15 (dn 5)~~	~~**T:** P16 15 (lf 5)~~	~~**U:** P16 15 (up 5)~~	~~**V:** P16 15 (rt 5)~~
~~**W:** P16 15 (dn 6)~~	~~**X:** P16 15 (lf 6)~~	~~**Y:** P16 15 (up 6)~~	~~**Z:** P16 15 (rt 6)~~
a: P16 15 (dn 7)	**b:** P16 15 (lf 7)	~~**c:** P16 15 (up 7)~~	~~**d:** P16 15 (rt 7)~~

Figure 7.20 A mapping of start positions for 15th order magic squares.

It is also instructive to compare the 9th order constructions with the 15th order constructions. For example, the F and G methods are still good (just as they were with the 5th, 7th, 11th, and 13th orders), and again, there is only one F and one G method that works. Similarly, the B and C methods both work for the 9th order and the 15th order, but the 9th order only has 3 of each that work, and the 15th order only has 5 that work.

We can also see that the N and P methods in the 9th order are analogous to the a and b methods in the 15th order. And in these cases when we look at the other methods that don't work, we also find that they don't work for the same reasons.

Mapping Multiple Construction Sequences
These mappings become very convenient for showing similarities between scores of magic squares (and hundreds of magic squares for higher orders). And, as mentioned earlier, they are very instructive when comparing differing order magic squares with similar properties. Before leaving this chapter, we will show the mappings of several 5th order and 7th order magic squares where there are several sequences that follow similar patterns. Here are the mappings:

A	CH	B	DE	
CH	B	DEF*		A
B	DE		AG*	CH
DE		A	CH	B
	A	CH	B	DE

P21 (rt 1, up 1)

Sequences:　1 2 3 4 5, or 1 2 3 5 4, or 1 3 5 2 4, or 1 2 5 3 4, or 1 5 2 4 3

A: P65 (dn 2)　　**B**: P65 (dn 1)　　**C**: P65 (lf 1)　　**D**: P65 (rt 2, dn 1)

E: P65 (lf 2)　　**F**: P65 (up 2)　　**G**: P65 (rt 2)　　**H**: P65 (rt 2, dn 2)

A	HKL	C	B	DJM	E	
HKL	C	B	DJM	E		A
C	B	DJM	EF*		A	HKL
B	DJM	E		AG*	HKL	C
DJM	E		A	HKL	C	B
E		A	HKL	C	B	DJM
	A	HKL	C	B	DJM	E

P21 (rt 1, up 1)

Sequences:　1 2 3 4 5 6 7, or 1 2 3 4 5 7 6, or 1 3 5 7 2 4 6, or 1 2 4 3 5 7 6, or 1 5 3 7 2 4 6

A: P87 (dn 2)　　**B**: P87 (dn 1)　　**C**: P87 (lf 1)　　**D**: P87 (rt 2, dn 1)

E: P87 (lf 2)　　**F**: P87 (up 2)　　**G**: P87 (rt 2)　　**H**: P87 (rt 2, dn 2)

J: P87 (dn 3)　　**K**: P87 (lf 3)　　**L**: P87 (up 3)　　**M**: P87 (rt 3)

Figure 7.21　Mappings of 5th and 7th order start positions along with multiple construction sequences. (*Not all sequences in the F and G positions will create a magic square.)

Other P21 Relations

The odd-order magic squares that we've looked at up to this point in this chapter use the same P21 relation of right one cell, and up one cell. This means that the position of the second element in the sequence (which is usually 2) is located to the right one cell, and up one cell from the position of the first element in the sequence (the number 1). We have abbreviated this so that we can quickly and concisely show it as P21 (rt 1, up 1). However, there are still many other arrangements that we can use that will build other unique magic squares. Another common relation is the "knight's move" positioning, which gets its name from the way a knight would move in chess. In this section we will describe the P21 parameter as moving right two cells and down 1 (a knight's move). Here are a few 7th order magic squares that use this construction method:

4	47	41	35	22	16	10
17	11	5	48	42	29	23
30	24	18	12	6	49	36
43	37	31	25	19	13	7
14	1	44	38	32	26	20
27	21	8	2	45	39	33
40	34	28	15	9	3	46

P1 (5, 2)
P21 (rt 2, dn 1)
P87 (rt 3, dn 2)
Seq 1 2 3 4 5 6 7

37	14	33	3	22	48	18
49	19	38	8	34	4	23
5	24	43	20	39	9	35
10	29	6	25	44	21	40
15	41	11	30	7	26	45
27	46	16	42	12	31	1
32	2	28	47	17	36	13

P1 (6, 7)
P21 (rt 2, dn 1)
P87 (lf 1, up 3)
Seq 1 2 3 4 5 6 7

39	12	34	7	22	44	17
45	18	40	13	35	1	23
2	24	46	19	41	14	29
8	30	3	25	47	20	42
21	36	9	31	4	26	48
27	49	15	37	10	32	5
33	6	28	43	16	38	11

P1 (5, 2)
P21 (rt 2, dn 1)
P87 (rt 4, up 4)
Seq 1 2 3 4 5 6 7

Figure 7.22   Three 7th order magic squares along with their construction methods where the P21 parameter uses a "knight's move" method.

The following is a mapping of the knight's move 7th order magic squares. As before, the letter shows where the number 1 can be placed to create a magic square such that the corresponding P87 will work with the current P21.

				ABCD		
K	E	F	BD		AC	
H		BD				AC
AC	BD	J				
BD	AC					
		AC	G			BD
			AC		BD	

P21 (rt 2, dn 1)
Sequences: 1 2 3 4 5 6 7

A: P87 (rt3, dn 2)   **B**: P87 (lf 1, up 3)   **C**: P87 (rt 4, up 4)   **D**: P87 (lf 2, up 2)

Pan-Diagonal Descriptions:
E: P87 (rt 3, up 3)   **F**: P87 (lf 2, dn 3)   **G**: P87 (up 3)   **H**: P87 (rt 1, up 2)
J: P87 (lf 2)   **K**: P87 (rt 1, dn 3)

Figure 7.23   A mapping of 7th order magic squares showing regular magic squares along with several pan-diagonal squares.

The above mapping shows both standard and pan-diagonal relationships for 7th order magic squares that use a knight's move relationship between numbers. Note that since the 1 can actually be anywhere in the array in a pan-diagonal magic square, the letters E thru K in the array above show where the 1 needs to be placed in order for the 25 (which is the middle number of the sequence 1 thru 49) to be in the center cell of the array.

And there are still many other P21 relationships. As a final example in this section, here is a 7th order magic square where the 2 is right two cells and up three cells from the 1:

31	9	36	21	48	26	4
16	43	28	6	33	11	38
1	35	13	40	18	45	23
42	20	47	25	3	30	8
27	5	32	10	37	15	49
12	39	17	44	22	7	34
46	24	2	29	14	41	19

P1 (3, 1)
P21 (rt 2, up 3)
P87 (rt 1, up 2)
Seq 1 2 3 4 5 6 7

Figure 7.24   A 7th order magic square along with yet another P21 description in its construction method.

Defining a P81 Relation Instead of the P87 Relation

Sometimes it is easier to construct the magic square by defining the position of the 8 with respect to the 1 instead of with respect to the 7. The following magic square is an excellent example of this. The numbers 1 thru 7 are placed to the right 2 cells, and down 1 cell from the previous number. Now, to define the position of the 8, we could say that it is 4 cells to the left of the 7, and down 2. However, it is much easier to say that the 8 is to the right 1 cell and down 1 cell from the 1. The numbers 8 thru 14 are then placed in the same manner as the numbers 1 thru 7 were; and then the 15 is right 1 cell and down 1 cell from the 8; and so on.

4	47	41	35	22	16	10
17	11	5	48	42	29	23
30	24	18	12	6	49	36
43	37	31	25	19	13	7
14	1	44	38	32	26	20
27	21	8	2	45	39	33
40	34	28	15	9	3	46

P1 (5, 2)
P21 (rt 2, dn 1)
P81 (rt 1, dn 1)
Seq 1 2 3 4 5 6 7

Figure 7.25   A 7th order magic square along with its construction method. In this example the P87 parameter is replaced with a P81 parameter.

Therefore, the magic square above can actually be described in either of these two ways:

P1 (5, 2)
P21 (rt 2, dn 1)
P87 (lf 4, dn2)
Seq 1 2 3 4 5 6 7

P1 (5, 2)
P21 (rt 2, dn 1)
P81 (rt 1, dn 1)
Seq 1 2 3 4 5 6 7

Therefore, it is important to keep in mind that the description of the construction method is a tool to help us see patterns, but it isn't an end all description of unique magic squares.

Through all of these examples, we have kept the sets within the sequence in the same order. For example, in a 5th order magic square there are 5 sets with 5 numbers each in the sequence. And any change that was made to the first set was then reflected through the other four sets. In other words, if we started with our normal sequence for the numbers 1 through 25, we have:

$$1, 2, 3, 4, 5, \quad 6, 7, 8, 9, 10, \quad 11, 12, 13, 14, 15, \quad 16, 17, 18, 19, 20, \quad 21, 22, 23, 24, 25$$

Then if we swapped the number 3 and 5 in the first set, we would also make a similar swap in the other sets, which would include swapping the 8 and 10, the 13 and 15, the 18 and 20, and the 23 and 25, in order to generate a new sequence.

$$1, 2, 5, 4, 3, \quad 6, 7, 10, 9, 8, \quad 11, 12, 15, 14, 13, \quad 16, 17, 20, 19, 18, \quad 21, 22, 25, 24, 23$$

The sequence above would be our new sequence, and we could now use it just as our standard sequence to generate a new magic square.

1	18	9	25	12
19	10	22	11	3
7	21	13	4	20
23	14	5	17	6
15	2	16	8	24

Figure 7.26   A 5th order magic square using an altered sequence.

And we can use the following description to define this magic square:

P1 (1, 1)
P21 (rt 1, up 1)
P65 (rt 2)
Seq 1 2 5 4 3

What is interesting here is that we can define the position of 6 in terms of the 3 (which is how we usually would do it because the 3 is the last number of the first set of 5), or we could define the 6 in terms of the 5 (which is the middle number of the first set of five numbers). Either method will work, as long as we are consistent.

And as we saw just a little earlier, using these different methods of describing magic squares can in some cases cause the same magic square to be listed more than once. Therefore, if you are trying to create a list of magic squares that are unique to each other, then only a single method should be used to list the magic squares (don't combine a method that shows a P61 with methods that show a P63 or a P65). But if you are comparing similarities between different orders with similar methods, then comparing a P61 with a P81 using this notation becomes another tool to help with the comparisons.

Rearranging Sets within a Sequence
Here's an interesting construction method where I not only rearrange the numbers within the sets, but also the sets themselves. Using the same sequence as in the previous section, the numbers in the sets are in the order: 1, 2, 5, 4, 3; and then I also changed the order of these sets to 1, 3, 5, 2, 4; which then gives me the overall following sequence:

1, 2, 5, 4, 3,   11, 12, 15, 14, 13,   21, 22, 25, 24, 23,   6, 7, 10, 9, 8,   16, 17, 20, 19, 18

This is the listing that we will use for our sequence. The second number in each set will be placed right one and up one from its preceding number. The first number of the next set will be down two from the last number of the preceding set. Using this construction method and the new sequence we just generated, we get the following semi-magic square:

1	8	14	20	22
9	15	17	21	3
12	16	23	4	10
18	24	5	7	11
25	2	6	13	19

Figure 7.27   A 5th order semi-magic square with an altered sequence generated from an altered set order.

Now, when we shift the magic square so that the number 13 is in the center cell, we then have a regular magic square.

16	23	4	10	12
24	5	7	11	18
2	6	13	19	25
8	14	20	22	1
15	17	21	3	9

Figure 7.28   A 5th order magic square with altered sets within the altered sequence.

The description I could use to define this magic square would include an extra line that would describe the order of the sets. It would look something like this:

P1 (4, 5)
P21 (rt 1, up 1)
P65 (dn 2)
Seq 1 2 5 4 3
Sets 1 3 5 2 4

Keep in mind that the P65 term is referring to the last number of each set in relation to the first number of the following set, and not the actual 6 and 5 elements within the array.

It is also interesting to note that even though the sequence appeared to be very convoluted due to our rearranging of the sets, if we now take a step back and look at the magic square itself, we find that we can also use the following simpler description to define this magic square:

P1 (4, 5)
P21 (rt 1, up 1)
P65 (rt 3, dn 3)
Seq 1 2 5 4 3

There is no need to rearrange the sets. Instead, we can say that the entire sequence would now be:

1, 2, 5, 4, 3,   6, 7, 10, 9, 8,   11, 12, 15, 14, 13,   16, 17, 20, 19, 18,   21, 22, 25, 24, 23

In other words, even though it is possible to have an extra line in the description which defines the order of the sets, I find that it usually isn't needed because there can be also an easier way to describe the magic square, as we saw in this current example.

This example just shows that not only are there a myriad of ways to create magic squares; there can also be several ways to describe the same a magic square, and therefore (as mentioned previously) we need to be careful when describing them. As such, these descriptions are a useful tool to help categorize magic squares, but they aren't the final answer to describing every unique odd-ordered magic square.

Chapter 8
The 8th Order and Double Even-Ordered Magic Squares

The 8th Order Magic Square

The first 4th order magic square we looked at had a very interesting construction. Half of the numbers in the square didn't move from their original position. And of the half that did, each number simply traded places with a corresponding number on the opposite side, and through the middle of the square. There are several 8th order magic squares that share this similar construction. Let's start with what has come to be our standard initial square (where one is located in the upper left corner, and the numbers continue in ascending order going from left to right and from top to bottom):

1	2	3	4	5	6	7	8
9	10	11	12	13	14	15	16
17	18	19	20	21	22	23	24
25	26	27	28	29	30	31	32
33	34	35	36	37	38	39	40
41	42	43	44	45	46	47	48
49	50	51	52	53	54	55	56
57	58	59	60	61	62	63	64

Figure 8.1   An 8th order initial square (or array) of numbers.

Now, consider this 8th order magic square. The type of construction for this magic square is very similar to the first 4th order magic square we looked at several chapters ago. Of the 64 numbers in this array, half of them swapped positions with their corresponding number on the opposite side and through the middle of the square, while the other half remained in their original position.

1	2	62	61	60	59	7	8
9	10	54	53	52	51	15	16
48	47	19	20	21	22	42	41
40	39	27	28	29	30	34	33
32	31	35	36	37	38	26	25
24	23	43	44	45	46	18	17
49	50	14	13	12	11	55	56
57	58	6	5	4	3	63	64

Figure 8.2   An 8th order swap-thru-the-center magic square. Shaded numbers did not move.

Specifically, each number is swapped through the middle, and these are the steps:
- Swap the four center numbers on the top row with the four center numbers on the eighth row.
- Swap the four center numbers on the second row with the four center numbers on the seventh row.
- Swap the four center numbers in the first column with the four center numbers in the eighth column.
- Swap the four center numbers in the second column with the four center numbers in the seventh column.

This series of operations transformed the initial square in figure 8.1 into the magic square in figure 8.2.

Other Various Swap-through-the-Center Methods

Let's return to the initial square in figure 8.1. This time we will use it to create a magic square that is similar to the magic square in figure 8.2. To create this magic square we again swap each number through the middle using these steps:
- Swap the four center numbers on the top row with the four center numbers on the seventh row (instead of the eighth).
- Swap the center four numbers from the second row with the center four numbers on the eighth row (instead of the seventh).
- Swap the four center numbers on the first column with the four center numbers on the seventh column.
- Swap the four center numbers on the second column with the four center numbers on the eighth column.
- Finally, take the 2 x 2 arrays in each corner and diagonally swap the numbers (so that in the upper left corner the 1 and 10 trade places as do the 2 and 9, etc.)

10	9	54	53	52	51	16	15
2	1	62	61	60	59	8	7
47	48	19	20	21	22	41	42
39	40	27	28	29	30	33	34
31	32	35	36	37	38	25	26
23	24	43	44	45	46	17	18
58	57	6	5	4	3	64	63
50	49	14	13	12	11	56	55

Figure 8.3   Another 8th order swap-thru-the-center magic square. Shaded numbers did not move.

The sixteen numbers in the 4 x 4 array in the center are the only numbers that did not move from their original locations.

The following magic square also uses the initial square from figure 8.1 as its starting point and has the numbers swap through the center:

1	63	3	61	60	6	58	8
56	10	54	12	13	51	15	49
17	47	19	45	44	22	42	24
40	26	38	28	29	35	31	33
32	34	30	36	37	27	39	25
41	23	43	21	20	46	18	48
16	50	14	52	53	11	55	9
57	7	59	5	4	62	2	64

Figure 8.4   Yet another 8th order swap-thru-the-center magic square. Again, the shaded numbers did not move.

The magic square above has half of the numbers trade places with their corresponding number through the center of the square. In this case, the numbers that traded places and those that remained in their original position create a checker-boarded pattern throughout the square.

And, there are still many other ways of using the initial square in figure 8.1 to create other unique magic squares. The following magic square has several variations. We'll look at two of them:

1	63	3	61	60	6	58	8
56	10	11	52	53	14	15	49
17	18	46	45	44	43	23	24
40	31	38	28	29	35	26	33
32	39	30	36	37	27	34	25
41	42	22	21	20	19	47	48
16	50	51	12	13	54	55	9
57	7	59	5	4	62	2	64

Figure 8.5   Still another 8th order swap-thru-the-center magic square.

Of the 64 numbers in this magic square, 28 of them did not move; 28 traded places with their corresponding number on the opposite side of the square through the middle, and the remaining 8 traded places with their corresponding numbers straight across (and not through the middle). Specifically, the numbers 12 and 52 traded straight across with each other, as did 13 with 53; 26 with 31; and 34 with 39.

However, these 8 numbers could have traded with their opposite side through the middle, and 8 others could have traded straight across, as we see in this next example.

1	63	3	60	61	6	58	8
56	10	11	53	52	14	15	49
17	18	46	45	44	43	23	24
32	39	38	28	29	35	34	25
40	31	30	36	37	27	26	33
41	42	22	21	20	19	47	48
16	50	51	13	12	54	55	9
57	7	59	4	5	62	2	64

Figure 8.6   And another 8th order swap-thru-the-center magic square.

This time, the numbers 4 and 60 traded straight across with each other, as did 5 with 61; 25 with 32; and 33 with 40. Other than the different straight across trades, the Magic Square in figure 8.6 is identical to the one in figure 8.5.

We have now seen five variations of creating 8th order magic squares by swapping numbers through the center of the square; and there are still several other methods that could be defined. Swapping numbers through the middle and swapping numbers straight across as well as combinations of both of these swaps is a powerful method that works very well with double even magic squares.

Swap-by-Number and Swap-by-Position
Returning again to the first magic square of this chapter, we can say that we have two ways of defining these swaps: one is a swap-by-position and the other is a swap-by-number. The following list shows the swap by number on the left, and the swap by position on the right:

3 with 62	(1, 3) with (8, 6)
4 with 61	(1, 4) with (8, 5)
5 with 60	(1, 5) with (8, 4)
6 with 59	(1, 6) with (8, 3)
11 with 54	(2, 3) with (7, 6)
12 with 53	(2, 4) with (7, 5)
13 with 52	(2, 5) with (7, 4)
14 with 51	(2, 6) with (7, 3)
17 with 48	(3, 1) with (6, 8)
25 with 40	(4, 1) with (5, 8)
33 with 32	(5, 1) with (4, 8)
41 with 24	(6, 1) with (3, 8)
18 with 47	(3, 2) with (6, 7)
26 with 39	(4, 2) with (5, 7)
34 with 31	(5, 2) with (4, 7)
42 with 23	(6, 2) with (3, 7)

Swap-by-Number	**Swap-by-Position**

We can apply either of these swap methods to other initial squares and create yet other unique magic squares. In a few cases, we can apply both swap methods to the same initial square and the result will create two unique magic squares. One such example uses the following initial square.

This next initial square is an 8 x 8 array that has been sub-divided into four 4 x 4 arrays, with the numbers one thru 16 in the upper left 4 x 4 array, the numbers 17 thru 32 in the upper right array, the numbers 33 thru 48 in the lower left array, and finally the numbers 49 thru 64 in the lower right 4 x 4 array.

1	2	3	4	17	18	19	20
5	6	7	8	21	22	23	24
9	10	11	12	25	26	27	28
13	14	15	16	29	30	31	32
33	34	35	36	49	50	51	52
37	38	39	40	53	54	55	56
41	42	43	44	57	58	59	60
45	46	47	48	61	62	63	64

Figure 8.7   Another 8th order initial square (or array) of numbers.

Now, we can perform both of the swap methods on this initial square to create two unique magic squares. Swapping by position, we get the following magic square:

1	2	62	61	48	47	19	20
5	6	58	57	44	43	23	24
56	55	11	12	25	26	38	37
52	51	15	16	29	30	34	33
32	31	35	36	49	50	14	13
28	27	39	40	53	54	10	9
41	42	22	21	8	7	59	60
45	46	18	17	4	3	63	64

Figure 8.8   An 8th order magic square created from the second initial square by using the swap-by-position method. The shaded numbers did not move from their original position in the modified initial square.

And, swapping by number gives us this magic square:

1	2	62	61	48	47	19	20
60	59	7	8	21	22	42	41
9	10	54	53	40	39	27	28
52	51	15	16	29	30	34	33
32	31	35	36	49	50	14	13
37	38	26	25	12	11	55	56
24	23	43	44	57	58	6	5
45	46	18	17	4	3	63	64

Figure 8.9   An 8th order magic square created from the second initial square by using the swap-by-number method. Again, the shaded numbers did not move from their original position in the modified initial square.

These two magic squares are unique to each other, yet they share some very interesting similarities. For example, the first, fourth, fifth and eighth rows are identical to each other in both magic squares. The second row as compared with the seventh row, as well as the third compared with the sixth rows read backwards from each other, and yet other than that, are also the same.

Other Initial Square Patterns
As we have shown in previous chapters throughout this book, there are two parts to creating a variety of magic squares. They are the initial sequence (or position) and the

method. The method defines how we move numbers from their original position to their new position (and there are many of these methods). And as we've seen before, the initial arrangement of numbers can be just as important as the method is used upon the initial arrangement. Here are four more initial squares that can be used to create a large variety of magic squares using the methods that we have already considered in this chapter. (The numbers 1, 2, 3, and 4 have been highlighted to help see their differences.)

<table>
<tr><td>1</td><td>2</td><td>5</td><td>6</td><td>9</td><td>10</td><td>13</td><td>14</td></tr>
<tr><td>3</td><td>4</td><td>7</td><td>8</td><td>11</td><td>12</td><td>15</td><td>16</td></tr>
<tr><td>17</td><td>18</td><td>21</td><td>22</td><td>25</td><td>26</td><td>29</td><td>30</td></tr>
<tr><td>19</td><td>20</td><td>23</td><td>24</td><td>27</td><td>28</td><td>31</td><td>32</td></tr>
<tr><td>33</td><td>34</td><td>37</td><td>38</td><td>41</td><td>42</td><td>45</td><td>46</td></tr>
<tr><td>35</td><td>36</td><td>39</td><td>40</td><td>43</td><td>44</td><td>47</td><td>48</td></tr>
<tr><td>49</td><td>50</td><td>53</td><td>54</td><td>57</td><td>58</td><td>61</td><td>62</td></tr>
<tr><td>51</td><td>52</td><td>55</td><td>56</td><td>59</td><td>60</td><td>63</td><td>64</td></tr>
</table>

<table>
<tr><td>1</td><td>3</td><td>5</td><td>7</td><td>9</td><td>11</td><td>13</td><td>15</td></tr>
<tr><td>2</td><td>4</td><td>6</td><td>8</td><td>10</td><td>12</td><td>14</td><td>16</td></tr>
<tr><td>17</td><td>19</td><td>21</td><td>23</td><td>25</td><td>27</td><td>29</td><td>31</td></tr>
<tr><td>18</td><td>20</td><td>22</td><td>24</td><td>26</td><td>28</td><td>30</td><td>32</td></tr>
<tr><td>33</td><td>35</td><td>37</td><td>39</td><td>41</td><td>43</td><td>45</td><td>47</td></tr>
<tr><td>34</td><td>36</td><td>38</td><td>40</td><td>42</td><td>44</td><td>46</td><td>48</td></tr>
<tr><td>49</td><td>51</td><td>53</td><td>55</td><td>57</td><td>59</td><td>61</td><td>63</td></tr>
<tr><td>50</td><td>52</td><td>54</td><td>56</td><td>58</td><td>60</td><td>62</td><td>64</td></tr>
</table>

<table>
<tr><td>1</td><td>17</td><td>2</td><td>18</td><td>3</td><td>19</td><td>4</td><td>20</td></tr>
<tr><td>33</td><td>49</td><td>34</td><td>50</td><td>35</td><td>51</td><td>36</td><td>52</td></tr>
<tr><td>5</td><td>21</td><td>6</td><td>22</td><td>7</td><td>23</td><td>8</td><td>24</td></tr>
<tr><td>37</td><td>53</td><td>38</td><td>54</td><td>39</td><td>55</td><td>40</td><td>56</td></tr>
<tr><td>9</td><td>25</td><td>10</td><td>26</td><td>11</td><td>27</td><td>12</td><td>28</td></tr>
<tr><td>41</td><td>57</td><td>42</td><td>58</td><td>43</td><td>59</td><td>44</td><td>60</td></tr>
<tr><td>13</td><td>29</td><td>14</td><td>30</td><td>15</td><td>31</td><td>16</td><td>32</td></tr>
<tr><td>45</td><td>61</td><td>46</td><td>62</td><td>47</td><td>63</td><td>48</td><td>64</td></tr>
</table>

<table>
<tr><td>1</td><td>5</td><td>9</td><td>13</td><td>2</td><td>6</td><td>10</td><td>14</td></tr>
<tr><td>17</td><td>21</td><td>25</td><td>29</td><td>18</td><td>22</td><td>26</td><td>30</td></tr>
<tr><td>33</td><td>37</td><td>41</td><td>45</td><td>34</td><td>38</td><td>42</td><td>46</td></tr>
<tr><td>49</td><td>53</td><td>57</td><td>61</td><td>50</td><td>54</td><td>58</td><td>62</td></tr>
<tr><td>3</td><td>7</td><td>11</td><td>15</td><td>4</td><td>8</td><td>12</td><td>16</td></tr>
<tr><td>19</td><td>23</td><td>27</td><td>31</td><td>20</td><td>24</td><td>28</td><td>32</td></tr>
<tr><td>35</td><td>39</td><td>43</td><td>47</td><td>36</td><td>40</td><td>44</td><td>48</td></tr>
<tr><td>51</td><td>55</td><td>59</td><td>63</td><td>52</td><td>56</td><td>60</td><td>64</td></tr>
</table>

Figure 8.10 Four additional 8th order initial squares.

And there are still other 8 x 8 array patterns that can be used as initial squares (some of which are variations of the patterns above). Each of these initial squares can be used with any of the methods listed above; and in each case would create unique magic squares.

The 12th Order Magic Square

In many ways, the 12th order magic square is similar to the 8th order magic square. And just as we did with the 8th order initial square, we'll start by defining a 12 x 12 initial array of numbers.

<table>
<tr><td>1</td><td>2</td><td>3</td><td>4</td><td>5</td><td>6</td><td>7</td><td>8</td><td>9</td><td>10</td><td>11</td><td>12</td></tr>
<tr><td>13</td><td>14</td><td>15</td><td>16</td><td>17</td><td>18</td><td>19</td><td>20</td><td>21</td><td>22</td><td>23</td><td>24</td></tr>
<tr><td>25</td><td>26</td><td>27</td><td>28</td><td>29</td><td>30</td><td>31</td><td>32</td><td>33</td><td>34</td><td>35</td><td>36</td></tr>
<tr><td>37</td><td>38</td><td>39</td><td>40</td><td>41</td><td>42</td><td>43</td><td>44</td><td>45</td><td>46</td><td>47</td><td>48</td></tr>
<tr><td>49</td><td>50</td><td>51</td><td>52</td><td>53</td><td>54</td><td>55</td><td>56</td><td>57</td><td>58</td><td>59</td><td>60</td></tr>
<tr><td>61</td><td>62</td><td>63</td><td>64</td><td>65</td><td>66</td><td>67</td><td>68</td><td>69</td><td>70</td><td>71</td><td>72</td></tr>
<tr><td>73</td><td>74</td><td>75</td><td>76</td><td>77</td><td>78</td><td>79</td><td>80</td><td>81</td><td>82</td><td>83</td><td>84</td></tr>
<tr><td>85</td><td>86</td><td>87</td><td>88</td><td>89</td><td>90</td><td>91</td><td>92</td><td>93</td><td>94</td><td>95</td><td>96</td></tr>
<tr><td>97</td><td>98</td><td>99</td><td>100</td><td>101</td><td>102</td><td>103</td><td>104</td><td>105</td><td>106</td><td>107</td><td>108</td></tr>
<tr><td>109</td><td>110</td><td>111</td><td>112</td><td>113</td><td>114</td><td>115</td><td>116</td><td>117</td><td>118</td><td>119</td><td>120</td></tr>
<tr><td>121</td><td>122</td><td>123</td><td>124</td><td>125</td><td>126</td><td>127</td><td>128</td><td>129</td><td>130</td><td>131</td><td>132</td></tr>
<tr><td>133</td><td>134</td><td>135</td><td>136</td><td>137</td><td>138</td><td>139</td><td>140</td><td>141</td><td>142</td><td>143</td><td>144</td></tr>
</table>

Figure 8.11 A 12th order initial square.

The array above is our "standard" initial array. We start with the 1 in the upper left corner and list all the numbers in ascending order from left to right and from top to bottom until we reach the highest number in the lower right corner.

The first 12th order magic square that we'll create is similar to the first 4th and 8th order magic squares we created (magic squares in figures 4.2 and 8.2).

1	2	3	141	140	139	138	137	136	10	11	12
13	14	15	129	128	127	126	125	124	22	23	24
25	26	27	117	116	115	114	113	112	34	35	36
108	107	106	40	41	42	43	44	45	99	98	97
96	95	94	52	53	54	55	56	57	87	86	85
84	83	82	64	65	66	67	68	69	75	74	73
72	71	70	76	77	78	79	80	81	63	62	61
60	59	58	88	89	90	91	92	93	51	50	49
48	47	46	100	101	102	103	104	105	39	38	37
109	110	111	33	32	31	30	29	28	118	119	120
121	122	123	21	20	19	18	17	16	130	131	132
133	134	135	9	8	7	6	5	4	142	143	144

Figure 8.12 A 12th order magic square.

Again, half of the numbers swap places with their corresponding number through the middle and on the opposite side of the square. The numbers that don't move are located in the 3 x 3 arrays in each of the corners as well as the 6 x 6 array in the center of the square. (These are the grayed-in squares.)

A Second Method

We can use variations of the methods used to create 8th order magic squares to also create other 12th order magic squares. Here is a variation of the second method from above that we are now using on the 12 x 12 initial array to create a 12th order magic square.

27	26	25	117	116	115	114	113	112	36	35	34
15	14	13	129	128	127	126	125	124	24	23	22
3	2	1	141	140	139	138	137	136	12	11	10
106	107	108	40	41	42	43	44	45	97	98	99
94	95	96	52	53	54	55	56	57	85	86	87
82	83	84	64	65	66	67	68	69	73	74	75
70	71	72	76	77	78	79	80	81	61	62	63
58	59	60	88	89	90	91	92	93	49	50	51
46	47	48	100	101	102	103	104	105	37	38	39
135	134	133	9	8	7	6	5	4	144	143	142
123	122	121	21	20	19	18	17	16	132	131	130
111	110	109	33	32	31	30	29	28	120	119	118

Figure 8.13 A second 12th order magic square.

In the magic square above, we took the center six numbers from the top row and swapped them through the middle with the center six numbers on the tenth row. We then performed the same swap with the second and eleventh rows and then with the third and twelfth rows. Next, we did the same thing with the center six numbers in the first and tenth columns, the second and eleventh columns and finally the third and twelfth columns.

The numbers in the center 6 x 6 array didn't move; however, to complete our magic square we took the 3 x 3 arrays in each corner and swapped the numbers through the middle within their respective arrays. This final operation completed the magic square.

Several Other Swap-Through-the-Middle Magic Squares
As the magic square gets larger and larger, the number of possible swap-through-the-middle magic squares also increases. All of these magic squares are some variation of each other. Case in point, the next three magic squares are 12th order variations of each other. Again, the shaded areas show those numbers that did not move from their original positions.

1	2	3	136	137	138	139	140	141	10	11	12
13	14	15	124	125	126	127	128	129	22	23	24
25	26	27	112	113	114	115	116	117	34	35	36
48	47	46	105	104	103	102	101	100	39	38	37
60	59	58	93	92	91	90	89	88	51	50	49
72	71	70	81	80	79	78	77	76	63	62	61
84	83	82	69	68	67	66	65	64	75	74	73
96	95	94	57	56	55	54	53	52	87	86	85
108	107	106	45	44	43	42	41	40	99	98	97
109	110	111	28	29	30	31	32	33	118	119	120
121	122	123	16	17	18	19	20	21	130	131	132
133	134	135	4	5	6	7	8	9	142	143	144

Figure 8.14 The first of three swap-through-the-middle 12th order magic squares.

The following is a variation of the magic square above.

1	2	3	136	140	139	138	137	141	10	11	12
13	14	15	124	128	127	126	125	129	22	23	24
25	26	27	112	116	115	114	113	117	34	35	36
48	47	46	105	101	102	103	104	100	39	38	37
96	95	94	57	53	54	55	56	52	87	86	85
84	83	82	69	65	66	67	68	64	75	74	73
72	71	70	81	77	78	79	80	76	63	62	61
60	59	58	93	89	90	91	92	88	51	50	49
108	107	106	45	41	42	43	44	40	99	98	97
109	110	111	28	32	31	30	29	33	118	119	120
121	122	123	16	20	19	18	17	21	130	131	132
133	134	135	4	8	7	6	5	9	142	143	144

Figure 8.15 The 2nd of three swap-through-the-middle 12th order magic squares.

And here is yet another magic square that is a variation of the two preceding magic squares.

1	2	3	141	137	138	139	140	136	10	11	12
13	14	15	129	125	126	127	128	124	22	23	24
25	26	27	117	116	115	114	113	112	34	35	36
108	107	106	40	41	42	43	44	45	99	98	97
60	59	94	52	92	91	90	89	57	87	50	49
72	71	82	64	80	79	78	77	69	75	62	61
84	83	70	76	68	67	66	65	81	63	74	73
96	95	58	88	56	55	54	53	93	51	86	85
48	47	46	100	101	102	103	104	105	39	38	37
109	110	111	33	32	31	30	29	28	118	119	120
121	122	123	21	17	18	19	20	16	130	131	132
133	134	135	9	5	6	7	8	4	142	143	144

Figure 8.16   The 3rd of three swap-through-the-middle 12th order magic squares. These three magic squares all share similar swap patterns.

We can still use other variations of the methods listed above (for the 8th order squares) to create even more 12th order magic squares. And again, we can also create variations to our initial square and then use various methods to either swap-by-position or swap-by-number and create even more magic squares.

The following two 12 x 12 arrays are variations to the initial square. The first initial square divides the 12 x 12 array into 3 x 3 cells. This gives us 16 of these cells with nine numbers in each cell. The 16 cells are arranged with the lowest numbers in the upper left cell, and they ascend going left-to-right and top-to-bottom. Likewise, the numbers in each cell are also arranged with the lowest number in the upper left corner, and the numbers ascend going from left-to-right and from top-to-bottom.

1	2	3	10	11	12	19	20	21	28	29	30
4	5	6	13	14	15	22	23	24	31	32	33
7	8	9	16	17	18	25	26	27	34	35	36
37	38	39	46	47	48	55	56	57	64	65	66
40	41	42	49	50	51	58	59	60	67	68	69
43	44	45	52	53	54	61	62	63	70	71	72
73	74	75	82	83	84	91	92	93	100	101	102
76	77	78	85	86	87	94	95	96	103	104	105
79	80	81	88	89	90	97	98	99	106	107	108
109	110	111	118	119	120	127	128	129	136	137	138
112	113	114	121	122	123	130	131	132	139	140	141
115	116	117	124	125	126	133	134	135	142	143	144

Figure 8.17   A second 12th order initial square.

This second initial square divides the 12 x 12 array into 4 x 4 cells. This gives us nine of these cells with 16 numbers in each cell. Again, the cells themselves, as well as the numbers within each cell are arranged with the lowest number in the upper left corner, and the largest number in the lower right corner.

1	2	3	4	17	18	19	20	33	34	35	36
5	6	7	8	21	22	23	24	37	38	39	40
9	10	11	12	25	26	27	28	41	42	43	44
13	14	15	16	29	30	31	32	45	46	47	48
49	50	51	52	65	66	67	68	81	82	83	84
53	54	55	56	69	70	71	72	85	86	87	88
57	58	59	60	73	74	75	76	89	90	91	92
61	62	63	64	77	78	79	80	93	94	95	96
97	98	99	100	113	114	115	116	129	130	131	132
101	102	103	104	117	118	119	120	133	134	135	136
105	106	107	108	121	122	123	124	137	138	139	140
109	110	111	112	125	126	127	128	141	142	143	144

Figure 8.18   A third 12th order initial square.

It is possible to use these initial squares along with the methods we've discussed to create a multitude of magic squares. For example, the following 12 x 12 array used the initial square above and the second swap-through-the-middle method (that we saw at the beginning of this chapter) to create this unique magic square. (The shaded numbers are those numbers that didn't move from their initial position in the square above.)

11	10	9	133	120	119	118	117	104	44	43	42
7	6	5	137	124	123	122	121	108	40	39	38
3	2	1	141	128	127	126	125	112	36	35	34
130	131	132	16	29	30	31	32	45	97	98	99
94	95	96	52	65	66	67	68	81	61	62	63
90	91	92	56	69	70	71	72	85	57	58	59
86	87	88	60	73	74	75	76	89	53	54	55
82	83	84	64	77	78	79	80	93	49	50	51
46	47	48	100	113	114	115	116	129	13	14	15
111	110	109	33	20	19	18	17	4	144	143	142
107	106	105	37	24	23	22	21	8	140	139	138
103	102	101	41	28	27	26	25	12	136	135	134

Figure 8.19   Another swap-through-the-middle 12th order magic squares using the second initial square pattern.

Higher Order Double-Even Magic Squares

In this chapter we have looked at swap-through-the-middle methods and initial squares for 8th order and 12th order magic squares. Again, these same patterns and methods can be used on 16th order and higher order double-even magic squares.

The following magic square is a 16th order swap-through-the-middle magic square. Again, we started with our standard initial square, and left the 4 x 4 arrays in each of the corners in place, as well as the center 8 x 8 array. The remaining numbers traded places with their counterparts through the center of the square.

1	2	3	4	252	251	250	249	248	247	246	245	13	14	15	16
17	18	19	20	236	235	234	233	232	231	230	229	29	30	31	32
33	34	35	36	220	219	218	217	216	215	214	213	45	46	47	48
49	50	51	52	204	203	202	201	200	199	198	197	61	62	63	64
192	191	190	189	69	70	71	72	73	74	75	76	180	179	178	177
176	175	174	173	85	86	87	88	89	90	91	92	164	163	162	161
160	159	158	157	101	102	103	104	105	106	107	108	148	147	146	145
144	143	142	141	117	118	119	120	121	122	123	124	132	131	130	129
128	127	126	125	133	134	135	136	137	138	139	140	116	115	114	113
112	111	110	109	149	150	151	152	153	154	155	156	100	99	98	97
96	95	94	93	165	166	167	168	169	170	171	172	84	83	82	81
80	79	78	77	181	182	183	184	185	186	187	188	68	67	66	65
193	194	195	196	60	59	58	57	56	55	54	53	205	206	207	208
209	210	211	212	44	43	42	41	40	39	38	37	221	222	223	224
225	226	227	228	28	27	26	25	24	23	22	21	237	238	239	240
241	242	243	244	12	11	10	9	8	7	6	5	253	254	255	256

Figure 8.20 A 16th order swap-through-the-middle magic square.

The final array we'll consider in this section also started with the standard 16 x 16 initial square; however this time the array was divided into 2 x 2 cells where half of these cells traded places with their counterparts through the center of the array, and the other half remained in their original position. This method creates a very interesting checkerboard pattern. And again, the shaded squares are those that remained in their original positions.

1	2	254	253	5	6	250	249	248	247	11	12	244	243	15	16
17	18	238	237	21	22	234	233	232	231	27	28	228	227	31	32
224	223	35	36	220	219	39	40	41	42	214	213	45	46	210	209
208	207	51	52	204	203	55	56	57	58	198	197	61	62	194	193
65	66	190	189	69	70	186	185	184	183	75	76	180	179	79	80
81	82	174	173	85	86	170	169	168	167	91	92	164	163	95	96
160	159	99	100	156	155	103	104	105	106	150	149	109	110	146	145
144	143	115	116	140	139	119	120	121	122	134	133	125	126	130	129
128	127	131	132	124	123	135	136	137	138	118	117	141	142	114	113
112	111	147	148	108	107	151	152	153	154	102	101	157	158	98	97
161	162	94	93	165	166	90	89	88	87	171	172	84	83	175	176
177	178	78	77	181	182	74	73	72	71	187	188	68	67	191	192
64	63	195	196	60	59	199	200	201	202	54	53	205	206	50	49
48	47	211	212	44	43	215	216	217	218	38	37	221	222	34	33
225	226	30	29	229	230	26	25	24	23	235	236	20	19	239	240
241	242	14	13	245	246	10	9	8	7	251	252	4	3	255	256

Figure 8.21 A second 16th order swap-through-the-middle magic square.

The Franklin Semi-Magic Square

Benjamin Franklin would probably roll over in his grave if all we talked about were swap-through-the-middle double-even magic squares. There are still other ways to create double-even magic squares. Benjamin Franklin developed one such method that he used to create an 8th order and a 16th order magic square. In this section we'll take a closer look at the 8th order semi-magic square that bears his name.

52	61	4	13	20	29	36	45
14	3	62	51	46	35	30	19
53	60	5	12	21	28	37	44
11	6	59	54	43	38	27	22
55	58	7	10	23	26	39	42
9	8	57	56	41	40	25	24
50	63	2	15	18	31	34	47
16	1	64	49	48	33	32	17

Figure 8.22 An 8th order semi-magic square developed by Benjamin Franklin.

The semi-magic square above is known as the Franklin Square, and it has an interesting construction. The numbers are placed in pairs, in a zigzag pattern throughout the square. The two arrays below make it easier to see how the pairs of numbers are placed within the square. Notice that the numbers 1 thru 8 and 57 thru 64 are in columns 2 and 3 while the numbers 17 thru 24 and 41 thru 48 are in columns 5 and 8 (see the array on the left in the figure below).

	61	4		20			45
	3	62		46			19
	60	5		21			44
	6	59		43			22
	58	7		23			42
	8	57		41			24
	63	2		18			47
	1	64		48			17

52			13		29	36	
14			51		35	30	
53			12		28	37	
11			54		38	27	
55			10		26	39	
9			56		40	25	
50			15		31	34	
16			49		33	32	

Figure 8.23 These two 8 x 8 arrays make it easier to see how the numbers are placed within the Franklin Square.

Continuing on, we see that the numbers 25 thru 40 are in columns 6 and 7 while the numbers 9 thru 16 and 49 thru 56 are in columns 1 and 4 (see the array on the right in the figure above). All of these numbers are placed within the array in sets of two or sets of four in such a way that their placement makes a zigzag pattern with respect to each other as well as with respect to preceding and following numbers.

There are critics who have pointed out that Franklin's square isn't that great because it is only a semi-magic square (after all, the main diagonals don't add up to the 8th order constant). However, Franklin's square does have some very unique properties of its own, and as we will see in just a minute, maybe Benjamin Franklin preferred it that way.

The most interesting property of a Franklin semi-magic square is that the right-angle diagonals all add up to the 8th order constant. And it doesn't matter which way we consider these right-angle diagonals. In other words, we can start anywhere on the left side, go diagonally up four cells, across one cell, and then diagonally down four cells, and the numbers within those cells add up to 260. Or we could have started anywhere on the left side and gone diagonally down four cells, across one cell, and then diagonally up four cells, and again the numbers within the cells add up to 260.

52	61	4	13	20	29	36	45
14	3	62	51	46	35	30	19
53	60	5	12	21	28	37	44
11	6	59	54	43	38	27	22
55	58	7	10	23	26	39	42
9	8	57	56	41	40	25	24
50	63	2	15	18	31	34	47
16	1	64	49	48	33	32	17

52	61	4	13	20	29	36	45
14	3	62	51	46	35	30	19
53	60	5	12	21	28	37	44
11	6	59	54	43	38	27	22
55	58	7	10	23	26	39	42
9	8	57	56	41	40	25	24
50	63	2	15	18	31	34	47
16	1	64	49	48	33	32	17

Figure 8.24   Two examples of horizontal bent diagonals within the Franklin Square.

And this still holds true if we take these diagonals and wrap them around the top or the bottom of the square (as shown below):

52	61	4	13	20	29	36	45
14	3	62	51	46	35	30	19
53	60	5	12	21	28	37	44
11	6	59	54	43	38	27	22
55	58	7	10	23	26	39	42
9	8	57	56	41	40	25	24
50	63	2	15	18	31	34	47
16	1	64	49	48	33	32	17

52	61	4	13	20	29	36	45
14	3	62	51	46	35	30	19
53	60	5	12	21	28	37	44
11	6	59	54	43	38	27	22
55	58	7	10	23	26	39	42
9	8	57	56	41	40	25	24
50	63	2	15	18	31	34	47
16	1	64	49	48	33	32	17

Figure 8.25   Two examples of horizontal bent diagonals that wrap around the top and bottom of the Franklin Square.

Similarly, we can start anywhere on the top row and go down diagonally either to the right or left for four cells, go straight down one cell and then continue diagonally back, and the sum of the numbers within those cells also adds up to 260. How cool is that?

52	61	4	13	20	29	36	45
14	3	62	51	46	35	30	19
53	60	5	12	21	28	37	44
11	6	59	54	43	38	27	22
55	58	7	10	23	26	39	42
9	8	57	56	41	40	25	24
50	63	2	15	18	31	34	47
16	1	64	49	48	33	32	17

52	61	4	13	20	29	36	45
14	3	62	51	46	35	30	19
53	60	5	12	21	28	37	44
11	6	59	54	43	38	27	22
55	58	7	10	23	26	39	42
9	8	57	56	41	40	25	24
50	63	2	15	18	31	34	47
16	1	64	49	48	33	32	17

Figure 8.26   Two examples of vertical bent diagonals within the Franklin Square.

And once again, we can take these vertical diagonals and wrap them around to the right or the left, and again they maintain their property of adding up to 260:

52	61	4	13	20	29	36	45
14	3	62	51	46	35	30	19
53	60	5	12	21	28	37	44
11	6	59	54	43	38	27	22
55	58	7	10	23	26	39	42
9	8	57	56	41	40	25	24
50	63	2	15	18	31	34	47
16	1	64	49	48	33	32	17

52	61	4	13	20	29	36	45
14	3	62	51	46	35	30	19
53	60	5	12	21	28	37	44
11	6	59	54	43	38	27	22
55	58	7	10	23	26	39	42
9	8	57	56	41	40	25	24
50	63	2	15	18	31	34	47
16	1	64	49	48	33	32	17

Figure 8.27 Two examples of vertical bent diagonals that wrap around the right and left sides within the Franklin Square.

In addition to this, Benjamin Franklin also noted that the following two sets of eight numbers also added to 260:

52	61	4	13	20	29	36	45
14	3	62	51	46	35	30	19
53	60	5	12	21	28	37	44
11	6	59	54	43	38	27	22
55	58	7	10	23	26	39	42
9	8	57	56	41	40	25	24
50	63	2	15	18	31	34	47
16	1	64	49	48	33	32	17

52	61	4	13	20	29	36	45
14	3	62	51	46	35	30	19
53	60	5	12	21	28	37	44
11	6	59	54	43	38	27	22
55	58	7	10	23	26	39	42
9	8	57	56	41	40	25	24
50	63	2	15	18	31	34	47
16	1	64	49	48	33	32	17

Figure 8.28 Two examples a set of eight cells that add to 260 within the Franklin Square.

These are very interesting properties for a magic square that is only considered semi-magic. And yet, maybe that is what Benjamin Franklin wanted. I say this because if we take the Franklin Square and consider the four right columns and flip them over. It becomes a regular magic square (but is that as cool as a square whose right angle diagonals all add up to the 8th order constant instead?).

52	61	4	13	48	33	32	17
14	3	62	51	18	31	34	47
53	60	5	12	41	40	25	24
11	6	59	54	23	26	39	42
55	58	7	10	43	38	27	22
9	8	57	56	21	28	37	44
50	63	2	15	46	35	30	19
16	1	64	49	20	29	36	45

Figure 8.29 A regular magic square created from the Franklin Square.

In addition to this, we can take a Franklin square and consider the four top rows and swap them end for end, which then causes the semi-magic square to become pan-diagonal (now that is pretty cool, and it just might be as cool as a square with right angle diagonals that add up to the magic square constant).

89

45	36	29	20	13	4	61	52
19	30	35	46	51	62	3	14
44	37	28	21	12	5	60	53
22	27	38	43	54	59	6	11
55	58	7	10	23	26	39	42
9	8	57	56	41	40	25	24
50	63	2	15	18	31	34	47
16	1	64	49	48	33	32	17

Figure 8.30   A pan-diagonal magic square created from the Franklin Square.

Benjamin Franklin may well have been aware of these properties within these variations of his magic square, and if he was aware, then he obviously chose to share his square in a way that was very unique to other contemporary magic squares of that era.

Other Zigzag Constructions
The Benjamin Franklin square uses a zigzag type of construction to create the magic square. There are variations of this method that we can use that will create still other magic squares. In this next example we start by placing the numbers in a sequence-of-four zigzag pattern throughout the square.

53	12	5	60	37	28	21	44
59	6	11	54	43	22	27	38
55	10	7	58	39	26	23	42
57	8	9	56	41	24	25	40
52	13	4	61	36	29	20	45
62	3	14	51	46	19	30	35
50	15	2	63	34	31	18	47
64	1	16	49	48	17	32	33

Figure 8.31   An initial 8 x 8 square using a Franklin-like construction method.

The array above shows the initial placement of the numbers within the 8 x 8 array. I've shaded the numbers 1 thru 8 in columns 2 and 3, and the numbers 17 thru 24 in columns 6 and 7 so that it is easier to see the zigzag pattern. Summing the numbers across each of the rows gives us 260 as the total for each row. Each of the rows already adds up to the 8th order constant; however, the columns do not. Swapping pairs of numbers on rows 2, 3, 6 and 7 will cause the columns to add up to 260, but the diagonals still don't. Here is the semi-magic square we get after the swaps:

53	12	5	60	37	28	21	44
6	59	54	11	22	43	38	27
10	55	58	7	26	39	42	23
57	8	9	56	41	24	25	40
52	13	4	61	36	29	20	45
3	62	51	14	19	46	35	30
15	50	63	2	31	34	47	18
64	1	16	49	48	17	32	33

Figure 8.32   A semi-magic square created from the Franklin-like initial square.

Now, if we take the square above and shift the members one column right or left, or one row up or down, we get a regular magic square. Here is the square from above after being shifted up one row.

6	59	54	11	22	43	38	27
10	55	58	7	26	39	42	23
57	8	9	56	41	24	25	40
52	13	4	61	36	29	20	45
3	62	51	14	19	46	35	30
15	50	63	2	31	34	47	18
64	1	16	49	48	17	32	33
53	12	5	60	37	28	21	44

Figure 8.33   The resulting magic square created from the previous semi-magic square.

Similarly, we can back up to our initial square (at the beginning of this section), and swap pairs of numbers within the first, fourth, fifth and eighth rows to create this semi-magic square.

12	53	60	5	28	37	44	21
59	6	11	54	43	22	27	38
55	10	7	58	39	26	23	42
8	57	56	9	24	41	40	25
13	52	61	4	29	36	45	20
62	3	14	51	46	19	30	35
50	15	2	63	34	31	18	47
1	64	49	16	17	48	33	32

Figure 8.34   A 2nd semi-magic square created from the Franklin-like square.

Then again, if we shift the members of the square one column right or left, or one row up or down, we get a regular magic square. Here is the magic square that we get when we take the square from above and shift the rows down one row.

1	64	49	16	17	48	33	32
12	53	60	5	28	37	44	21
59	6	11	54	43	22	27	38
55	10	7	58	39	26	23	42
8	57	56	9	24	41	40	25
13	52	61	4	29	36	45	20
62	3	14	51	46	19	30	35
50	15	2	63	34	31	18	47

Figure 8.35   A 2nd magic square created from the 2nd semi-magic square above.

And there are still other variations to the zigzag creation method that can be used on doubly-even magic squares. As before and as always, imagination is the key that unlocks the door to these and other methods.

Chapter 9
Nested Magic Squares and Magic Rings

Odd-Ordered Magic Rings
It is possible to have one or more magic squares within a magic square. The next chapter looks at magic square templates and uses arrays of magic squares to build larger magic squares. In this chapter we will look at a square array of numbers surrounded by a ring of numbers where the square array is a magic square, and then with the ring around the magic square, it creates another, larger magic square. These magic squares are concentric to each other, and therefore are called concentric, or border magic squares.

We'll start by going through an example. We'll create a 5th order magic square, by taking out the center 3 x 3 section of the 5 x 5 array, and use it to create a 3rd order magic square. We can do this because the center section contains 3 sets of 3 numbers that are evenly spaced. We will then use the remaining ring of numbers to create several 5th order magic squares surrounding the center 3 x 3 magic square. This will give us a 5th order magic square with a 3rd order magic square nested inside of it.

1	2	3	4	5
6	7	8	9	10
11	12	13	14	15
16	17	18	19	20
21	22	23	24	25

$\rightarrow$

7	8	9
12	13	14
17	18	19

$\rightarrow$

18	7	14
9	13	17
12	19	8

Initial Array　　　　**Center Removed**　　　**Magic Square**

Figure 9.1   A 5 x 5 array has the center 3 x 3 array removed and made into a 3rd order magic square.

Note that the rows, columns, and diagonals of the 3 x 3 magic square above add to 39. Now, with the center section removed, we are left with this ring of numbers below.

1	2	3	4	5
6				10
11				15
16				20
21	22	23	24	25

Figure 9.2   The 5 x 5 ring after the center 3 x 3 array has been removed.

We can rearrange the numbers in this ring, so that when we place the 3 x 3 magic square back in the ring, we will have a 5th order magic square. Here are four possible ways that the ring can be rearranged:

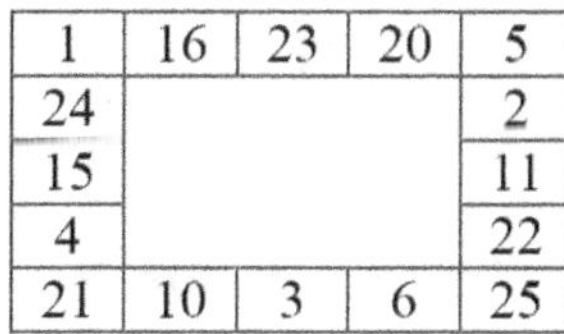

1	16	23	20	5
24				2
15				11
4				22
21	10	3	6	25

1	20	23	16	5
4				22
15				11
24				2
21	6	3	10	25

1	16	23	20	5
4				22
15				11
24				2
21	10	3	6	25

1	20	23	16	5
24				2
15				11
4				22
21	6	3	10	25

Figure 9.3   A set of four distinct 5 x 5 rings that will now create 5th order magic squares when the 3rd order square is returned to the center.

In each of these rings, the pairs of numbers that make up the inside rows, columns, and diagonals add to 26; while the outside rows and columns (around the perimeter) add to 65. Now, when we put the 3rd order magic square (whose constant is equal to 39) back inside the rings, we get this set of 5th order magic squares:

1	16	23	20	5
24	14	7	18	2
15	17	13	9	11
4	8	19	12	22
21	10	3	6	25

1	20	23	16	5
4	14	7	18	22
15	17	13	9	11
24	8	19	12	2
21	6	3	10	25

1	16	23	20	5
4	14	7	18	22
15	17	13	9	11
24	8	19	12	2
21	10	3	6	25

1	20	23	16	5
24	14	7	18	2
15	17	13	9	11
4	8	19	12	22
21	6	3	10	25

Figure 9.4   Four distinct 5th order magic squares created from combining the 3rd order square in the center and the 5 x 5 ring.

And since the center 3 x 3 section is a magic square, we can actually rotate, and/or mirror it within its magic ring to get still other unique magic squares. It should be noted that with the many variations of the 3 x 3 magic squares within the several types of magic rings, some rotation/mirroring combinations will produce identical magic squares; however, the majority of these 5th order magic squares will be unique to each other.

We can do the same thing with higher odd-order magic squares. Let's look at a 7 x 7 square of numbers.

1	2	3	4	5	6	7
8	9	10	11	12	13	14
15	16	17	18	19	20	21
22	23	24	25	26	27	28
29	30	31	32	33	34	35
36	37	38	39	40	41	42
43	44	45	46	47	48	49

9	10	11	12	13
16	17	18	19	20
23	24	25	26	27
30	31	32	33	34
37	38	39	40	41

Initial Array   **Center Section Removed**

Figure 9.5   A 7 x 7 array with the center 5 x 5 array removed.

By taking out the 5 x 5 square in the middle, we have 5 sets of 5 evenly spaced numbers which can be used to create a 5th order magic square. We then use the remaining 7 x 7 ring of numbers to create a magic ring. This magic ring can then circumscribe any of the 5th order magic squares to create a 7th order magic square with a nested 5th order magic square. Here is one possible 7 x 7 magic ring:

1	8	21	46	48	44	7
5						45
47						3
28						22
15						35
36						14
43	42	29	4	2	6	49

Figure 9.6   The 7 x 7 ring after the center 5 x 5 array has been removed.

With this initial magic ring, it is possible to create magic ring variations by rearranging the columns and/or rows. The only stipulation is that we must keep the column pairs and row pairs together. For example, we can swap the 8 column with the 48 column in the ring, but we must keep the 42 with the 8, and the 2 with the 48. Here are some variations to our 7 x 7 magic ring.

1	8	21	46	48	44	7
5						45
15						35
47						3
28						22
36						14
43	42	29	4	2	6	49

1	48	21	46	8	44	7
5						45
15						35
47						3
28						22
36						14
43	2	29	4	42	6	49

Figure 9.7   Two distinct 7 x 7 magic rings created from the original magic ring.

As one can easily infer, there are quite a few more 7 x 7 magic ring variations. Also, keep in mind that we can easily create a large number of 5th order magic squares that can be placed within each one of these 7 x 7 magic rings. Altogether, these various combinations can literally give us thousands of magic squares.

With what we've just discussed, we can now create a double-nested magic square. The following 7th order magic square has a nested 5th order magic square within it, and within the 5th order magic square is a nested 3rd order magic square.

1	48	21	46	8	44	7
5	9	30	39	34	13	45
15	40	32	17	26	10	35
47	27	19	25	31	23	3
28	12	24	33	18	38	22
36	37	20	11	16	41	14
43	2	29	4	42	6	49

Figure 9.8   A 7th order magic square with nested 5th and 3rd order magic squares.

The nested 3rd order magic square has a constant equal to 75. The constant for the nested 5th order magic square is 125; while the constant for the 7th order magic square is 175.

A Construction Method for Odd-Ordered Magic Ringed Magic Squares
The following diagram illustrates a method of building an odd-ordered magic square where if each outer ring is removed, the remaining square is still a magic square; and this process can be continued all the way to the center 3 x 3 array. To build this magic ringed magic square, the square array is divided into regions A through M (excluding I) as shown in the figure below:

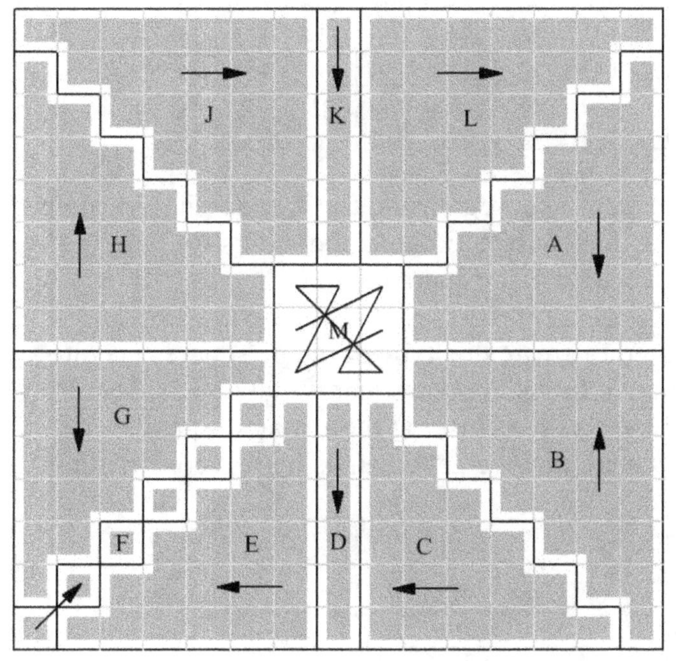

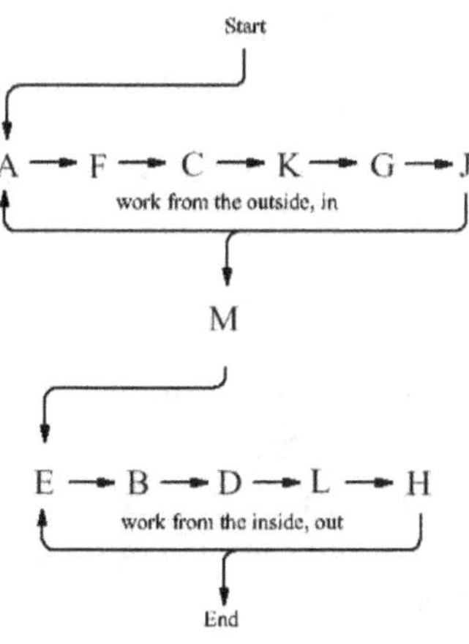

Figure 9.9   A construction method for any odd-ordered magic ringed magic square.

The way this method works is you start with 1 and go through all the numbers consecutively, starting in the outside column of region A moving down the column. Then move to region F and place one number, and then move to region C and place numbers along the bottom row, moving from right to left. Place the next number in region K, and then move to region G and place the next sequence of numbers in the outside column and moving down. Then move to region J and place the sequence of numbers in the top row, moving from left to right. At this point return to region A and repeat the cycle, moving towards the inside of the array one column or one row. This iterative cycle is repeated until you've entered only two numbers on the inside row of region J. At this point you move to the middle 3 x 3 array (region M) and enter the next 9 numbers as a 3rd ordered magic square.

Now, for the remaining regions, start at the inner-most row or column and continue placing the numbers sequentially as you move outward starting in region E, then moving to regions B, D, L and H (in that order). The iterative process is continued until you reach the outer rows and columns and the square is complete. The following is a 13th order magic square with regions A, C, G, and J shaded:

19	20	21	22	23	24	13	158	159	160	161	162	163
169	40	41	42	43	44	35	136	137	138	139	140	1
168	145	57	58	59	60	53	118	119	120	121	25	2
167	144	125	70	71	72	67	104	105	106	45	26	3
166	143	124	109	79	80	77	94	95	61	46	27	4
165	142	123	108	97	84	83	88	73	62	47	28	5
164	141	122	107	96	89	85	81	74	63	48	29	6
14	36	54	68	78	82	87	86	92	102	116	134	156
15	37	55	69	75	90	93	76	91	101	115	133	155
16	38	56	64	99	98	103	66	65	100	114	132	154
17	39	49	112	111	110	117	52	51	50	113	131	153
18	30	129	128	127	126	135	34	33	32	31	130	152
7	150	149	148	147	146	157	12	11	10	9	8	151

Figure 9.10 A 13th order magic square with regions A, C, G, and J shaded so that it is easier to see how this construction method works for building odd-ordered magic ringed magic squares.

Even though the construction method used to create this type of magic squares is quite unique, it is the property that they are magic ringed, magic squares that makes them truly amazing. Let's take a few minutes and look a little closer at these magic squares. The following is the 9th order nested magic square that we get when using this construction method.

13	14	15	16	9	74	75	76	77
81	26	27	28	23	60	61	62	1
80	65	35	36	33	50	51	17	2
79	64	53	40	39	44	29	18	3
78	63	52	45	41	37	30	19	4
10	24	34	38	43	42	48	58	72
11	25	31	46	49	32	47	57	71
12	20	55	54	59	22	21	56	70
5	68	67	66	73	8	7	6	69

Figure 9.11 The 9th order magic ringed, magic square created with the construction method detailed in this section.

This 9th order magic square has a center 3 x 3 array, and three rings surrounding the center. Looking at the outside ring, we see that all the pairs of numbers that are opposite each other on the rows, columns, and diagonals add to 82. We would naturally expect the pairs of numbers within the outside ring to all add to the same number. However, if we now remove the outside ring of this magic square and look at the new outside ring of the remaining square, we see that the pairs of numbers again add to 82. And now, if we do it once more, we find that the third outside ring also has the pairs of numbers that again add to 82. Even when we consider the inside 3 x 3 array, we again see that the outside ring of opposite numbers on the rows columns and diagonals also add to 82, with the center number equal to 41 (which is half of 82).

The following is the 11th order magic ringed, magic square. The sum of the numbers that are opposite of each other on the rows, columns and diagonals add to 122. And this remains true for all the numbers within the same corresponding ring all the way into the center where the center is equal to 61, which is half of 122.

16	17	18	19	20	11	112	113	114	115	116
121	33	34	35	36	29	94	95	96	97	1
120	101	46	47	48	43	80	81	82	21	2
119	100	85	55	56	53	70	71	37	22	3
118	99	84	73	60	59	64	49	38	23	4
117	98	83	72	65	61	57	50	39	24	5
12	30	44	54	58	63	62	68	78	92	110
13	31	45	51	66	69	52	67	77	91	109
14	32	40	75	74	79	42	41	76	90	108
15	25	88	87	86	93	28	27	26	89	107
6	105	104	103	102	111	10	9	8	7	106

Figure 9.12 The 11th order magic ringed, magic square created with the construction method detailed in this section.

And here is the 13th order magic ringed, magic square (that we also saw earlier when we examined how it was constructed).

19	20	21	22	23	24	13	158	159	160	161	162	163
169	40	41	42	43	44	35	136	137	138	139	140	1
168	145	57	58	59	60	53	118	119	120	121	25	2
167	144	125	70	71	72	67	104	105	106	45	26	3
166	143	124	109	79	80	77	94	95	61	46	27	4
165	142	123	108	97	84	83	88	73	62	47	28	5
164	141	122	107	96	89	85	81	74	63	48	29	6
14	36	54	68	78	82	87	86	92	102	116	134	156
15	37	55	69	75	90	93	76	91	101	115	133	155
16	38	56	64	99	98	103	66	65	100	114	132	154
17	39	49	112	111	110	117	52	51	50	113	131	153
18	30	129	128	127	126	135	34	33	32	31	130	152
7	150	149	148	147	146	157	12	11	10	9	8	151

Figure 9.13 The 13th order magic ringed, magic square created with the construction method detailed in this section.

The 13th order magic ringed, magic square again has this same property. The sum of the numbers that are opposite of each other on the rows, columns and diagonals add to 170. And this remains true for all the numbers within the same corresponding ring all the way into the center where the center is equal to 85, which is half of 170. Notice also that 170 is equal to $13^2 + 1$; likewise, 122 is equal to $11^2 + 1$; and 82 is equal to $9^2 + 1$.

This property where each ring adds to the same number is unique to all odd-ordered magic ringed, magic squares where the magic rings extend from the outside ring all the way to the inside 3 x 3 magic square. And in such cases, the center number will be half the value of the sum of the pairs of corresponding numbers in each ring.

Even- Ordered Magic Rings

I have not found a way to place a 6 x 6 ring around a 4th order magic square so that it creates a 6th order magic square with a nested 4th order magic square inside. You might think that using the Devedec method that we saw in chapter 6 might work, but it doesn't. The two types of even-ordered magic squares have different types of symmetries; therefore, it appears to be impossible to create such a type of even-ordered concentric magic square; however, far be it from me to actually say that it is completely impossible. However, it is possible to create a 4th order magic square within an 8 x 8 ring such that the 4th order magic square is nested and centered within an 8th order magic square. Let's look at how this could be done. We'll start with a standard 8 x 8 initial square and remove the center 4 x 4 array of numbers.

1	2	3	4	5	6	7	8
9	10	11	12	13	14	15	16
17	18	19	20	21	22	23	24
25	26	27	28	29	30	31	32
33	34	35	36	37	38	39	40
41	42	43	44	45	46	47	48
49	50	51	52	53	54	55	56
57	58	59	60	61	62	63	64

$\rightarrow$

19	20	21	22
27	28	29	30
35	36	37	38
43	44	45	46

Figure 9.14 An initial 8 x 8 array of numbers with the center 4 x 4 array removed.

We now use the numbers in this 4 x 4 array to create a 4th order magic square. We can use any method we want to create the 4th order magic square. In this case, the constant of this particular square will be equal to 130. Here is our 4th order magic square:

19	45	44	22
38	28	29	35
30	36	37	27
43	21	20	46

Figure 9.15 A 4th Order Magic Square made from the 4 x 4 array.

Next, we take the outside ring, which is two cells wide, and use it to make a magic ring. Here is our original ring on the left and a magic ring on the right:

1	2	3	4	5	6	7	8
9	10	11	12	13	14	15	16
17	18					23	24
25	26					31	32
33	34					39	40
41	42					47	48
49	50	51	52	53	54	55	56
57	58	59	60	61	62	63	64

$\rightarrow$

1	63	62	4	5	59	58	8
56	55	11	13	12	14	50	49
48	18					23	41
25	34					39	32
33	26					31	40
24	42					47	17
16	15	51	53	52	54	10	9
57	7	6	60	61	3	2	64

Figure 9.16   The initial 8 x 8 ring on the left is arranged into one of many possible 8th order magic rings on the right.

Now by placing the 4th order magic square within the 8 x 8 magic ring, we have an 8th order magic square with a nested 4th order magic square inside.

1	63	62	4	5	59	58	8
56	55	11	13	12	14	50	49
48	18	19	45	44	22	23	41
25	34	38	28	29	35	39	32
33	26	30	36	37	27	31	40
24	42	43	21	20	46	47	17
16	15	51	53	52	54	10	9
57	7	6	60	61	3	2	64

Figure 9.17   A Nested 4th Order Magic Square within an 8th Order Magic Square

As mentioned earlier in this chapter, we can keep the outside ring the same, and rearrange the center 4 x 4 cell into any 4th order magic square and thus create a new and unique nested 8th order magic square. Even an operation as simple as mirroring the original 4th order magic square will create another unique 8th order magic square when it is placed back within the magic ring.

19	45	44	22
38	28	29	35
30	36	37	27
43	21	20	46

$\rightarrow$

22	44	45	19
35	29	28	38
27	37	36	30
46	20	21	43

Figure 9.18   The original 4th order magic square on the left is mirrored on the right.

The following is another unique magic square using the same magic ring as in the previous example along with the mirrored 4th order magic square from the previous example.

1	63	62	4	5	59	58	8
56	55	11	13	12	14	50	49
48	18	22	44	45	19	23	41
25	34	35	29	28	38	39	32
33	26	27	37	36	30	31	40
24	42	46	20	21	43	47	17
16	15	51	53	52	54	10	9
57	7	6	60	61	3	2	64

Figure 9.19   A previous 4th order magic square is mirrored and placed again within a previous 8th order magic ring to create a unique magic square.

Alternate Initial Square Patterns

An interesting property that we see time and again with magic squares is that we can take an alternate initial square, apply the same steps, or method, on this new initial square, and in this way create a new and unique magic square. It is no different with these nested magic squares. For example, here is an alternate 8 x 8 initial array:

1	2	3	4	17	18	19	20
5	6	7	8	21	22	23	24
9	10	11	12	25	26	27	28
13	14	15	16	29	30	31	32
33	34	35	36	49	50	51	52
37	38	39	40	53	54	55	56
41	42	43	44	57	58	59	60
45	46	47	48	61	62	63	64

Figure 9.20   An alternate initial 8 x 8 array.

First, we remove the center 4 x 4 array of numbers and make that into a magic square.

11	12	25	26
15	16	29	30
35	36	49	50
39	40	53	54

→

11	53	40	26
50	16	29	35
30	36	49	15
39	25	12	54

Figure 9.21   The center 4 x 4 array is removed and made into a 4th order magic square.

The constant to the magic square above is equal to 130. Now we create the 8 x 8 magic ring, and place the 4th order magic square back within the magic ring to arrive at our 8th order magic square with the nested 4th order magic square. Here is the final product:

1	63	62	4	17	47	46	20
60	59	7	21	8	22	42	41
56	10	11	53	40	26	27	37
13	34	50	16	29	35	51	32
33	14	30	36	49	15	31	52
28	38	39	25	12	54	55	9
24	23	43	57	44	58	6	5
45	19	18	48	61	3	2	64

Figure 9.22   A nested 4th order magic square within an 8th order magic square created from an alternate initial 8 x 8 array.

And just to show that this isn't some kind of abstract fluke of nature, here is a second example where we start with yet another initial 8 x 8 array on the left, and use it to create a concentric 4th order magic square within the 8th order magic square on the right.

1	2	5	6	9	10	13	14
3	4	7	8	11	12	15	16
17	18	21	22	25	26	29	30
19	20	23	24	27	28	31	32
33	34	37	38	41	42	45	46
35	36	39	40	43	44	47	48
49	50	53	54	57	58	61	62
51	52	55	56	59	60	63	64

$\rightarrow$

1	63	60	6	9	55	52	14
62	61	7	11	8	12	50	49
48	18	21	43	40	26	29	35
19	34	42	24	27	37	45	32
33	20	28	38	41	23	31	46
30	36	39	25	22	44	47	17
16	15	53	57	54	58	4	3
51	13	10	56	59	5	2	64

Figure 9.23   A second initial 8 x 8 array on the left is used to create another unique 8th order magic square with a nested 4th order magic square on the right.

Finally, here is a 4th order square nested inside of an 8th order square which is nested inside of a 12th order magic square:

1	143	142	141	5	7	6	8	136	135	134	12
132	131	15	16	20	126	127	17	21	22	122	121
120	26	27	117	116	30	31	113	112	34	35	109
108	38	106	105	41	43	42	44	100	99	47	97
49	86	94	52	53	91	90	56	57	87	95	60
73	71	63	76	80	66	67	77	81	70	62	84
61	83	75	64	68	78	79	65	69	82	74	72
85	50	58	88	89	55	54	92	93	51	59	96
48	98	46	45	101	103	102	104	40	39	107	37
36	110	111	33	32	114	115	29	28	118	119	25
24	23	123	124	128	18	19	125	129	130	14	13
133	11	10	9	137	139	138	140	4	3	2	144

Figure 9.24   A 12th order magic square with a nested 8th order magic square within, and a nested 4th order magic square within it.

Due to the nature of these even-ordered concentric magic squares (where each magic ring is two cells wide), we can swap columns 1 and 2 with each other and then swap columns 11 and 12 with each other, and still have our 4th order nested inside an 8th order which is nested inside of a 12th order magic square. The following is the resulting magic square after these column swaps have been performed.

143	1	142	141	5	7	6	8	136	135	12	134
131	132	15	16	20	126	127	17	21	22	121	122
26	120	27	117	116	30	31	113	112	34	109	35
38	108	106	105	41	43	42	44	100	99	97	47
86	49	94	52	53	91	90	56	57	87	60	95
71	73	63	76	80	66	67	77	81	70	84	62
83	61	75	64	68	78	79	65	69	82	72	74
50	85	58	88	89	55	54	92	93	51	96	59
98	48	46	45	101	103	102	104	40	39	37	107
110	36	111	33	32	114	115	29	28	118	25	119
23	24	123	124	128	18	19	125	129	130	13	14
11	133	10	9	137	139	138	140	4	3	144	2

Figure 9.25   The previous concentric 12th order magic square with column 1 swapped with 2 and column 11 swapped with 12.

In fact, there are several pairs of swaps that we could perform. We could have also swapped columns 1 and 12 with each other; or just swapped column 2 with column 11; or we could have performed both of these swaps, each of which would still give us our concentric magic squares, where each would also still be unique from the others. And, we could have left the columns alone, and instead, performed these same swaps with the rows.

The swaps we just discussed focused on the outer ring of numbers. There are also pairs of columns and rows within this magic square that can be swapped with each other that will also maintain the overall concentricity of the magic square. These pairs of swaps include the following:

Column Swaps
        Swap columns 1 and 2, and swap columns 11 and 12
        Swap columns 1 and 12, and/or swap columns 2 and 11
        Swap columns 3 and 4, and swap columns 9 and 10
        Swap columns 3 and 10, and/or swap columns 4 and 9
        Swap columns 5 and 6, and swap columns 7 and 8
        Swap columns 5 and 8, and/or swap columns 6 and 7

Row Swaps
        Swap rows 1 and 2, and swap rows 11 and 12
        Swap rows 1 and 12, and/or swap rows 2 and 11
        Swap rows 3 and 4, and swap rows 9 and 10
        Swap rows 3 and 10, and/or swap rows 4 and 9
        Swap rows 5 and 6, and swap rows 7 and 8
        Swap rows 5 and 8, and/or swap rows 6 and 7

By performing any of these swaps or any combination of these swaps on many of the concentric 12th order magic squares, it is possible to have magic squares that will still maintain their concentric property.

As a final note, if we take our concentric magic square from above (the 12th order square with a nested 8th order and a nested 4th order magic square), and swap the two center columns, and the two center rows, and then highlight the numbers that are in their original positions, we find that the numbers in their original positions create concentric circles within our concentrically nested magic square:

1	143	142	141	5	6	7	8	136	135	134	12
132	131	15	16	20	127	126	17	21	22	122	121
120	26	27	117	116	31	30	113	112	34	35	109
108	38	106	105	41	42	43	44	100	99	47	97
49	86	94	52	53	90	91	56	57	87	95	60
61	83	75	64	68	79	78	65	69	82	74	72
73	71	63	76	80	66	67	77	81	70	62	84
85	50	58	88	89	54	55	92	93	51	59	96
48	98	46	45	101	102	103	104	40	39	107	37
36	110	111	33	32	115	114	29	28	118	119	25
24	23	123	124	128	19	18	125	129	130	14	13
133	11	10	9	137	138	139	140	4	3	2	144

Figure 9.26   The shaded numbers are the numbers that are in their original positions within this 12th order magic square which, by the way, has a nested 8th order magic square within, and within it is a nested 4th order magic square.

In the next chapter we'll use magic square templates to build larger magic squares from smaller ones. They won't have the so called "magic-ring" property, but they will still have magic squares nested within magic squares.

Chapter 10
Magic Square Templates

We'll continue our look at nested magic squares; however, in this chapter we'll go one step further and use magic squares within magic squares to build larger magic squares; and we'll do this by using a smaller order magic square as a template to build a larger order magic square.

Using the 3rd Order Method to Create a 9th Order Magic Square
We'll begin by creating a 9th order magic square from nine 3 x 3 magic squares arranged as a larger 3 x 3 magic square. To do this, we start with a 9 x 9 array and list the numbers 1 through 81 in our usual initial order (with values increasing from left to right, and from top to bottom). Then we create 9 separate 3 x 3 arrays. Then, to create the magic square, we treat each 3 x 3 array as a single cell and arranged these nine cells as we would a large 3 x 3 magic square. Finally, each of these cells will be treated as a 3 x 3 square and made into a magic square. The result will be a 9 x 9 magic square. Now that we have described what we are going to do, let's do it.

First we list the numbers 1 through 81 in a 9 x 9 array, and then we divide this array into nine 3 x 3 arrays:

<table>
<tr>
<td>

1	2	3
10	11	12
19	20	21

</td>
<td>

4	5	6
13	14	15
22	23	24

</td>
<td>

7	8	9
16	17	18
25	26	27

</td>
</tr>
<tr>
<td>

28	29	30
37	38	39
46	47	48

</td>
<td>

31	32	33
40	41	42
49	50	51

</td>
<td>

34	35	36
43	44	45
52	53	54

</td>
</tr>
<tr>
<td>

55	56	57
64	65	66
73	74	75

</td>
<td>

58	59	60
67	68	69
76	77	78

</td>
<td>

61	62	63
70	71	72
79	80	81

</td>
</tr>
</table>

Figure 10.1   An initial 9 x 9 array divided into nine 3 x 3 arrays.

This gives us nine sets of nine numbers. Each set of nine numbers is a 3 x 3 array. Next, we consider each 3 x 3 array as a single cell, and move them to where they would go for a 3rd order magic square solution. Here is the 3rd order magic square that we can use as a reference:

8	1	6
3	5	7
4	9	2

Figure 10.2 The 3rd order magic square.

And here is our 9 x 9 square with each 3 x 3 cell moved to its new location (where the new position corresponds with its position in the 3rd order magic square):

58	59	60
67	68	69
76	77	78

1	2	3
10	11	12
19	20	21

34	35	36
43	44	45
52	53	54

7	8	9
16	17	18
25	26	27

31	32	33
40	41	42
49	50	51

55	56	57
64	65	66
73	74	75

28	29	30
37	38	39
46	47	48

61	62	63
70	71	72
79	80	81

4	5	6
13	14	15
22	23	24

Figure 10.3 The nine 3 x 3 arrays are arranged as a 3rd order magic square.

Finally, each individual 3 x 3 array of numbers is arranged within itself, so that it creates a 3rd order magic square. Here is the magic square (with the spacing between the 3 x 3 arrays still in place):

77	58	69
60	68	76
67	78	59

20	1	12
3	11	19
10	21	2

53	34	45
36	44	52
43	54	35

26	7	18
9	17	25
16	27	8

50	31	42
33	41	49
40	51	32

74	55	66
57	65	73
64	75	56

47	28	39
30	38	46
37	48	29

80	61	72
63	71	79
70	81	62

23	4	15
6	14	22
13	24	5

Figure 10.4 Each 3 x 3 array is made into a 3rd order magic square, thus altogether making a 9th order magic square.

Now that we have the nine 3 x 3 arrays arranged as magic squares, we now have a 9th order magic square. But that's not all. Each 3 x 3 array can be rotated and/or mirrored within its own position while the other cells around it can either be held in their same orientation, or also rotated and/or mirrored. These combinations of rotations and mirrorings alone can give us hundreds of unique magic squares all related to this original magic square. For example, here is the magic square from above, except that the lower left cell has been rotated 90 degrees (which makes it unique from the square above):

77	58	69	20	1	12	53	34	45
60	68	76	3	11	19	36	44	52
67	78	59	10	21	2	43	54	35
26	7	18	50	31	42	74	55	66
9	17	25	33	41	49	57	65	73
16	27	8	40	51	32	64	75	56
39	46	29	80	61	72	23	4	15
28	38	48	63	71	79	6	14	22
47	30	37	70	81	62	13	24	5

Figure 10.5   A variation of the previous 9th order magic square where the lower left 3rd order square is rotated 90 degrees.

Of course, we could use this same method to create 16th order magic squares by using any of our 4th order magic squares as a template. We would simply divide the 16 x 16 array of into 16 cells, where each cell is a 4 x 4 array of numbers. Then arrange these 16 cells as a 4th order magic square, and then within each cell arrange those 16 numbers as a 4th order magic square. Using any combination of 4th order methods on each cell, as well as mirrorings and/or rotations, we could literally create thousands of distinct 16th order magic squares – all from this single template. The possibilities become mind boggling when we look at doing the same thing with the 25th order, and other higher order magic squares.

Also, the order doesn't need to be a perfect square like a 16th or 25th order for this method to work. We can also use templates to create other orders of squares, like a 12th, 15th, 18th and 20th order magic square. Here is a combination of a 3rd order method with 4th order arrays to create a 12th order magic square.

1	2	3	4		5	6	7	8		9	10	11	12
13	14	15	16		17	18	19	20		21	22	23	24
25	26	27	28		29	30	31	32		33	34	35	36
37	38	39	40		41	42	43	44		45	46	47	48

49	50	51	52		53	54	55	56		57	58	59	60
61	62	63	64		65	66	67	68		69	70	71	72
73	74	75	76		77	78	79	80		81	82	83	84
85	86	87	88		89	90	91	92		93	94	95	96

97	98	99	100		101	102	103	104		105	106	107	108
109	110	111	112		113	114	115	116		117	118	119	120
121	122	123	124		125	126	127	128		129	130	131	132
133	134	135	136		137	138	139	140		141	142	143	144

Figure 10.6   An initial 12 x 12 array is divided into nine 4 x 4 arrays.

We start by dividing this 12 x 12 array into nine 4 x 4 arrays. Now, treat each 4 x 4 array as a single cell. This gives us 9 cells which are arranged as a 3 x 3 square. Next, we arrange these 9 cells as a magic square to get:

101	102	103	104
113	114	115	116
125	126	127	128
137	138	139	140

1	2	3	4
13	14	15	16
25	26	27	28
37	38	39	40

57	58	59	60
69	70	71	72
81	82	83	84
93	94	95	96

9	10	11	12
21	22	23	24
33	34	35	36
45	46	47	48

53	54	55	56
65	66	67	68
77	78	79	80
89	90	91	92

97	98	99	100
109	110	111	112
121	122	123	124
133	134	135	136

49	50	51	52
61	62	63	64
73	74	75	76
85	86	87	88

105	106	107	108
117	118	119	120
129	130	131	132
141	142	143	144

5	6	7	8
17	18	19	20
29	30	31	32
41	42	43	44

Figure 10.7   The nine 4 x 4 arrays are arranged as a 3rd order magic square.

Then, we look at each 4 x 4 cell, and treat it as a 4 x 4 array, and we rearrange each of these arrays so that they are magic squares. There are hundreds of ways to create a 4th order magic square, and if wanted to, we could use a different method on each of the nine 4 x 4 squares; however, in this case, we will use the same method on each array. The following is the 12th order magic square (with the spacing between the 4 x 4 arrays still in place):

101	139	138	104
128	114	115	125
116	126	127	113
137	102	103	140

1	39	38	4
28	14	15	25
16	26	27	13
37	3	2	40

57	95	94	60
84	70	71	81
72	82	83	69
93	59	58	96

9	47	46	12
36	22	23	33
24	34	35	21
45	11	10	48

53	91	90	56
80	66	67	77
68	78	79	65
89	55	54	92

97	135	134	100
124	110	111	121
112	122	123	109
133	99	98	136

49	87	86	52
76	62	63	73
64	74	75	61
85	51	50	88

105	143	142	108
132	118	119	129
120	130	131	117
141	107	106	144

5	43	42	8
32	18	19	29
20	30	31	17
41	7	6	44

Figure 10.8   Each 4 x 4 array is made into a 4th order magic square, thus altogether making a 12th order magic square.

Each of the 4 x 4 arrays above is a magic square. Again, we can literally create thousands of 12th order magic squares by using different methods of creating 4th order magic squares within each 4 x 4 array. Then we can rotate and/or mirror each of these individual 4th order squares to produce large combinations of 12th order squares. The 12th order magic square below is identical to the 12th order magic square above except that the upper-left cell is a mirror of its counterpart in the previous magic square:

104	138	139	101	1	39	38	4	57	95	94	60
125	115	114	128	28	14	15	25	84	70	71	81
113	127	126	116	16	26	27	13	72	82	83	69
140	103	102	137	37	3	2	40	93	59	58	96
9	47	46	12	53	91	90	56	97	135	134	100
36	22	23	33	80	66	67	77	124	110	111	121
24	34	35	21	68	78	79	65	112	122	123	109
45	11	10	48	89	55	54	92	133	99	98	136
49	87	86	52	105	143	142	108	5	43	42	8
76	62	63	73	132	118	119	129	32	18	19	29
64	74	75	61	120	130	131	117	20	30	31	17
85	51	50	88	141	107	106	144	41	7	6	44

Figure 10.9   A variation of the previous 12th order magic square where the upper left 4th order square is mirrored from the previous magic square.

In the previous example, we used a combination of the 3rd order method and 4th order arrays to create a 12th order magic square. Alternately, we could have divided the original 12 x 12 array into sixteen 3 x 3 cells and created several other 12th order magic squares from that arrangement. Of course the possibilities become limitless when we consider even larger orders of magic squares.

Alternate Initial Squares

We can also start with alternate initial squares, use this method, and still create more unique magic squares. Let's take a look at an alternate initial 9 x 9 array of numbers.

1	4	7
28	31	34
55	58	61

2	5	8
29	32	35
56	59	62

3	6	9
30	33	36
57	60	63

10	13	16
37	40	43
64	67	70

11	14	17
38	41	44
65	68	71

12	15	18
39	42	45
66	69	72

19	22	25
46	49	52
73	76	79

20	23	26
47	50	53
74	77	80

21	24	27
48	51	54
75	78	81

Figure 10.10 An alternate initial 9 x 9 array divided into nine 3 x 3 arrays.

To create the initial square above, we refer to Initial Square 10.1. We take the numbers from the number one position in their respective cells and use them for the first cell in this initial square. The first cell consists of the numbers 1, 4, 7, 28, ... Likewise, the second cell is made from the numbers in the 2 position (the numbers 2, 5, 8, 29, …), and so on.

Again, we treat these nine cells as a 3 x 3 square, and make them into a magic square (while keeping the numbers within each cell in its same position).

20	23	26
47	50	53
74	77	80

1	4	7
28	31	34
55	58	61

12	15	18
39	42	45
66	69	72

3	6	9
30	33	36
57	60	63

11	14	17
38	41	44
65	68	71

19	22	25
46	49	52
73	76	79

10	13	16
37	40	43
64	67	70

21	24	27
48	51	54
75	78	81

2	5	8
29	32	35
56	59	62

Figure 10.11 The nine 3 x 3 arrays are first arranged as a 3rd order magic square.

Finally, each of the nine 3 x 3 cells is treated as a magic square. The result is a 9th order magic square. This is the 9th order magic square still divided into its nine 3 x 3 cells:

77	20	53
26	50	74
47	80	23

58	1	34
7	31	55
28	61	4

69	12	45
18	42	66
39	72	15

60	3	36
9	33	57
30	63	6

68	11	44
17	41	65
38	71	14

76	19	52
25	49	73
46	79	22

67	10	43
16	40	64
37	70	13

78	21	54
27	51	75
48	81	24

59	2	35
8	32	56
29	62	5

Figure 10.12 Each 3 x 3 array is made into a 3rd order magic square, thus making a unique 9th order magic square from the alternate initial square.

Using a Template and a 2 x 2 Cell
There are no 2nd order magic squares; however, in this example we will take a 6 x 6 array and divide it into nine 2 x 2 cells. We will then take these nine cells and arrange them as a third order magic square, and then finally, we will make a pair of swaps within each cell (except the center one) plus three more swaps, and create a magic square. Here is our initial 6 x 6 array of numbers:

1	2
3	4

5	6
7	8

9	10
11	12

13	14
15	16

17	18
19	20

21	22
23	24

25	26
27	28

29	30
31	32

33	34
35	36

Figure 10.13 The initial 6 x 6 array divided into nine cells.

Now we use our 3rd order magic square as a template to arrange the nine cells:

8	1	6
3	5	7
4	9	2

29	30	1	2	21	22
31	32	3	4	23	24
9	10	17	18	25	26
11	12	19	20	27	28
13	14	33	34	5	6
15	16	35	36	7	8

Figure 10.14 The 3rd order template is on the left and it is used as a guide to arrange the nine 2 x 2 arrays on the right.

Now we swap a pair of numbers in each cell, except the center cell.

29	31	1	4	24	22
30	32	3	2	23	21
10	9	17	18	25	26
11	12	19	20	28	27
16	14	35	34	5	7
15	13	33	36	6	8

Figure 10.15 Pairs of numbers are swapped in the four corner cells to bring us one step closer to the 6th order magic square.

The numbers in the four corner cells that were switched are the 30 with the 31, the 21 with the 24, the 13 with the 16, and the 6 with the 7. The numbers in the sides and the top and bottom cells that were switched are the 2 with the 4, the 35 with the 33, the 9 with the 10, and the 28 with the 27. With the sides and top and bottom cells, we could have also swapped the opposite pairs in each cell. Now to finish our magic square, we only have to make three more swaps.

29	31	4	1	24	22
30	32	3	2	23	21
10	12	17	18	28	26
11	9	19	20	25	27
16	14	35	34	5	7
15	13	33	36	6	8

Figure 10.16 Several asymmetric swaps are made in the center cells (along the top, bottom and sides), to finally create the 6th order magic square.

Finally, in the figure above, we swap the 9 and 12, the 25 and 28 and the 1 and 4; and now we have a magic square.

The following figure shows how the swapping works within the 2 x 2 cells. The figure below represents a 6th order magic square where the 36 elements are divided into nine 2 x 2 cells. It is within the four corner cells that we swap two of the elements diagonally while leaving the other two in place. The lines in the figure show which of the two elements trade places within each cell. The five 2 x 2 cells that make up the center section of the array will have their own swap method.

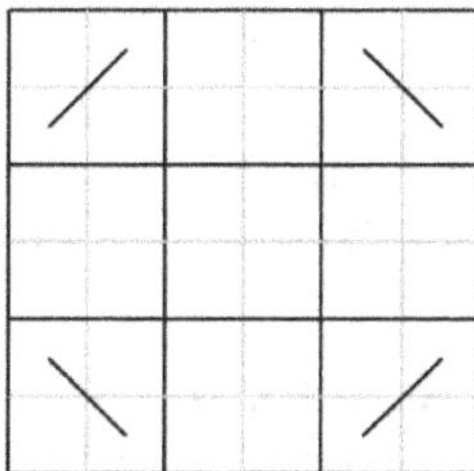

Figure 10.17 Swap pattern of the 2 x 2 cells within a 6th order magic square.

This pattern of swaps may not seem like very much; however, this pattern becomes even more pronounced as we look at higher orders of magic squares in just a bit. For now, let's look at it one more time within a 6th order magic square.

As in previous examples, we saw that we could use a different initial square and the same method to create other magic squares. Again, we treat the initial 6 x 6 array as a 3rd order square made up of 2 x 2 cells. Therefore, there are 9 cells with 4 numbers in each cell. In this case, the numbers in each cell have a difference of 9. For example, the first cell includes the numbers 1, 10, 19, 28 (each number is 9 more than the preceding number). The second cell includes the numbers 2, 11, 20, 29, etc. In this way, we evenly distribute the 36 numbers between the 9 cells. This is what our initial arrangement looks like:

1	10		2	11		3	12
19	28		20	29		21	30
4	13		5	14		6	15
22	31		23	32		24	33
7	16		8	17		9	18
25	34		26	35		27	36

Figure 10.18 An alternate initial 6 x 6 array divided into nine 2 x 2 arrays.

Now we take these 9 cells, and arrange them as we would in a 3rd order magic square. Doing so give us:

<table>
<tr><td>8</td><td>1</td><td>6</td></tr>
<tr><td>3</td><td>5</td><td>7</td></tr>
<tr><td>4</td><td>9</td><td>2</td></tr>
</table>

8	17
26	35

1	10
19	28

6	15
24	33

3	12
21	30

5	14
23	32

7	16
25	34

4	13
22	31

9	18
27	36

2	11
20	29

Figure 10.19 Again, the 3rd order template is on the left and it is used as a guide to arrange the nine 2 x 2 arrays on the right.

Next, we perform the same swaps as we did in the previous example. (Note that we are swapping by position, not by number). Specifically, the 17 and 26 trade places, as do the 6 and 33, the 11 and 20 and the 4 and 31 in each of the corner cells. Likewise, for the center cells on the top, bottom and sides, the 10 and 28 swap places, the 3 and 12, the 25 and 34, and the 9 and 27. These swaps give us the following square of numbers.

8	26
17	35

1	28
19	10

33	15
24	6

12	3
21	30

5	14
23	32

7	16
34	25

31	13
22	4

27	18
9	36

2	20
11	29

Figure 10.20 Pairs of numbers are swapped in the four corner cells and the four center cells along the top, bottom and sides.

Again, we still don't quite have a magic square; but after making the next three swaps, we will. The 1 and 28 trade places, as do the 3 and 30, and the 7 and the 34. (Or we could have swapped the 9 and 36 instead of the 1 and the 28.) Removing the spaces, we have our finished 6th order magic square:

8	26	28	1	33	15
17	35	19	10	24	6
12	30	5	14	34	16
21	3	23	32	7	25
31	13	27	18	2	20
22	4	9	36	11	29

Figure 10.21 A final set of three swaps creates the desired 6th order magic square.

Interestingly enough, this method does not appear to work with all types of initial squares (such as our usual initial square where we place our one in the upper left corner and list the numbers sequentially in ascending order from left to right and from top to bottom).

2 x 2 Cells in 10th and 14th Order Magic Squares

Now let's try the same thing with a 10th order square of numbers. We'll take a 10 x 10 array and divide it into twenty-five 2 x 2 cells. We then take these twenty-five cells, arrange them as a fifth order magic square, and then finally, make several swaps, and in this way create a 10th order magic square. Here is our initial 10 x 10 array of numbers:

1	2		5	6		9	10		13	14		17	18
3	4		7	8		11	12		15	16		19	20

21	22		25	26		29	30		33	34		37	38
23	24		27	28		31	32		35	36		39	40

41	42		45	46		49	50		53	54		57	58
43	44		47	48		51	52		55	56		59	60

61	62		65	66		69	70		73	74		77	78
63	64		67	68		71	72		75	76		79	80

81	82		85	86		89	90		93	94		97	98
83	84		87	88		91	92		95	96		99	100

Figure 10.22 The initial 10 x 10 array divided into 25 cells.

Now we will take a 5th order magic square and use it as a template on our initial square.

17	24	1	8	15
23	5	7	14	16
4	6	13	20	22
10	12	19	21	3
11	18	25	2	9

Figure 10.23 The 5th order magic square that will be used as a template.

Using the template, we get the following intermediate square within our 10 x 10 array:

65	66		93	94		1	2		29	30		57	58
67	68		95	96		3	4		31	32		59	60

89	90		17	18		25	26		53	54		61	62
91	92		19	20		27	28		55	56		63	64

13	14		21	22		49	50		77	78		85	86
15	16		23	24		51	52		79	80		87	88

37	38		45	46		73	74		81	82		9	10
39	40		47	48		75	76		83	84		11	12

41	42		69	70		97	98		5	6		33	34
43	44		71	72		99	100		7	8		35	36

Figure 10.24 The template is used as a guide to arrange the 2 x 2 arrays.

Finally, to create a magic square, we must perform a diagonal swap in every 2 x 2 cell except the cells in the center column and center row. Many of those cells will require three of the numbers to trade places. Without belaboring the detail, here is our 10th order magic square:

65	67	93	95	2	3	32	30	60	58
66	68	94	96	4	1	31	29	59	57
89	91	17	19	27	26	56	54	64	62
90	92	18	20	25	28	55	53	63	61
16	14	24	22	49	50	78	80	87	85
13	15	21	23	51	52	79	77	86	88
40	38	48	46	73	76	81	83	9	11
39	37	47	45	75	74	82	84	10	12
44	42	72	70	100	97	5	7	33	35
43	41	71	69	99	98	6	8	34	36

Figure 10.25 The resulting 10th order magic square created from the template.

Now when we compare the swap pattern of this magic square with that of the 6th order magic square, we begin to see many similarities. Again, the corner 2 x 2 cells only swap two diagonal members within their cells while the center column and row has its own swap pattern. The following figure shows the swap pattern that this method uses on 10th order magic squares. (There are multiple swap patterns within the center column and center row, each of which would create a unique magic square, and therefore are not shown in the figure below.)

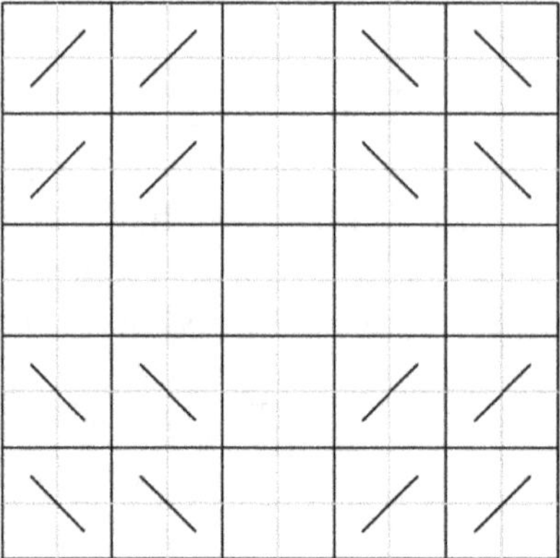

Figure 10.26 The swap pattern used within the 2 x 2 cells for 10th order magic squares.

It appears that we could have used any 5th order magic square as a template and then by making the same or similar swaps within each cell we could then create other unique 10th order magic squares. Likewise, there are alternate initial squares that we could use and then with the same or similar swaps within each cell we could create even more unique 10th order magic squares. One example of an alternate initial square would use the numbers 1, 26, 51 and 76 in the first 2 x 2 cell, and the numbers 2, 27, 52 and 77 in the second 2 x 2 cell, and so on.

The possibilities become mind boggling. It appears that this method works for any singly even-order magic square. As a final example in this section, we'll take a 14th order array of numbers and use a 7th order magic square as a temple to create a 14th order magic square. We begin by creating an initial square similar to the one used as initial square 10.13 and 10.21. (The first 2 x 2 cell has the numbers 1, 2, 3 and 4; the second 2 x 2 cell has the numbers 5, 6, 7 and 8; and so on.) We then select a seventh order magic square as the template.

30	39	48	1	10	19	28
38	47	7	9	18	27	29
46	6	8	17	26	35	37
5	14	16	25	34	36	45
13	15	24	33	42	44	4
21	23	32	41	43	3	12
22	31	40	49	2	11	20

Figure 10.27 The 7th order magic square that will be used as a template to create a 14th order magic square.

Using the template, we create an intermediate 14th order square:

117	118	153	154	189	190	1	2	37	38	73	74	109	110
119	120	155	156	191	192	3	4	39	40	75	76	111	112
149	150	185	186	25	26	33	34	69	70	105	106	113	114
151	152	187	188	27	28	35	36	71	72	107	108	115	116
181	182	21	22	29	30	65	66	101	102	137	138	145	146
183	184	23	24	31	32	67	68	103	104	139	140	147	148
17	18	53	54	61	62	97	98	133	134	141	142	177	178
19	20	55	56	63	64	99	100	135	136	143	144	179	180
49	50	57	58	93	94	129	130	165	166	173	174	13	14
51	52	59	60	95	96	131	132	167	168	175	176	15	16
81	82	89	90	125	126	161	162	169	170	9	10	45	46
83	84	91	92	127	128	163	164	171	172	11	12	47	48
85	86	121	122	157	158	193	194	5	6	41	42	77	78
87	88	123	124	159	160	195	196	7	8	43	44	79	80

Figure 10.28 Again, the template is used as a guide to arrange the 2 x 2 arrays.

Then finally, we create a magic square by performing a diagonal swap in every cell except the cells in the center column and center row. Many of the cells in the center column and center row again require three of the numbers to trade places. And again without belaboring the detail, here is our 14th order magic square:

117	119	153	155	189	191	2	3	40	38	76	74	112	110
118	120	154	156	190	192	4	1	39	37	75	73	111	109
149	151	185	187	25	27	35	34	72	70	108	106	116	114
150	152	186	188	26	28	36	33	71	69	107	105	115	113
181	183	21	23	29	31	67	66	104	102	140	138	148	146
182	184	22	24	30	32	65	68	103	101	139	137	147	145
20	18	56	54	64	62	97	98	134	136	143	141	179	177
17	19	53	55	61	63	99	100	135	133	142	144	178	180
52	50	60	58	96	94	129	132	165	167	173	175	13	15
51	49	59	57	95	93	131	130	166	168	174	176	14	16
84	82	92	90	128	126	164	161	169	171	9	11	45	47
83	81	91	89	127	125	163	162	170	172	10	12	46	48
88	86	124	122	160	158	193	196	5	7	41	43	77	79
87	85	123	121	159	157	194	195	6	8	42	44	78	80

Figure 10.29 The resulting 14th order magic square created from the template.

The following figure shows the swap pattern that was used to create the 14th order magic square above. Again we see that the corner 2 x 2 cells diagonally swap two of their members within each cell. And again, the 2 x 2 cells that makeup the center row and center column have their own swap patterns (which are not shown in the figure below). And actually, there are several different possibilities and each one will generate a unique magic square (while at the same time the values in the corner cells are kept in their same positions).

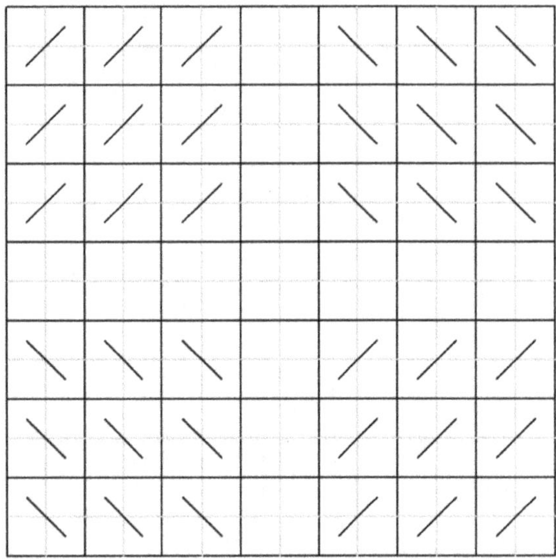

Figure 10.30 The diagonal lines show the swap pattern used within the 2 x 2 cells for 14th order magic squares.

2 x 2 Cells in 8th Order Magic Squares

We can also use a variation of this method to create an 8th order magic square. We do this by creating an initial 8 x 8 square that is broken up into sixteen 2 x 2 cells. We then populate the cells in a manner similar to our previous initial squares. Then by using a 4th order magic square as a template, we use a similar method to transform the 8 x 8 array into an 8th order magic square.

We begin by creating 16 cells made of 2 x 2 arrays. The 4 numbers in each cell will have a difference of 16. Therefore, the first cell is made of the numbers 1, 17, 33 and 49. The second cell is made of the numbers 2, 18, 34 and 50; and so on. Here is the initial 8 x 8 square of numbers grouped into the 16 cells:

1	17		2	18		3	19		4	20
33	49		34	50		35	51		36	52

5	21		6	22		7	23		8	24
37	53		38	54		39	55		40	56

9	25		10	26		11	27		12	28
41	57		42	58		43	59		44	60

13	29		14	30		15	31		16	32
45	61		46	62		47	63		48	64

Figure 10.31 The initial 8 x 8 array divided into sixteen 2 x 2 arrays.

Next, we treat these 16 cells as a 4th order square, and move them to their magic square positions. We'll use our first 4th order magic square as a template so that we can see where to place the cells in the 8 x 8 array. From the template, we see that cells 2 and 15 trade places, as well as 3 and 14, 5 and 12, and 8 and 9.

1	15	14	4
12	6	7	9
8	10	11	5
13	3	2	16

1	17		15	31		14	30		4	20
33	49		47	63		46	62		36	52

12	28		6	22		7	23		9	25
44	60		38	54		39	55		41	57

8	24		10	26		11	27		5	21
40	56		42	58		43	59		37	53

13	29		3	19		2	18		16	32
45	61		35	51		34	50		48	64

Figure 10.32 A 4th order template is on the left and it is used as a guide to arrange the sixteen 2 x 2 arrays on the right.

We don't have a magic square yet. To get a magic square, two of the numbers in each cell must trade places. As we move from left to right, within the first row of cells, the pairs of numbers that swap places are:

```
First Row:    33 and 17, 47 and 31, 14 and 62,  4 and 52
Second Row:   44 and 28,  6 and 54, 39 and 23,  9 and 57
Third Row:     8 and 56, 42 and 26, 11 and 59, 37 and 21
Fourth Row:   13 and 61,  3 and 51, 34 and 18, 48 and 32
```

Doing this, and removing the spaces, gives us the following 8th order magic square:

1	33	15	47	62	30	52	20
17	49	31	63	46	14	36	4
12	44	54	22	7	39	57	25
28	60	38	6	23	55	41	9
56	24	10	42	59	27	5	37
40	8	26	58	43	11	21	53
61	29	51	19	2	34	16	48
45	13	35	3	18	50	32	64

Figure 10.33 The resulting 8th order magic square created from the template.

No other swaps than the diagonal swaps in each cell were required to create this magic square. The following figure shows the swaps within each of the 2 x 2 cells. Notice that there are no center cells within the rows or columns that require their own type of swap. Instead, this swap pattern takes us from our initial pattern directly to a magic square.

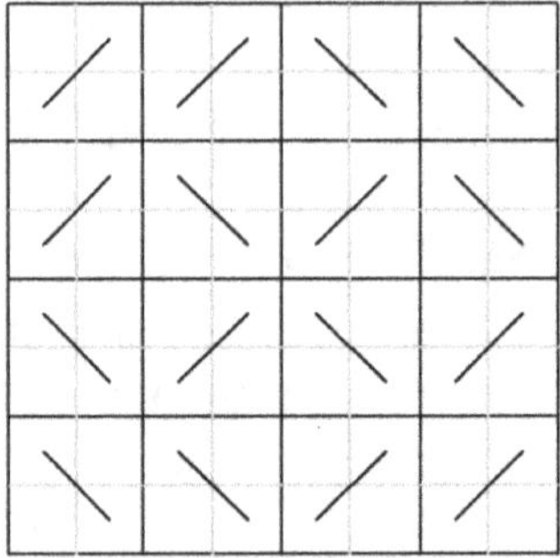

Figure 10.34 The diagonal lines show the swap pattern used within the 2 x 2 cells for 8th order magic squares.

So, what if we use a different 4th order template, can we still get an 8th order magic square? In this next example, we will use initial square 10.30 and a diabolical 4th order magic square as a template to place our cells in the 8 x 8 array:

1	15	4	14
8	10	5	11
13	3	16	2
12	6	9	7

1	17		15	31		4	20		14	30
33	49		47	63		36	52		46	62

8	24		10	26		5	21		11	27
40	56		42	58		37	53		43	59

13	29		3	19		16	32		2	18
45	61		35	51		48	64		34	50

12	28		6	22		9	25		7	23
44	60		38	54		41	57		39	55

Figure 10.35 Another 4th order template is used to arrange the sixteen 2 x 2 arrays on the right.

Again, we don't have a magic square yet, but if each cell (by location) performs the same swapping that was done in the previous example, we will. Here is the 8th order magic square after the swapping has been performed and the spaces have been removed:

1	33	15	47	52	20	62	30
17	49	31	63	36	4	46	14
8	40	58	26	5	37	59	27
24	56	42	10	21	53	43	11
61	29	3	35	64	32	2	34
45	13	19	51	48	16	18	50
60	28	54	22	9	41	7	39
44	12	38	6	25	57	23	55

Figure 10.36 The resulting 8th order magic square created from the new template.

There are also at least three other swap patterns that we can use on these initial squares to generate 8th order magic squares. These swap patterns look like this:

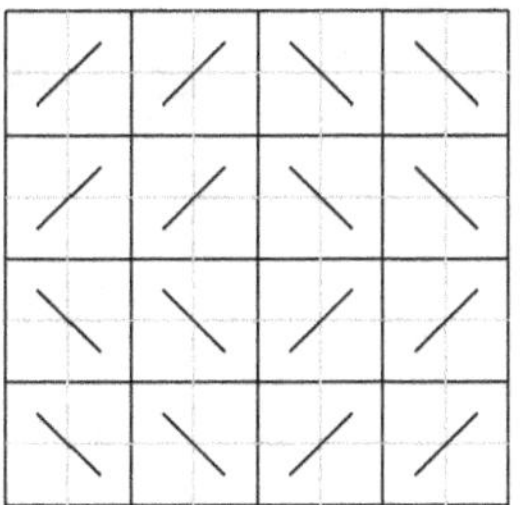 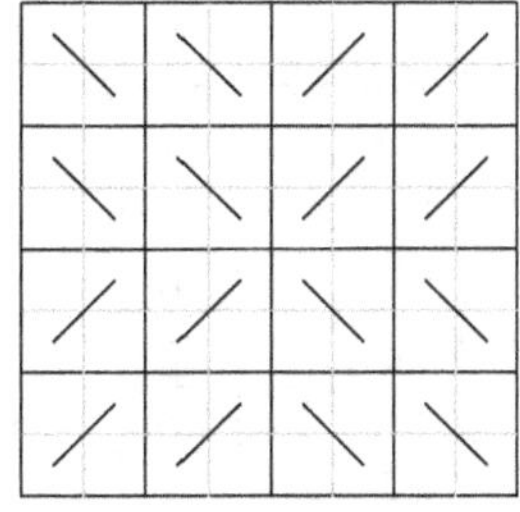 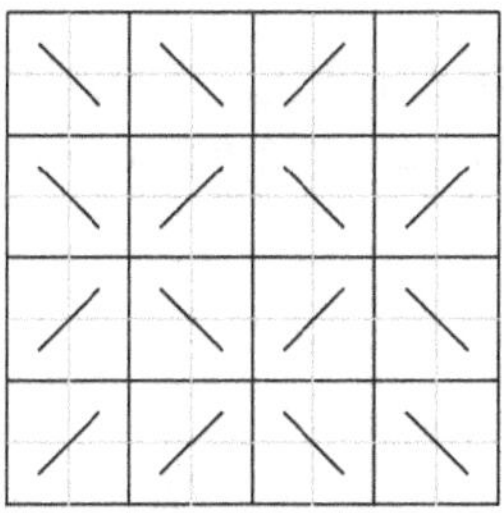

Figure 10.37 An alternate 2 x 2 swap patterns that can be used on 8th order arrays.

While these swap methods don't appear to work with every initial square pattern, it does work with the following pattern where we put the numbers 1, 2, 3, 4 in the first 2 x 2 cell, the numbers 5, 6, 7, 8 in the second cell, and so on:

1	2		5	6		9	10		13	14
3	4		7	8		11	12		15	16

17	18		21	22		25	26		29	30
19	20		23	24		27	28		31	32

33	34		37	38		41	42		45	46
35	36		39	40		43	44		47	48

49	50		53	54		57	58		61	62
51	52		55	56		59	60		63	64

Figure 10.38 An alternate initial 8 x 8 array divided into sixteen 2 x 2 arrays.

Again, we treat these 16 cells as a 4th order square, and move them to their magic square positions. From the template, we see that cells 2 and 15 trade places, as well as 3 and 14, 5 and 12, and 8 and 9.

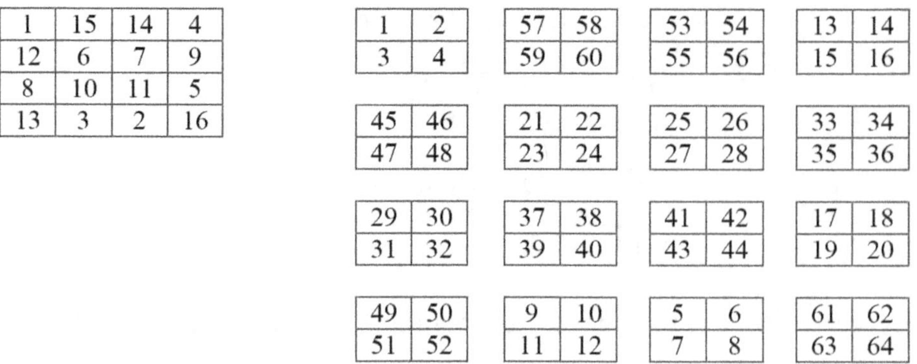

Figure 10.39 Again, the 4th order template on the left is used as a guide to arrange the sixteen 2 x 2 arrays on the right.

And once again, after we perform the diagonal swaps within the 2 x 2 cells, we have our 8th order magic square:

1	3	57	59	56	54	16	14
2	4	58	60	55	53	15	13
45	47	21	23	28	26	36	34
46	48	22	24	27	25	35	33
32	30	40	38	41	43	17	19
31	29	39	37	42	44	18	20
52	50	12	10	8	6	61	63
51	49	11	9	7	5	62	64

Figure 10.40 The resulting 8th order magic square created from the new template.

This method also appears to work with many types of higher doubly even-order magic squares. In other words, 12th order and 16th order magic squares follow this same type of method. And as mentioned earlier with the singly even-order magic squares, this type also appears to work with any type of valid template, and will work on several types of initial squares.

Designer Magic Squares
In the previous chapter we looked at nested magic squares and magic rings. In this chapter we used smaller ordered magic squares as templates to build higher order magic squares. We can use combinations of these two techniques (templates and magic rings) to build larger order magic squares that have a combination of these properties. I call these "designer" magic squares.

The following magic square was created from a 15 x 15 array of numbers. This larger array was then divided into nine 5 x 5 arrays where the numbers 1 through 25 are in the first array, 26 through 50 are in the second array, and so on. The nine 5 x 5 arrays are then arranged as a 3rd order magic square (where each cell of the 3 x 3 magic square is actually a 5 x 5 array) Each 5 x 5 array is then arranged as a 5 x 5 magic ring with a nested third order magic square. Altogether, this pattern creates a 15th order designer magic square with a combination of template and magic ring properties.

To help visualize what I am trying to say, here is a single 5 x 5 magic ring on the left, along with the internal 3 x 3 magic square on the right. These are used as a template for the nine 5 x 5 cells within the 15th order magic square below:

1	16	23	20	5
24				2
15				11
4				22
21	10	3	26	25

$\rightarrow$

18	7	14
15	9	17
12	19	8

176	191	198	195	180	1	16	23	20	5	126	141	148	145	130
199	193	182	189	177	24	18	7	14	2	149	143	132	139	127
190	184	188	192	186	15	9	13	17	11	140	134	138	142	136
179	187	194	183	197	4	12	19	8	22	129	137	144	133	147
196	185	178	181	200	21	10	3	6	25	146	135	128	131	150
51	66	73	70	55	101	116	123	120	105	151	166	173	170	155
74	68	57	64	52	124	118	107	114	102	174	168	157	164	152
65	59	63	67	61	115	109	113	117	111	165	159	163	167	161
54	62	69	58	72	104	112	119	108	122	154	162	169	158	172
71	60	53	56	75	121	110	103	106	125	171	160	153	156	175
76	91	98	95	80	201	216	223	220	205	26	41	48	45	30
99	93	82	89	77	224	218	207	214	202	49	43	32	39	27
90	84	88	92	86	215	209	213	217	211	40	34	38	42	36
79	87	94	83	97	204	212	219	208	222	29	37	44	33	47
96	85	78	81	100	221	210	203	206	225	46	35	28	31	50

Figure 10.41 A 15th order magic square made from 5th order nested magic squares (where each has a 3rd order magic square inside); and where the nine 5th order magic squares are arranged as a large 3rd order magic square to make the 15th order magic square.

The following is another designer magic square where I use a 4th order pan-diagonal magic square as a template for a 16th order magic square. The 16 x 16 array is divided into sixteen 4 x 4 arrays where the sixteen numbers within each 4 x 4 array is arranged as a pan-diagonal magic square. These sixteen arrays are in turn also arranged as a 4th order pan-diagonal magic square, thus giving us a 16th order pan-diagonal magic square.

222	219	213	212	174	171	165	164	78	75	69	68	62	59	53	52
209	216	218	223	161	168	170	175	65	72	74	79	49	56	58	63
220	221	211	214	172	173	163	166	76	77	67	70	60	61	51	54
215	210	224	217	167	162	176	169	71	66	80	73	55	50	64	57
14	11	5	4	126	123	117	116	158	155	149	148	238	235	229	228
1	8	10	15	113	120	122	127	145	152	154	159	225	232	234	239
12	13	3	6	124	125	115	118	156	157	147	150	236	237	227	230
7	2	16	9	119	114	128	121	151	146	160	153	231	226	240	233
190	187	181	180	206	203	197	196	46	43	37	36	94	91	85	84
177	184	186	191	193	200	202	207	33	40	42	47	81	88	90	95
188	189	179	182	204	205	195	198	44	45	35	38	92	93	83	86
183	178	192	185	199	194	208	201	39	34	48	41	87	82	96	89
110	107	101	100	30	27	21	20	254	251	245	244	142	139	133	132
97	104	106	111	17	24	26	31	241	248	250	255	129	136	138	143
108	109	99	102	28	29	19	22	252	253	243	246	140	141	131	134
103	98	112	105	23	18	32	25	247	242	256	249	135	130	144	137

Figure 10.42 A 16th order pan-diagonal magic square made from sixteen 4th order pan-diagonal magic squares all arranged as a larger 4th order pan-diagonal magic square.

And since each 4 x 4 cell is pan-diagonal, each of them can be shifted, rotated and/or mirrored in place so that any of these combinations will still create a 16th order pan-diagonal magic square. This will literally give us thousands of 16th order pan-diagonal magic squares, all derived from an initial 4th order pan-diagonal magic square.

Chapter 11
Building Double Even-Ordered Pan-Diagonal Magic Squares

Building 4th Order Pan-Diagonal Magic Squares

As mentioned earlier, a pan-diagonal magic square is one where not only the rows, columns and main diagonals all add up to the same constant, but so do all of the broken diagonals. In this section we will see a method that will create a 4th order magic square from scratch. We will start with an initial square, swap a couple of columns, and a couple of rows, and then finally swap a few numbers to arrive at the pan-diagonal magic square. First, we have the initial square on the left below, and on the right, we've taken the initial square, and swapped columns three and four with each other, and then swapped rows three and four with each other.

1	2	3	4
5	6	7	8
9	10	11	12
13	14	15	16

1	2	4	3
5	6	8	7
13	14	16	15
9	10	12	11

Figure 11.1   Start with the initial square on the left and then swap columns 3 and 4, and then swap rows 3 and 4 to create the intermediate square on the right.

Note that at this point we don't even have a magic square yet; however, after the following number swaps, we will have a pan-diagonal magic square. Using the above square on the right, first, swap numbers 2 and 14, and numbers 3 and 15; and then swap numbers 5 and 8, and numbers 9 and 12. Finally, swap number 6 and 11, and numbers 7 and 10. This gives us the following 4th order pan-diagonal magic square.

1	14	4	15
8	11	5	10
13	2	16	3
12	7	9	6

Figure 11.2   After completing the described number swaps, we have this 4th order pan-diagonal magic square.

We also could have started with either of the initial squares below:

1	2	5	6
3	4	7	8
9	10	13	14
11	12	15	16

1	2	9	10
5	6	13	14
3	4	11	12
7	8	15	16

Figure 11.3   Two other initial squares that can be used to create pan-diagonal magic squares.

Then after swapping columns 3 and 4 with each other, and then rows 3 and 4 with each other we get these two intermediate squares (the square below on the left corresponds with the initial square above on the left, and the square below on the right corresponds with the initial square above and on the right).

1	2	6	5
3	4	8	7
11	12	16	15
9	10	14	13

1	2	10	9
5	6	14	13
7	8	16	15
3	4	12	11

Figure 11.4   The two intermediate squares created from the two initial squares from the previous figure.

Finally, we perform the six pairs of number swaps within each square (these are swap-by-position exchanges even though we are identifying the positions by the numbers in those position). Specifically, on the left we swap the numbers 2 and 12, the numbers 5 and 15, the numbers 3 and 8, the numbers 9 and 14, the numbers 7 and 10, and then the numbers 4 and 13. This gives us the pan-diagonal magic square below, and on the left.

Using the intermediate square above and on the right we swap the numbers in the same positions, specifically the 2 and 8, the numbers 9 and 15, the numbers 5 and 14, the numbers 3 and 12, the numbers 4 and 13, and then the numbers 6 and 11. This gives us the pan-diagonal magic square below, and on the right.

1	12	6	15
8	13	3	10
11	2	16	5
14	7	9	4

1	8	10	15
14	11	5	4
7	2	16	9
12	13	3	6

Figure 11.5   These are the resulting pan-diagonal magic squares derived from the initial squares from the previous figure.

A close inspection of the three pan-diagonal magic squares created in this section shows that they are all unique and independent of each other.

8th Order Pan-Diagonal Magic Squares

In this section we are going to take several kinds of basic magic squares and use them as a starting point to create pan-diagonal magic squares. We'll start with a regular 8th order magic square:

1	2	62	61	60	59	7	8
9	10	54	53	52	51	15	16
48	47	19	20	21	22	42	41
40	39	27	28	29	30	34	33
32	31	35	36	37	38	26	25
24	23	43	44	45	46	18	17
49	50	14	13	12	11	55	56
57	58	6	5	4	3	63	64

Figure 11.6   A regular 8th order magic square from chapter 8.

This is the first magic square that we saw in chapter 8. This is a symmetrical magic square, and the operations we'll perform on this magic square will be symmetrical so that our resulting pan-diagonal square will also be symmetrical. We begin by swapping columns 2 and 4 with each other, and then swap column 5 with 7. Next we swap rows 2 and 4 with each other, followed by swapping row 5 with 7. Then we swap column 2 with 3, and column 6 with 7. Finally we swap rows 2 and 3 with each other and row 6 with 7. The resulting magic square is pan-diagonal.

1	62	61	2	7	60	59	8
48	19	20	47	42	21	22	41
40	27	28	39	34	29	30	33
9	54	53	10	15	52	51	16
49	14	13	50	55	12	11	56
32	35	36	31	26	37	38	25
24	43	44	23	18	45	46	17
57	6	5	58	63	4	3	64

Figure 11.7   A pan-diagonal square generated from a regular magic square.

This method doesn't create a pan-diagonal magic square from just any 8th order magic square, but it does appears to work on specific magic squares where we started with a certain kind of initial square, and then created a magic square using the swap-by-position method that we used to create Magic Square 8.2 (See chapter 8). This method of creating the magic square swaps the middle four numbers from the first and second rows with the middle four numbers from the bottom and next-to-the-bottom rows. And then swaps middle four numbers from the first and second columns with the middle four numbers from the last and next-to-the-last columns. In the example below, we start with another initial square (on the left), and the resulting magic square is on the right.

1	2	3	4	17	18	19	20
5	6	7	8	21	22	23	24
9	10	11	12	25	26	27	28
13	14	15	16	29	30	31	32
33	34	35	36	49	50	51	52
37	38	39	40	53	54	55	56
41	42	43	44	57	58	59	60
45	46	47	48	61	62	63	64

1	2	62	61	48	47	19	20
5	6	58	57	44	43	23	24
56	55	11	12	25	26	38	37
52	51	15	16	29	30	34	33
32	31	35	36	49	50	14	13
28	27	39	40	53	54	10	9
41	42	22	21	8	7	59	60
45	46	18	17	4	3	63	64

Figure 11.8   An alternate initial square on the left used to generate a regular 8th order magic square on the right (shaded numbers didn't move).

Now, to create the pan-diagonal magic square, we start with the magic square (on the right), and swap columns 2 with 4 and 5 with 7. Next, we swap rows 2 with 4 and 5 with 7. Finally, we swap columns 2 with 3 and 6 with 7, and then swap rows 2 with 3 and 6 with 7. Resulting with our pan-diagonal magic square:

1	62	61	2	19	48	47	20
56	11	12	55	38	25	26	37
52	15	16	51	34	29	30	33
5	58	57	6	23	44	43	24
41	22	21	42	59	8	7	60
32	35	36	31	14	49	50	13
28	39	40	27	10	53	54	9
45	18	17	46	63	4	3	64

Figure 11.9   The regular magic square from above is now used to generate a pan-diagonal 8th order magic square in this figure.

Below are a couple of other examples of this whole process. In each case, we use a unique initial 8 x 8 square, and follow the same method to create a regular the 8th order magic square. Then after the magic square has been created, we swap the same rows and columns (in the same order) on the magic square to create the 8th order pan-diagonal magic square.

Since each of the initial squares is unique, the magic squares are also unique, and because of that, the resulting pan-diagonal magic squares are also unique to each other. (Again, keep in mind that the initial squares aren't magic; they are simply symmetrical arrays that we use as our starting point.)

1	2	5	6	9	10	13	14
3	4	7	8	11	12	15	16
17	18	21	22	25	26	29	30
19	20	23	24	27	28	31	32
33	34	37	38	41	42	45	46
35	36	39	40	43	44	47	48
49	50	53	54	57	58	61	62
51	52	55	56	59	60	63	64

Initial Square

1	2	60	59	56	55	13	14
3	4	58	57	54	53	15	16
48	47	21	22	25	26	36	35
46	45	23	24	27	28	34	33
32	31	37	38	41	42	20	19
30	29	39	40	43	44	18	17
49	50	12	11	8	7	61	62
51	52	10	9	6	5	63	64

Magic Square

1	60	59	2	13	56	55	14
48	21	22	47	36	25	26	35
46	23	24	45	34	27	28	33
3	58	57	4	15	54	53	16
49	12	11	50	61	8	7	62
32	37	38	31	20	41	42	19
30	39	40	29	18	43	44	17
51	10	9	52	63	6	5	64

Pan-diagonal Magic Square

1	17	2	18	3	19	4	20
33	49	34	50	35	51	36	52
5	21	6	22	7	23	8	24
37	53	38	54	39	55	40	56
9	25	10	26	11	27	12	28
41	57	42	58	43	59	44	60
13	29	14	30	15	31	16	32
45	61	46	62	47	63	48	64

Initial Square

1	17	63	47	62	46	4	20
33	49	31	15	30	14	36	52
60	44	6	22	7	23	57	41
28	12	38	54	39	55	25	9
56	40	10	26	11	27	53	37
24	8	42	58	43	59	21	5
13	29	51	35	50	34	16	32
45	61	19	3	18	2	48	64

Magic Square

1	63	47	17	4	62	46	20
60	6	22	44	57	7	23	41
28	38	54	12	25	39	55	9
33	31	15	49	36	30	14	52
13	51	35	29	16	50	34	32
56	10	26	40	53	11	27	37
24	42	58	8	21	43	59	5
45	19	3	61	48	18	2	64

Pan-diagonal Magic Square

Figure 11.10 The same sequence is followed in these two examples. An initial square is at the upper-left, the regular magic square is at the upper-right and just below is the pan-diagonal 8th order magic square.

It is interesting to note that we can also create an 8th order pan-diagonal magic square by starting with the second magic square from chapter 8 and swapping columns 2 with 4 and then 5 with 7; and then finally swapping rows 2 with 4 and 5 with 7. The resulting pan-diagonal magic square is related to the first pan-diagonal magic square created in this chapter, but it was created with fewer swaps. In the figure below, the initial magic square is on the left, and the pan-diagonal magic square is on the right:

10	9	54	53	52	51	16	15
2	1	62	61	60	59	8	7
47	48	19	20	21	22	41	42
39	40	27	28	29	30	33	34
31	32	35	36	37	38	25	26
23	24	43	44	45	46	17	18
58	57	6	5	4	3	64	63
50	49	14	13	12	11	56	55

10	53	54	9	16	51	52	15
39	28	27	40	33	30	29	34
47	20	19	48	41	22	21	42
2	61	62	1	8	59	60	7
58	5	6	57	64	3	4	63
23	44	43	24	17	46	45	18
31	36	35	32	25	38	37	26
50	13	14	49	56	11	12	55

Figure 11.11 The regular magic square is on the left and the resulting pan-diagonal magic square is on the right.

Just as the first two magic squares from chapter 8 are related to each other, so are their resulting pan-diagonal magic squares. In fact, it is possible to go from this latest pan-diagonal magic square back to the first magic square in this chapter by simply swapping columns 1 with 4 and then 5 with 8. Then swapping rows 1 with 4 and then 5 with 8; and finally swapping columns 2 with 3 and then 6 with 7 and swapping rows 2 with 3 and then 6 with 7.

Using a 4 x 4 Pan-Diagonal Template to Create an 8 x 8 Pan-Diagonal Magic Square
We can also use 2 x 2 cells and a 4th order pan-diagonal magic square template to create an 8th order pan-diagonal magic square. We'll start with the pan-diagonal (diabolical) magic square that we saw towards the end of chapter 4; and use it as a template to fill our 8 x 8 array (which we have conveniently broken into sixteen 2 x 2 cells).

14	11	5	4
1	8	10	15
12	13	3	6
7	2	16	9

53	54
55	56

41	42
43	44

17	18
19	20

13	14
15	16

1	2
3	4

29	30
31	32

37	38
39	40

57	58
59	60

45	46
47	48

49	50
51	52

9	10
11	12

21	22
23	24

25	26
27	28

5	6
7	8

61	62
63	64

33	34
35	36

Figure 11.12 A 4th order pan-diagonal magic square on the left is used as a template to place 2 x 2 cells within the 8x8 array on the right.

And as before, we don't have a magic square yet; however, if we swap the diagonals within each 2 x 2 cell as we did before, we get the following magic square.

53	55	41	43	20	18	16	14
54	56	42	44	19	17	15	13
1	3	29	31	40	38	60	58
2	4	30	32	39	37	59	57
48	46	52	50	9	11	21	23
47	45	51	49	10	12	22	24
28	26	8	6	61	63	33	35
27	25	7	5	62	64	34	36

Figure 11.13 The resulting 8th order magic square after performing diagonal swaps within the 2 x 2 cells in the previous 8 x 8 array.

The 8 x 8 array above is a magic square; but it isn't pan-diagonal, at least not yet. To create a pan-diagonal magic square we must swap columns 1 and 2 with each other and columns 7 and 8 with each other. Then we must swap rows 1 and 2 with each other and rows 7 and 8 with each other. That now gives us the following pan-diagonal magic square.

56	54	42	44	19	17	13	15
55	53	41	43	20	18	14	16
3	1	29	31	40	38	58	60
4	2	30	32	39	37	57	59
46	48	52	50	9	11	23	21
45	47	51	49	10	12	24	22
25	27	7	5	62	64	36	34
26	28	8	6	61	63	35	33

Figure 11.14 The resulting 8th order pan-diagonal magic square.

We also could have created another pan-diagonal magic square by taking our magic square from above and swapping columns 3 and 4 with each other and columns 5 and 6 with each other; and then by swapping rows 3 and 4 with each other and rows 5 and 6 with each other. Doing so creates the following pan-diagonal magic square.

53	55	43	41	18	20	16	14
54	56	44	42	17	19	15	13
2	4	32	30	37	39	59	57
1	3	31	29	38	40	60	58
47	45	49	51	12	10	22	24
48	46	50	52	11	9	21	23
28	26	6	8	63	61	33	35
27	25	5	7	64	62	34	36

Figure 11.15 A second 8th order pan-diagonal magic square from our original magic square.

And the madness doesn't stop there either. We could have just as easily started with our original magic square and swapped columns 3 and 4 with each other and columns 5 and 6 with each other; and then swapped rows 1 and 2 with each other and rows 7 and 8 with each other to create this unique pan-diagonal magic square.

54	56	44	42	17	19	15	13
53	55	43	41	18	20	16	14
1	3	31	29	38	40	60	58
2	4	32	30	37	39	59	57
48	46	50	52	11	9	21	23
47	45	49	51	12	10	22	24
27	25	5	7	64	62	34	36
28	26	6	8	63	61	33	35

Figure 11.16 A third 8th order pan-diagonal magic square from the original.

Or, we could have swapped columns 1 and 2 with each other and columns 7 and 8 with each other; and then swapped rows 3 and 4 with each other and rows 5 and 6 with each other to create this pan-diagonal magic square:

55	53	41	43	20	18	14	16
56	54	42	44	19	17	13	15
4	2	30	32	39	37	57	59
3	1	29	31	40	38	58	60
45	47	51	49	10	12	24	22
46	48	52	50	9	11	23	21
26	28	8	6	61	63	35	33
25	27	7	5	62	64	36	34

Figure 11.17 A fourth 8th order pan-diagonal magic square from our original.

In each case we started with a single 8th order magic square (that was created from the template of a 4th order pan-diagonal magic square); and then we used that 8th order magic square to create four unique pan-diagonal magic squares.

In an earlier chapter we saw that we could fill the 2 x 2 cells with numbers that had a difference of 16 so that the first cell was made up of the numbers 1, 17, 33, and 49. The second cell is made up of the numbers 2, 18, 34, and 50; and so on continue with the remaining cells.

14	11	5	4
1	8	10	15
12	13	3	6
7	2	16	9

14	30		11	27		5	21		4	20
46	62		43	59		37	53		36	52

1	17		8	24		10	26		15	31
33	49		40	56		42	58		47	63

12	28		13	29		3	19		6	22
44	60		45	61		35	51		38	54

7	23		2	18		16	32		9	25
39	55		34	50		48	64		41	57

Figure 11.18 The 4th order pan-diagonal magic square on the left is used as a template to place the alternate initial values within the 2 x 2 cells (which are within the 8x8 array) on the right.

After using the 2 x 2 cell swap method, the partial magic square above becomes the magic square below.

14	46	11	43	53	21	52	20
30	62	27	59	37	5	36	4
1	33	8	40	58	26	63	31
17	49	24	56	42	10	47	15
60	28	61	29	3	35	6	38
44	12	45	13	19	51	22	54
55	23	50	18	16	48	9	41
39	7	34	2	32	64	25	57

Figure 11.19 The resulting 8th order magic square.

And since we used a pan-diagonal magic square for our template, we can use the same column and row swaps that we saw above to create several 8th order pan-diagonal magic squares.

The Doubling Function and Double Even-Ordered Pan-Diagonal Magic Squares
It is possible to take a magic square, perform certain operations on that magic square, and still end up with a magic square. Some operations are fairly consistent in that they will always (or almost always) create a magic square from a magic square. Other operations are very sporadic. Sometimes they work, and sometimes they don't. One such sporadic operation is a doubling function.

Let's look at an 8th order magic square to see how the doubling function works. We start by multiplying the value in each cell by two. If the resulting value is greater than 64, we subtract 65 to get a new result between 1 and 63. Notice that doubling the numbers from 1 to 32 gives us the even numbers from 2 to 64 inclusive. And, doubling the numbers 33 to 64, and then subtracting 65, gives us the odd numbers between 1 and 63 inclusive.

This series of operations is closed in that it transforms every number from 1 to 64 into another number between 1 and 64, while at the same time keeping every number unique within the set. However, just because we start with a magic square doesn't mean that we will have another magic square after performing this operation. In many cases we do not.

On the other hand, there are many interesting patterns that become evident when this doubling function is applied to pan-diagonal magic squares. An example can be seen in the following set of 8th order pan-diagonal magic squares. Here is an 8th order pan-diagonal magic square.

1	62	44	23	7	60	46	17
9	54	36	31	15	52	38	25
24	43	61	2	18	45	59	8
63	4	22	41	57	6	20	47
34	29	11	56	40	27	13	50
42	21	3	64	48	19	5	58
55	12	30	33	49	14	28	39
32	35	53	10	26	37	51	16

Figure 11.20 An initial 8th order pan-diagonal magic square.

Now, if we double each cell (using the algorithm described above), we get another pan-diagonal magic square that is unique from the first one.

2	59	23	46	14	55	27	34
18	43	7	62	30	39	11	50
48	21	57	4	36	25	53	16
61	8	44	17	49	12	40	29
3	58	22	47	15	54	26	35
19	42	6	63	31	38	10	51
45	24	60	1	33	28	56	13
64	5	41	20	52	9	37	32

Figure 11.21 The pan-diagonal magic square after the first doubling iteration.

And if we do it again, we get still another pan-diagonal magic square that is unique from the previous two. Each doubling iteration produces another unique pan-diagonal magic square. Here is the third one in this series:

4	53	46	27	28	45	54	3
36	21	14	59	60	13	22	35
31	42	49	8	7	50	41	32
57	16	23	34	33	24	15	58
6	51	44	29	30	43	52	5
38	19	12	61	62	11	20	37
25	48	55	2	1	56	47	26
63	10	17	40	39	18	9	64

Figure 11.22 The pan-diagonal magic square after the second doubling iteration.

Altogether, if we continue this doubling function, we will cycle through 12 pan-diagonal magic squares before the cycle repeats itself. And even more remarkable is the fact that it appears that if we take any 8th order pan-diagonal magic square and perform this doubling function on it repeatedly, we will create twelve pan-diagonal magic squares before cycling back to our original magic square in its original orientation.

The following is a pan-diagonal bi-magic magic square. Remember that to be bi-magic, the magic square is not only a regular magic square, but it is also a magic square when each cell is squared.

16	41	36	5	27	62	55	18
26	63	54	19	13	44	33	8
1	40	45	12	22	51	58	31
23	50	59	30	4	37	48	9
38	3	10	47	49	24	29	60
52	21	32	57	39	2	11	46
43	14	7	34	64	25	20	53
61	28	17	56	42	15	6	35

Figure 11.23 An 8th order pan-diagonal bi-magic square.

We can also perform the doubling function on the above magic square and it also will cycle through 12 pan-diagonal magic squares before returning back to this original pan-diagonal magic square. In addition to this, every third magic square in this cycle will also be a bi-magic pan-diagonal magic square. Here is the third pan-diagonal magic square which is also the next bi-magic magic square in this cycle.

63	3	28	40	21	41	50	14
13	49	42	22	39	27	4	64
8	60	35	31	46	18	9	53
54	10	17	45	32	36	59	7
44	24	15	51	2	62	37	25
26	38	61	1	52	16	23	43
19	47	56	12	57	5	30	34
33	29	6	58	11	55	48	20

Figure 11.24 The pan-diagonal bi-magic square after the third doubling iteration.

This property of doubling the values and creating another pan-diagonal magic square appears to exist with many (or possibly all) 8th order pan-diagonal magic squares.

It was with the 8th order pan-diagonal magic squares that I first discovered this property; however, it appears that all doubly-even-ordered pan-diagonal magic squares share this property to at least some degree. For example, we can take a 4th order pan-diagonal magic square, perform the doubling function on it, and create another 4th order pan-diagonal magic square. (Of course, the doubling function is a little different in this case. Here we double the numbers 1 through 16, and those numbers whose result is greater than 16, we subtract 17.) The 4th order magic square below and on the left is pan-diagonal. The 4th order pan-diagonal magic square on the right was created from the left square by applying the doubling function.

1	12	6	15
8	13	3	10
11	2	16	5
14	7	9	4

2	7	12	13
16	9	6	3
5	4	15	10
11	14	1	8

Figure 11.25 A 4th order pan-diagonal magic square on the left and its resulting pan-diagonal magic square on the right after the first doubling iteration.

The 4th order pan-diagonal magic squares iterate through 8 cycles before returning to the original square in its original orientation.

To perform a doubling function on a 16th order magic square, we take the numbers 1 to 256 and double them. And then for any number larger than 256, we subtract 257. Performing this operation on every number between 1 and 256 will give us all of the numbers from 1 to 256. The following pan-diagonal magic square can be doubled four times before returning to a rotation of the original. After 16 doubling cycles, we return to the original pan-diagonal magic square in its original orientation. Here is my initial magic square from that cycle:

1	17	251	235	234	250	20	4	13	29	247	231	230	246	32	16
2	18	252	236	233	249	19	3	14	30	248	232	229	245	31	15
176	192	86	70	71	87	189	173	164	180	90	74	75	91	177	161
175	191	85	69	72	88	190	174	163	179	89	73	76	92	178	162
159	143	101	117	120	104	142	158	147	131	105	121	124	108	130	146
160	144	102	118	119	103	141	157	148	132	106	122	123	107	129	145
50	34	204	220	217	201	35	51	62	46	200	216	213	197	47	63
49	33	203	219	218	202	36	52	61	45	199	215	214	198	48	64
193	209	59	43	42	58	212	196	205	221	55	39	38	54	224	208
194	210	60	44	41	57	211	195	206	222	56	40	37	53	223	207
112	128	150	134	135	151	125	109	100	116	154	138	139	155	113	97
111	127	149	133	136	152	126	110	99	115	153	137	140	156	114	98
95	79	165	181	184	168	78	94	83	67	169	185	188	172	66	82
96	80	166	182	183	167	77	93	84	68	170	186	187	171	65	81
242	226	12	28	25	9	227	243	254	238	8	24	21	5	239	255
241	225	11	27	26	10	228	244	253	237	7	23	22	6	240	256

Figure 11.26 A 16th order pan-diagonal magic square.

For pan-diagonal magic squares that are 4th, 8th, and 16th order, we find that every iteration of the doubling function will create another pan-diagonal magic square; and after a fixed number of iterations, you return again to the original square in its original orientation. The 4th order squares take 8 iterations to return to the original square in its original orientation. The 8th order squares require 12 iterations and 16th order magic squares require 16 iterations to return to the original square in its original orientation. The following table sums up the properties of the doubling function on double even pan-diagonal magic squares from the 4th order through the 16th order.

Order	Iteration (Cycle) Cnt	# of Pan-Diag	Spacing between Mag Sqs
4	8	8	1
8	12	12	1
12	28	12	1 or 7
16	16	16	1

Table 11.1   This table shows the effect of the doubling function on pan-diagonal magic squares for doubly even-orders of magic squares.

Notice that the 12th order pan-diagonal magic squares have an interesting cycle. It takes 28 iterations to complete the cycle, and within this cycle there will be six pan-diagonal magic squares in a row followed by one semi-magic square, and then there will be seven non-magic squares in a row, and then the pattern repeats one more time in order to complete the cycle of 28 iterations.

The table does not show any singly even magic squares (6th, 10th, and 14th order magic squares). That is because it is believed that it isn't possible to have pan-diagonal magic squares within these orders (and my experience with magic squares substantiates that claim).

Row Swapping within a Pan-Diagonal Magic Square
Once we have a pan-diagonal magic square, we can shift the rows or columns, and the magic square will still be pan-diagonal. (To shift a row we take a row from the bottom and place it on the top, or take it from the top and place it on the bottom. Similarly, to shift a column, we take a column from the left, and place it on the right, or take it from the right and place it on the left.)

If we were to take a regular magic square and shift a row or column, the main diagonals probably wouldn't add up to the constant anymore. The magic square would become a semi-magic square. However, with a pan-diagonal magic square this isn't a problem since all the broken diagonals also add up to the constant; therefore, shifting the rows or columns of the magic square simply causes the other broken diagonals to become the main diagonals.

However, if we were to take a pan-diagonal magic square and symmetrically swap the columns or rows, we would still have a magic square, but in most cases it wouldn't be pan-diagonal anymore. (For example, in the case of a 16th order magic square, swapping the 2nd and 15th columns is a symmetrical swap, so is swapping the 3rd and 14th rows, and so on.) With this understanding, we see that the following pan-diagonal magic square has this additional interesting property of being able to swap symmetrical columns and rows, and still being pan-diagonal. We start with the 16th order pan-diagonal magic square, perform some symmetrical column swaps, and our result will be another pan-diagonal magic square. Here is our initial pan-diagonal magic square:

8	7	254	253	255	256	5	6	11	12	241	242	244	243	10	9
24	23	238	237	239	240	21	22	27	28	225	226	228	227	26	25
169	170	83	84	82	81	172	171	166	165	96	95	93	94	167	168
185	186	67	68	66	65	188	187	182	181	80	79	77	78	183	184
153	154	99	100	98	97	156	155	150	149	112	111	109	110	151	152
137	138	115	116	114	113	140	139	134	133	128	127	125	126	135	136
56	55	206	205	207	208	53	54	59	60	193	194	196	195	58	57
40	39	222	221	223	224	37	38	43	44	209	210	212	211	42	41
216	215	46	45	47	48	213	214	219	220	33	34	36	35	218	217
200	199	62	61	63	64	197	198	203	204	49	50	52	51	202	201
121	122	131	132	130	129	124	123	118	117	144	143	141	142	119	120
105	106	147	148	146	145	108	107	102	101	160	159	157	158	103	104
73	74	179	180	178	177	76	75	70	69	192	191	189	190	71	72
89	90	163	164	162	161	92	91	86	85	176	175	173	174	87	88
232	231	30	29	31	32	229	230	235	236	17	18	20	19	234	233
248	247	14	13	15	16	245	246	251	252	1	2	4	3	250	249

Figure 11.27 A 16th order pan-diagonal magic square.

Now, we swap columns 5 and 6; swap columns 7 and 8; swap columns 9 and 10; and finally swap columns 11 and 12, and our resulting magic square is still pan-diagonal.

8	7	254	253	256	255	6	5	12	11	242	241	244	243	10	9
24	23	238	237	240	239	22	21	28	27	226	225	228	227	26	25
169	170	83	84	81	82	171	172	165	166	95	96	93	94	167	168
185	186	67	68	65	66	187	188	181	182	79	80	77	78	183	184
153	154	99	100	97	98	155	156	149	150	111	112	109	110	151	152
137	138	115	116	113	114	139	140	133	134	127	128	125	126	135	136
56	55	206	205	208	207	54	53	60	59	194	193	196	195	58	57
40	39	222	221	224	223	38	37	44	43	210	209	212	211	42	41
216	215	46	45	48	47	214	213	220	219	34	33	36	35	218	217
200	199	62	61	64	63	198	197	204	203	50	49	52	51	202	201
121	122	131	132	129	130	123	124	117	118	143	144	141	142	119	120
105	106	147	148	145	146	107	108	101	102	159	160	157	158	103	104
73	74	179	180	177	178	75	76	69	70	191	192	189	190	71	72
89	90	163	164	161	162	91	92	85	86	175	176	173	174	87	88
232	231	30	29	32	31	230	229	236	235	18	17	20	19	234	233
248	247	14	13	16	15	246	245	252	251	2	1	4	3	250	249

Figure 11.28 The resulting 16th order pan-diagonal magic square after four pairs of columns have been swapped.

We can also do the same thing with the similar rows. We could have started with either our original pan-diagonal magic square, or the one above; then we could have swapped the same numbered rows, and we would again have a pan-diagonal magic square. In this case, we'll take the magic square above and swap rows 5 and 6; swap rows 7 and 8; swap rows 9 and 10 and finally swap rows 11 and 12, and we will create another unique pan-diagonal magic square. Here it is:

8	7	254	253	256	255	6	5	12	11	242	241	244	243	10	9
24	23	238	237	240	239	22	21	28	27	226	225	228	227	26	25
169	170	83	84	81	82	171	172	165	166	95	96	93	94	167	168
185	186	67	68	65	66	187	188	181	182	79	80	77	78	183	184
137	138	115	116	113	114	139	140	133	134	127	128	125	126	135	136
153	154	99	100	97	98	155	156	149	150	111	112	109	110	151	152
40	39	222	221	224	223	38	37	44	43	210	209	212	211	42	41
56	55	206	205	208	207	54	53	60	59	194	193	196	195	58	57
200	199	62	61	64	63	198	197	204	203	50	49	52	51	202	201
216	215	46	45	48	47	214	213	220	219	34	33	36	35	218	217
105	106	147	148	145	146	107	108	101	102	159	160	157	158	103	104
121	122	131	132	129	130	123	124	117	118	143	144	141	142	119	120
73	74	179	180	177	178	75	76	69	70	191	192	189	190	71	72
89	90	163	164	161	162	91	92	85	86	175	176	173	174	87	88
232	231	30	29	32	31	230	229	236	235	18	17	20	19	234	233
248	247	14	13	16	15	246	245	252	251	2	1	4	3	250	249

Figure 11.29 The resulting 16th order pan-diagonal magic square after four pairs of rows have been swapped.

Now, if we swap columns 5 and 6; swap columns 7 and 8; swap columns 9 and 10; and finally swap columns 11 and 12; we will have our fourth unique pan-diagonal magic square, and then swapping the rows one more time will bring us back to the original pan-diagonal magic square that we started with.

Keep in mind that these row and column swapping examples appear to be exceptions to the rule. Almost always whenever we have a pan-diagonal magic square and swap symmetrical rows and/or columns within the magic square, the resulting magic square will not be pan-diagonal any more (although if we perform symmetrical row and column swaps, the resulting square will still be a regular magic square).

Odd/Even Number Swapping of a Pan-Diagonal Magic Square
In the second chapter where we discussed magic square operations, it was briefly mentioned that performing an odd/even number swap on a pan-diagonal magic square will produce another pan-diagonal magic square. Briefly, an odd/even number swap is where the 1 is swapped with the 2, the 3 is swapped with the 4, the 5 with the 6, etc. In other words, an odd number is swapped with its subsequent even number. Let's look at a few pan-diagonal examples:

55	53	41	43	20	18	14	16
56	54	42	44	19	17	13	15
4	2	30	32	39	37	57	59
3	1	29	31	40	38	58	60
45	47	51	49	10	12	24	22
46	48	52	50	9	11	23	21
26	28	8	6	61	63	35	33
25	27	7	5	62	64	36	34

56	54	42	44	19	17	13	15
55	53	41	43	20	18	14	16
3	1	29	31	40	38	58	60
4	2	30	32	39	37	57	59
46	48	52	50	9	11	23	21
45	47	51	49	10	12	24	22
25	27	7	5	62	64	36	34
26	28	8	6	61	63	35	33

Figure 11.30 The upper 8th order array is the initial pan-diagonal magic square and the lower array is the resulting pan-diagonal magic square after performing the odd/even number swap.

An odd/even number swap in the pan-diagonal square above is the same in this case as simply swapping pairs of rows. Notice that by swapping row 1 with row 2, row 3 with row 4, and so forth has the same effect as performing an odd/even swap. Either way, it produces another unique pan-diagonal magic square. Here is another example, where the upper array is the initial 8th order pan-diagonal magic square, and the lower array is the resulting pan-diagonal magic square after performing the odd/even swap.

55	12	11	56	49	14	13	50
26	37	38	25	32	35	36	31
18	45	46	17	24	43	44	23
63	4	3	64	57	6	5	58
7	60	59	8	1	62	61	2
42	21	22	41	48	19	20	47
34	29	30	33	40	27	28	39
15	52	51	16	9	54	53	10

56	11	12	55	50	13	14	49
25	38	37	26	31	36	35	32
17	46	45	18	23	44	43	24
64	3	4	63	58	5	6	57
8	59	60	7	2	61	62	1
41	22	21	42	47	20	19	48
33	30	29	34	39	28	27	40
16	51	52	15	10	53	54	9

Figure 11.31 The initial and resulting 8th order pan-diagonal magic squares after performing the odd/even swap.

An interesting characteristic of the pan-diagonal magic square shown above is that if you manually perform the swaps (as opposed as to having a software routine do it), you find that halfway through the swap process, you have yet another pan-diagonal magic square. The shaded areas in the figure below show the numbers where the odd/even swap was performed. Notice that the 4 x 4 arrays in each of the quadrants either have congruent or complimentary shaded patterns.

55	11	12	56	49	13	14	50
25	37	38	26	31	35	36	32
17	45	46	18	23	43	44	24
63	3	4	64	57	5	6	58
8	60	59	7	2	62	61	1
42	22	21	41	48	20	19	47
34	30	29	33	40	28	27	39
16	52	51	15	10	54	53	9

Figure 11.32 The resulting 8th order pan-diagonal magic squares after performing only half of the odd/even swaps on the original pan-diagonal magic square.

Finally, here is one more set of 8th order pan-diagonal magic squares. Again, the upper array is the initial magic square, and the lower array is the resulting pan-diagonal magic square after performing the odd/even swap.

1	63	31	33	4	62	30	36
56	10	42	24	53	11	43	21
52	14	46	20	49	15	47	17
5	59	27	37	8	58	26	40
25	39	7	57	28	38	6	60
48	18	50	16	45	19	51	13
44	22	54	12	41	23	55	9
29	35	3	61	32	34	2	64

2	64	32	34	3	61	29	35
55	9	41	23	54	12	44	22
51	13	45	19	50	16	48	18
6	60	28	38	7	57	25	39
26	40	8	58	27	37	5	59
47	17	49	15	46	20	52	14
43	21	53	11	42	24	56	10
30	36	4	62	31	33	1	63

Figure 11.33 Yet another set of initial and resulting 8th order pan-diagonal magic squares after performing the odd/even swap.

Chapter 12
Building Odd-Ordered Pan-Diagonal Magic Squares

Odd-Ordered Pan-Diagonal Magic Squares
Now, we'll look at some odd-ordered magic squares and convert them into pan-diagonal magic squares. We'll start with a 5th order magic square.

23	6	19	2	15		P1 (2, 3)
10	18	1	14	22		P21 (rt 1, up 1)
17	5	13	21	9		P65 (up 2)
4	12	25	8	16		Seq 1 2 3 4 5
11	24	7	20	3		

Figure 12.1   A regular 5th order magic square.

The magic square above can be described using our notation (which is also shown above). We can now use this magic square to create a 5th order pan-diagonal magic square. All we need to do is swap column 1 with column 2, swap column 4 with column 5, and then finally swap column 2 with column 4. This gives us the following pan-diagonal magic square:

6	15	19	23	2
18	22	1	10	14
5	9	13	17	21
12	16	25	4	8
24	3	7	11	20

Figure 12.2   The resulting pan-diagonal magic square.

We also could have performed the same swaps on Magic Square from figure 12.1, except at the end; instead of swapping column 2 with column 4, we could have swapped column 1 with column 5. It gives us the following pan-diagonal magic square; however, if we look closely, we see that it is the mirror of the previous pan-diagonal magic square.

2	23	19	15	6
14	10	1	22	18
21	17	13	9	5
8	4	25	16	12
20	11	7	3	24

Figure 12.3   The second pan-diagonal magic square (which is actually a mirror of the previous pan-diagonal magic square).

It is interesting to note that if we start with the magic square from figure 12.1 again, but this time we perform the same swaps on the rows instead of the columns, we get the following pan-diagonal magic square. Note also that this magic square is unique from the previous pan-diagonal magic square.

4	12	25	8	16
23	6	19	2	15
17	5	13	21	9
11	24	7	20	3
10	18	1	14	22

Figure 12.4   A third pan-diagonal magic square.

The curious reader can use the following description to create a magic square, and then use the same swapping algorithm on that magic square and from it create two more pan-diagonal magic squares by either performing the swaps on the columns or on the rows. This magic square is the same as that listed in figure 12.1, except that in this case, P1 is at (3, 2) instead of (2, 3) and P65 is (rt 2) instead of (up 2).

P1 (3, 2)
P21 (rt 1, up 1)
P65 (rt 2)
Seq 1 2 3 4 5

And if that weren't enough, we can change the sequences and use either of the methods along with their variations to create even more pan-diagonal magic squares. In the following example, we'll use the description above with the sequence 1 3 5 2 4. This description will create a semi-magic square, as shown below (however, this semi-magic square becomes a regular magic square if the rows and columns are shifted so that the 13 is placed in the center cell):

5	16	7	23	14
19	10	21	12	3
8	24	15	1	17
22	13	4	20	6
11	2	18	9	25

P1 (3, 4)
P21 (rt 1, up 1)
P65 (rt 2)
Seq 1 3 5 2 4

Figure 12.5   A 5th order semi-magic square.

Then by using the swap algorithm described above on this semi-magic square, we get the following pan-diagonal magic square:

23	5	7	14	16
12	19	21	3	10
1	8	15	17	24
20	22	4	6	13
9	11	18	25	2

Figure 12.6   The resulting pan-diagonal magic square.

As a final note, we'll refer back to the mapping we did in chapter 7. (The mapping is reproduced below). In that mapping, we created a 5 x 5 table, and in it we were able to show the construction of 32 different and unique magic squares. We determined that there were five A magic squares, and likewise five B, C, D, E and H magic squares making for a total of 30 of the 32 magic squares described by the map. However, the F and G magic squares only appeared once in that mapping. Yet, it is from the F and G magic squares that we created our pan-diagonal magic squares.

A	CH	B	DE	
CH	B	DEF		A
B	DE		AG	CH
DE		A	CH	B
	A	CH	B	DE

P21 (rt 1, up 1)
Seq 1 2 3 4 5

A P1 (see chart) **B** P1 (see chart) **C** P1 (see chart) **D** P1 (see chart)
   P65 (dn 2)   P65 (dn 1)   P65 (lf 1)   P65 (rt 2, dn 1)

E P1 (see chart) **F** P1 (see chart) **G** P1 (see chart) **H** P1 (see chart)
   P65 (lf 2)   P65 (up 2)   P65 (rt 2)   P65 (rt 2, dn 2)

Figure 12.7   A mapping of 5th order magic squares (from Chapter 7)

In other words, if we use the swap algorithm above on the A, B, C, D, E or H magic squares, they won't create pan-diagonal magic squares. Only the F and G magic squares appear to have this property.

The 7th Order Pan-Diagonal Magic Square
In several ways the 7th order, 11th order and 13th order magic squares share similar characteristics with regard to building pan-diagonal magic squares and to each other. We'll look at one of those characteristics within the next few sections. We'll begin by taking a 7th order magic square, swap a few columns and create a pan-diagonal magic square. We'll start with a magic square we saw earlier where the number one is right above the center cell. The resulting magic square is below and on the left, and the corresponding construction description is on the right.

46	15	40	9	34	3	28
21	39	8	33	2	27	45
38	14	32	1	26	44	20
13	31	7	25	43	19	37
30	6	24	49	18	36	12
5	23	48	17	42	11	29
22	47	16	41	10	35	4

P1 (3, 4)
P21 (rt 1, up 1)
P87 (up 2)
Seq 1 2 3 4 5 6 7

Figure 12.8   A regular 7th order magic square from Chapter 7.

To create this pan-diagonal magic square, we begin by swapping columns 1 and 3 with each other, and then columns 5 and 7. Next we swap columns 2 and 3 with each other and columns 5 and 6. Finally we swap columns 3 and 5, and our resulting magic square is pan-diagonal.

40	46	3	9	15	28	34
8	21	27	33	39	45	2
32	38	44	1	14	20	26
7	13	19	25	31	37	43
24	30	36	49	6	12	18
48	5	11	17	23	29	42
16	22	35	41	47	4	10

Figure 12.9   The 7th order pan-diagonal magic square after the column swaps.

There is an added dimension that comes with the 7th order magic square. If we perform our sequence of operations in a reverse order on the original magic square, we get another pan-diagonal magic square that is unique from the one above. In the pan-diagonal magic square below, we started with the magic square at the beginning of this section, but this time we swap columns 3 and 5 first. Then we swap columns 2 and 3 and columns 5 and 6; and finally, we swap columns 1 and 3 and columns 5 and 7. Here is the resulting pan-diagonal magic square:

15	34	46	9	28	40	3
39	2	21	33	45	8	27
14	26	38	1	20	32	44
31	43	13	25	37	7	19
6	18	30	49	12	24	36
23	42	5	17	29	48	11
47	10	22	41	4	16	35

Figure 12.10 The second pan-diagonal magic square after alternate column swaps.

To create these pan-diagonal magic squares, we only swapped columns. Again, we could have just as easily swapped the corresponding rows instead of columns and created still other pan-diagonal magic squares. Again, this method doesn't work for just any 7th order magic square, but it does appear to work in the same way for the F and G magic squares (refer to the 7th order mapping in chapter 7). And it also appears to work with various sequences (like 1 3 5 7 2 4 6, as well as 1 2 3 4 5 6 7).

There is also another interesting pattern that we see between these two swap algorithms that shows us how they are related to each other. Let's take our initial magic square again, and this time we'll number the columns, starting with 1 on the left and going to 7 on the right:

1	2	3	4	5	6	7
46	15	40	9	34	3	28
21	39	8	33	2	27	45
38	14	32	1	26	44	20
13	31	7	25	43	19	37
30	6	24	49	18	36	12
5	23	48	17	42	11	29
22	47	16	41	10	35	4

Figure 12.11 The initial 7th order magic square from the beginning of this section.

Now, we'll perform our first swap sequence, but we will only keep track of the column numbers (and not the entire column). We started our swap algorithm by swapping columns 1 and 3 with each other, then swapping columns 5 and 7 with each other.

```
1   2   3   4   5   6   7
3       1       7       5
```

Then we swapped columns 2 and 3 with each other and columns 5 and 6 with each other:

```
1   2   3   4   5   6   7
3       1       7       5
    1   2       6   7
```

And finally we swapped columns 3 and 5 with each other:

```
1   2   3   4   5   6   7
3       1       7       5
    1   2       6   7
        6       2
```

Therefore, if we look at just the column numbers themselves we see that

```
we started with:   1   2   3   4   5   6   7
and we ended with: 3   1   6   4   2   7   5
```

Now, if we do the same things with our second swap algorithm (where we performed the sequence of operations in reverse order), we find that

```
we started with:   1   2   3   4   5   6   7
and we ended with: 2   5   1   4   7   3   6
```

As we consider the swapped columns, we find that there is an interesting relation here. To generate this first sequence, we start with the 3 and add 5 to generate the next number (however, if the sum is over 7, then we subtract 7 to get the next number). Therefore, the next number is a 1. Again, we add 5 to get a 6. We again add 5 (to get 11) but since it is greater than 7, we subtract seven to give us 4. We continue adding 5, and subtracting 7 when necessary and we end up with the following sequence: 3, 1, 6, 4, 2, 7, 5.

Similarly, the second sequence above starts with the number 2, and adds 3 each time (while subtracting 7 when necessary), thus generating the sequence: 2, 5, 1, 4, 7, 3, 6.

What we have here is a method to take a certain kind of regular magic square, and by rearranging the columns in either of these two ways, we can generate a pan-diagonal magic square. And what we just did is very instructive because we will see this same type of relationship again with the other odd-order pan-diagonal magic squares.

11th and 13th Order Pan-Diagonal Magic Squares

The 11th order magic squares share these same characteristics as the 7th order magic squares. Let's consider some examples. Again, we'll start with a magic square that has the 1 right above the center cell. Here is the magic square along with its description:

116	45	106	35	96	25	86	15	76	5	66
55	105	34	95	24	85	14	75	4	65	115
104	44	94	23	84	13	74	3	64	114	54
43	93	33	83	12	73	2	63	113	53	103
92	32	82	22	72	1	62	112	52	102	42
31	81	21	71	11	61	111	51	101	41	91
80	20	70	10	60	121	50	100	40	90	30
19	69	9	59	120	49	110	39	89	29	79
68	8	58	119	48	109	38	99	28	78	18
7	57	118	47	108	37	98	27	88	17	67
56	117	46	107	36	97	26	87	16	77	6

P1 (5, 6)
P21 (rt 1, up 1)
P12 11 (up 2)
Seq 1 2 3 4 5 6 7 8 9 10 11

Figure 12.12 A regular 11th order magic square.

Now, we will use the following swap algorithm to create a pan-diagonal magic square from this magic square. These are the steps we'll use:

    Swap column 5 with column 10, and column 7 with column 2
    Swap column 8 with 9, and column 3 with 4
    Swap column 4 with 7 and column 5 with 8
    Then swap column 8 with 11, and column1 with 4
    Finally, swap column 1 with 11

The result of these nine column swaps is the following pan-diagonal magic square.

5	86	35	116	76	25	106	66	15	96	45
65	14	95	55	4	85	34	115	75	24	105
114	74	23	104	64	13	94	54	3	84	44
53	2	83	43	113	73	33	103	63	12	93
102	62	22	92	52	1	82	42	112	72	32
41	111	71	31	101	61	21	91	51	11	81
90	50	10	80	40	121	70	30	100	60	20
29	110	59	19	89	49	9	79	39	120	69
78	38	119	68	28	109	58	18	99	48	8
17	98	47	7	88	37	118	67	27	108	57
77	26	107	56	16	97	46	6	87	36	117

Figure 12.13 The resulting 11th order pan-diagonal magic square.

As we saw with the 7th order magic squares, if go through our swap algorithm in reverse order, we can create another magic square which is unique from the one we just created. Here is the algorithm in reverse:

> Swap column 1 with 11
> Then swap columns 8 with 11, and 1 with 4
> Swap columns 4 with 7, and 5 with 8
> Swap columns 8 with 9, and 3 with 4
> And finally, swap columns 5 with 10, and 7 with 2

And here is the pan-diagonal magic square that it creates:

35	66	86	106	5	25	45	76	96	116	15
95	115	14	34	65	85	105	4	24	55	75
23	54	74	94	114	13	44	64	84	104	3
83	103	2	33	53	73	93	113	12	43	63
22	42	62	82	102	1	32	52	72	92	112
71	91	111	21	41	61	81	101	11	31	51
10	30	50	70	90	121	20	40	60	80	100
59	79	110	9	29	49	69	89	120	19	39
119	18	38	58	78	109	8	28	48	68	99
47	67	98	118	17	37	57	88	108	7	27
107	6	26	46	77	97	117	16	36	56	87

Figure 12.14 A second 11th order pan-diagonal magic square.

Now, if we look at the actual column positions and compare the positions of the columns in the original magic square with the positions of those same columns in the pan-diagonal squares, we see some interesting relationships. For the first 11th order magic square we have:

we started with:	1	2	3	4	5	6	7	8	9	10	11
and we ended with:	10	7	4	1	9	6	3	11	8	5	2

And for the second 11th order magic square we get:

we started with:	1	2	3	4	5	6	7	8	9	10	11
and we ended with:	4	11	7	3	10	6	2	9	5	1	8

In both cases the starting square was a regular magic square, and the resulting square was a pan-diagonal magic square.

Now this is where it gets interesting. If we use a list of numbers from 1 to 11, and start with the number 10, then take every eighth number (going from left to right), and we get the sequence 10 7 4 1 9 6 3 11 8 5 2. Now, when we arrange the columns from the original magic square in this order, it gives us a pan-diagonal magic square. Then with the second sequence, start with the number 4 and take every seventh number (again, going from left to right) which gives us the sequence 4 11 7 3 10 6 2 9 5 1 8, which also gives us a pan-diagonal magic square when arranging the columns in that order. But, there are still a few more options available when we have 11 columns to work with. For example, we can take the initial sequence, start with the 5 and take every ninth number (going from left to right) and generate the sequence 5 3 1 10 8 6 4 2 11 9 7; or we can start with the 9 in the initial sequence and take every sixth number and create the sequence 9 4 10 5 11 6 1 7 2 8 3. Again, arranging the columns in either case will also generate pan-diagonal magic squares from our initial, regular, run-of-the-mil, standard magic square. And then, in addition to all of this, we could do the same with the rows (instead of the columns), and generate even more pan-diagonal magic squares from our original magic square.

And it doesn't stop there, many of these original magic squares can use different sequences (as shown in the description), and create other regular magic squares, and then by using the preceding column (or row) swap algorithm, we can create still other pan-diagonal magic squares. The following semi-magic square was created using a different sequence in the description. The magic square and its construction method are below:

121	45	101	36	92	27	83	18	74	9	65
54	110	34	90	25	81	16	72	7	63	119
108	43	99	23	79	14	70	5	61	117	52
41	97	32	88	12	68	3	59	115	50	106
95	30	86	21	77	1	57	113	48	104	39
28	84	19	75	10	66	111	46	102	37	93
82	17	73	8	64	120	55	100	35	91	26
15	71	6	62	118	53	109	44	89	24	80
69	4	60	116	51	107	42	98	33	78	13
2	58	114	49	105	40	96	31	87	22	67
56	112	47	103	38	94	29	85	20	76	11

P1 (5, 6)
P21 (rt 1, up 1)
P12 11 (up 2)
Seq 1 3 5 7 9 11 2 4 6 8 10

Figure 12.15 An 11th order semi-magic square.

The semi-magic square above (it is semi-magic because one of the main diagonals does not add up to 671) was used to create the pan-diagonal magic square below. The swap algorithm that was used is:

Swap column 1 with 11
Then swap columns 8 with 11, and 1 with 4
Swap columns 4 with 7, and 5 with 8
Swap columns 8 with 9, and 3 with 4
And finally, swap columns 5 with 10, and 7 with 2

Which also means that the order of the columns

started as:	1	2	3	4	5	6	7	8	9	10	11
and ended as:	4	11	7	3	10	6	2	9	5	1	8

In this case, even though we started with a semi-magic square, the algorithm above transformed the semi-magic square into the pan-diagonal magic square below:

36	65	83	101	9	27	45	74	92	121	18
90	119	16	34	63	81	110	7	25	54	72
23	52	70	99	117	14	43	61	79	108	5
88	106	3	32	50	68	97	115	12	41	59
21	39	57	86	104	1	30	48	77	95	113
75	93	111	19	37	66	84	102	10	28	46
8	26	55	73	91	120	17	35	64	82	100
62	80	109	6	24	53	71	89	118	15	44
116	13	42	60	78	107	4	33	51	69	98
49	67	96	114	22	40	58	87	105	2	31
103	11	29	47	76	94	112	20	38	56	85

Figure 12.16 The resulting 11th order pan-diagonal magic square.

We can also take the semi-magic square from figure 12.15, and shift it diagonally down and left three times (until the 61 is in the center cell). Doing so, we find that this operation converts the semi-magic square into a regular magic square. Then if we perform the same column swapping operations as above, we get the following pan-diagonal magic square:

42	60	78	107	4	33	51	69	98	116	13
96	114	22	40	58	87	105	2	31	49	67
29	47	76	94	112	20	38	56	85	103	11
83	101	9	27	45	74	92	121	18	36	65
16	34	63	81	110	7	25	54	72	90	119
70	99	117	14	43	61	79	108	5	23	52
3	32	50	68	97	115	12	41	59	88	106
57	86	104	1	30	48	77	95	113	21	39
111	19	37	66	84	102	10	28	46	75	93
55	73	91	120	17	35	64	82	100	8	26
109	6	24	53	71	89	118	15	44	62	80

Figure 12.17 A second 11th order pan-diagonal magic square.

These two pan-diagonal magic squares (from the previous two figures) are essentially the same magic square except that the second square is shifted down three cells, and then shifted left two cells from the first square. However, according to our definition of unique magic squares, they are unique to each other because neither is a rotation and/or mirroring of the other.

These same methods can be extended to 13th order magic squares and beyond. If the curious reader would like to create a 13th order magic square using the following description, they could then use the swap algorithm listed below to create a 13th order pan-diagonal magic square. Here is a description of the construction method:

P1 (6, 7)
P21 (rt 1, up 1)
P14 13 (up 2)
Seq 1 2 3 4 5 6 7 8 9 10 11 12 13

And here is the swap algorithm:

> Swap columns 8 with 12, and 6 with 2
> Then swap columns 9 with 10, and 5 with 4
> Swap columns 11 with 13, and 1 with 3
> Swap column 5 with 9
> Swap column 3 with 11
> And finally, swap column 2 with 12

The swap algorithm can also be written showing that the order of the columns

started as:	1	2	3	4	5	6	7	8	9	10	11	12	13
and ended as:	3	8	13	5	10	2	7	12	4	9	1	6	11

And just as we saw with previous pan-diagonal magic squares, there are still several other column arrangements that will create 13th order pan-diagonal magic squares. Here are two more:

column ordering A:	6	4	2	13	11	9	7	5	3	1	12	10	8
column ordering B:	12	9	6	3	13	10	7	4	1	11	8	5	2

And there are still many other sequences that will work with this order of magic square. This method of finding pan-diagonal magic squares is very successful with 7th order, 11th order and 13th order magic squares; however, I have been unsuccessful to get this method to work with 9th and 15th order magic squares. And this raises the question that maybe this method will only work with orders of magic squares that are prime numbers. Who knows? More research is still needed, and there is still much to learn about larger orders of magic squares.

Re-Ordering the Columns and Rows of Pan-Diagonal Magic Squares
By taking regular, specific odd-order magic squares, it is possible to reorder either the rows or the columns and create a pan-diagonal magic square. But it also appears that if we take any pan-diagonal magic square and perform the same type of reordering operation on it; in most cases it will create another pan-diagonal magic square. And this re-ordering can be used with both odd and even-ordered pan-diagonal magic squares.

In other words, the way that this works is that we start with a pan-diagonal magic square, and then rearrange either the columns or rows, and in most cases the result will be yet another unique pan-diagonal magic square. For example, here is a row/column reordering for an 8th order pan-diagonal magic square.

The 8th Order columns or rows

started as:	1	2	3	4	5	6	7	8
and ended as:	1	4	7	2	5	8	3	6

We will take a pan-diagonal magic square on the left, and then rearrange the order of the columns to create a new pan-diagonal magic square on the right.

Initial Pan-Diagonal Square

1	32	24	9	49	48	40	57
60	37	45	52	12	21	29	4
59	38	46	51	11	22	30	3
2	31	23	10	50	47	39	58
7	26	18	15	55	42	34	63
62	35	43	54	14	19	27	6
61	36	44	53	13	20	28	5
8	25	17	16	56	41	33	64

Resulting Pan-Diagonal Magic Square

1	9	40	32	49	57	24	48
60	52	29	37	12	4	45	21
59	51	30	38	11	3	46	22
2	10	39	31	50	58	23	47
7	15	34	26	55	63	18	42
62	54	27	35	14	6	43	19
61	53	28	36	13	5	44	20
8	16	33	25	56	64	17	41

Figure 12.18 Converting an 8th order pan-diagonal magic square into another.

In this example, to go from the initial pan-diagonal square (on the left) to the resulting square (on the right):

> first column stays the same
> fourth column becomes the second column
> seventh column becomes the third column
> second column becomes the fourth column
> fifth column stays the same
> eighth column becomes the sixth column
> third column becomes the seventh column, and
> sixth column becomes the eighth column.

And thus we have a new and unique pan-diagonal magic square from our original pan-diagonal magic square. Note that the rows could have been rearranged instead of the columns, and we would have gotten yet another unique 8th order pan-diagonal magic square. In this particular case, if we apply the ordering again on the resulting square, we will end up with our original pan-diagonal magic square. The following is a listing of other column/row re-orderings for 12th order and 16th order pan-diagonal magic squares.

The 12th Order columns or rows

started as:	1	2	3	4	5	6	7	8	9	10	11	12
and ended as:	1	6	11	4	9	2	7	12	5	10	3	8

The 16th Order columns or rows

started as:	1	2	3	4	5	6	7	8	9	10	11	12	13	14	15	16
and ended as:	1	4	7	10	13	16	3	6	9	12	15	2	5	8	11	14

started as:	1	2	3	4	5	6	7	8	9	10	11	12	13	14	15	16
and ended as:	1	6	11	16	5	10	15	4	9	14	3	8	13	2	7	12

started as:	1	2	3	4	5	6	7	8	9	10	11	12	13	14	15	16
and ended as:	1	8	15	6	13	4	11	2	9	16	7	14	5	12	3	10

My experience has shown that in most cases rearranging the columns or rows as shown above will result in another unique pan-diagonal magic square.

The Doubling Function and Odd-Ordered Pan-Diagonal Magic Squares
In the previous chapter we looked at a doubling function that we used on even-ordered pan-diagonal magic squares. There is also a doubling function that works with odd-ordered pan-diagonal magic squares; however, with the odd-ordered magic squares it works a little different, and we don't always get another pan-diagonal magic square.

Let's look at a 7th order magic square to see how the doubling function works with odd-ordered magic squares. Again, we start by multiplying the value in each cell by two. If the resulting value is greater than 49, we subtract 49 to get a new result between 1 and 49. Notice that doubling the numbers from 1 to 24 gives us the even numbers from 2 to 48 inclusive. And, doubling the numbers 25 to 49, and then subtracting 49, gives us the odd numbers between 1 and 49 inclusive. One thing that we can quickly see with this operation is that the number 49 maps to itself. In other words, applying the doubling function on a 7th order magic square will never move the 49 to another cell within the square. All the other numbers, from 1 to 48, will move to another location in a predictable manner.

Again, this series of operations is closed in that it transforms every number from 1 to 49 into another number between 1 and 49, while at the same time keeping every number unique within the set. However, just because we start with a magic square doesn't mean that we will have another magic square after performing this operation. In many cases we do not. However, by repeatedly performing this operation, the array of numbers will cycle through a series of squares and return again to the original square. A 7th order pan-diagonal magic square requires 21 doubling iterations before it returns to the original square in its original orientation. And within this series of squares, there will be three unique pan-diagonal magic squares that will be created and a couple of semi-magic squares.

Similarly, the 11th order pan-diagonal magic square requires 110 doubling iterations to return to the original square at the original orientation; and along the way there will be seven other pan-diagonal magic squares created, and two other semi-magic squares. The following table sums up a few of the properties of the doubling function for odd-ordered pan-diagonal magic squares from the 5th order through the 13th order.

Order	Iteration (Cycle) Count	# of Pan-Diag Magic Squares	# of Semi-Magic	Spacing between Mag Sqs
5	20	1	2	4 or 8
7	21	3	2	3 or 6
9	54	1	0	54
11	110	7	2	10 or 20
13	156	9	2	12 or 24

Table 12.1   This table shows the effect of the doubling function on pan-diagonal magic squares for odd-orders of magic squares.

It should also be noted that when you start with a pan-diagonal magic square and then repeatedly perform the doubling operation on that square, the semi-magic squares that are created become regular magic squares when the center number is placed in the center cell. With a 7th order magic square, the center number is 25, and by shifting the rows and columns of the semi-magic square, it becomes a regular magic square when the 25 is moved to the center cell of the square.

Here is a 7th order pan-diagonal magic square. We'll use it as an example to show how the doubling operation works with odd-ordered magic squares.

3	40	28	9	46	34	15
27	8	45	33	21	2	39
44	32	20	1	38	26	14
19	7	37	25	13	43	31
36	24	12	49	30	18	6
11	48	29	17	5	42	23
35	16	4	41	22	10	47

Figure 12.19 The initial pan-diagonal magic square that we will use as an example for the doubling operation.

After six iterations of the doubling function, we get a semi-magic square (which can be made into a regular magic square by simply shifting the rows and columns until the 25 is located within the center cell of the array). Then after three more iterations, we get the following pan-diagonal magic square:

17	47	28	2	32	13	36
6	29	10	40	21	44	25
37	18	48	22	3	33	14
26	7	30	11	41	15	45
8	38	19	49	23	4	34
46	27	1	31	12	42	16
35	9	39	20	43	24	5

Figure 12.20 The pan-diagonal magic square that we get after nine iterations of the doubling operation.

After three more iterations, we get another semi-magic square, and then after another three iterations of the doubling operation we get this pan-diagonal magic square:

10	19	28	30	39	48	1
41	43	3	12	21	23	32
16	25	34	36	45	5	14
47	7	9	18	27	29	38
22	31	40	49	2	11	20
4	13	15	24	33	42	44
35	37	46	6	8	17	26

Figure 12.21 The pan-diagonal magic square that we get after six more iterations of the doubling operation.

Then, after six more iterations, we arrive back at the original pan-diagonal magic square that we started with, making a total of 21 iterations of the doubling operation before cycling back to our initial pan-diagonal magic square. It is also interesting to note that the numbers 7, 14, 21, 28, 35, 42 as well as the 49 all remained within their same locations in the array if we consider just the pan-diagonal magic squares.

Odd-Ordered Pan-Diagonal Magic Squares from Scratch

So far in this chapter we have built pan-diagonal magic squares from existing magic squares. Here is a method that will build an odd-ordered pan-diagonal magic square from scratch. Variations of this method appear to work on 5th order, 7th order, 11th order and 13th order magic squares. Here is the pan-diagonal magic square along with its description:

<table>
<tr><td>1</td><td>10</td><td>14</td><td>18</td><td>22</td></tr>
<tr><td>19</td><td>23</td><td>2</td><td>6</td><td>15</td></tr>
<tr><td>7</td><td>11</td><td>20</td><td>24</td><td>3</td></tr>
<tr><td>25</td><td>4</td><td>8</td><td>12</td><td>16</td></tr>
<tr><td>13</td><td>17</td><td>21</td><td>5</td><td>9</td></tr>
</table>

P1 (1, 1)
P21 (rt 2, dn 1)
P65 (dn 2)
Seq 1 2 3 4 5

Figure 12.22 A 5th order pan-diagonal magic square.

The key to making this magic square pan-diagonal is the placement of the sixth element in relation to the fifth. Actually, the sixth element is placed to the right of the second element, but our description considers where the sixth element is in relation to the fifth element, and that is how we describe it; but as we look at higher order magic squares, we will see that we place the first number of the following set to the right of the second number of the preceding set. Again, we can see this in the 7th order magic square. In this case, the eighth element is placed to the right of the second element. Here is the 7th order pan-diagonal magic square along with its description:

<table>
<tr><td>1</td><td>14</td><td>20</td><td>26</td><td>32</td><td>38</td><td>44</td></tr>
<tr><td>39</td><td>45</td><td>2</td><td>8</td><td>21</td><td>27</td><td>33</td></tr>
<tr><td>28</td><td>34</td><td>40</td><td>46</td><td>3</td><td>9</td><td>15</td></tr>
<tr><td>10</td><td>16</td><td>22</td><td>35</td><td>41</td><td>47</td><td>4</td></tr>
<tr><td>48</td><td>5</td><td>11</td><td>17</td><td>23</td><td>29</td><td>42</td></tr>
<tr><td>30</td><td>36</td><td>49</td><td>6</td><td>12</td><td>18</td><td>24</td></tr>
<tr><td>19</td><td>25</td><td>31</td><td>37</td><td>43</td><td>7</td><td>13</td></tr>
</table>

P1 (1, 1)
P21 (rt 2, dn 1)
P87 (dn 2, lf 2)
Seq 1 2 3 4 5 6 7

Figure 12.23 A 7th order pan-diagonal magic square.

In the next example the pattern continues with the 11th order pan-diagonal magic square.

1	22	32	42	52	62	72	82	92	102	112
103	113	2	12	33	43	53	63	73	83	93
84	94	104	114	3	13	23	44	54	64	74
65	75	85	95	105	115	4	14	24	34	55
35	45	66	76	86	96	106	116	5	15	25
16	26	36	46	56	77	87	97	107	117	6
118	7	17	27	37	47	57	67	88	98	108
99	109	119	8	18	28	38	48	58	68	78
69	79	89	110	120	9	19	29	39	49	59
50	60	70	80	90	100	121	10	20	30	40
31	41	51	61	71	81	91	101	111	11	21

P1 (1, 1)
P21 (rt 2, dn 1)
P12 11 (dn 2, lf 6)
Seq 1 2 3 4 5 6 7 8 9 10 11

Figure 12.24 An 11th order pan-diagonal magic square.

Returning to the 5th and 7th order magic squares we see that they are still pan-diagonal
even if we keep everything else the same in the description, and only change the
sequence. Here is the same magic square as that of magic square 12.22 except with a
slightly different sequence:

1	9	15	18	22
20	23	2	6	14
7	11	19	25	3
24	5	8	12	16
13	17	21	4	10

P1 (1, 1)
P21 (rt 2, dn 1)
P65 (dn 2)
Seq 1 2 3 5 4

Figure 12.25 A 5th order pan-diagonal magic square with an alternate sequence.

And here are two variations of magic square from figure 12.23. Again, the only
difference between these magic squares is the sequence used to build the magic square;
yet in each case, the magic squares are still pan-diagonal.

1	13	21	26	32	38	44
39	45	2	8	20	28	33
27	35	40	46	3	9	15
10	16	22	34	42	47	4
49	5	11	17	23	29	41
30	36	48	7	12	18	24
19	25	31	37	43	6	14

P1 (1, 1)
P21 (rt 2, dn 1)
P87 (dn 2, lf 2)
Seq 1 2 3 4 5 7 6

1	13	21	26	32	37	45
39	44	3	8	20	28	33
27	35	40	46	2	10	15
9	17	22	34	42	47	4
49	5	11	16	24	29	41
31	36	48	7	12	18	23
19	25	30	38	43	6	14

P1 (1, 1)
P21 (rt 2, dn 1)
P87 (dn 2, lf 2)
Seq 1 3 2 4 5 7 6

Figure 12.26 Two 7th order pan-diagonal magic squares with similar construction
methods, yet different sequences.

Comparing Odd-Order Pan-Diagonal Construction Descriptions
Let's return again to our 5th, 7th, and 11th order magic squares and look at the construction descriptions of these pan-diagonal magic squares.

5th Order	7th Order		11th Order	
P1 (1, 1)	P1 (1, 1)		P1 (1, 1)	
P21 (rt 2, dn 1)	P21 (rt 2, dn 1)		P21 (rt 2, dn 1)	
P65 (dn 2, lf 0)	P87 (dn 2, lf 2)		P12 11 (dn 2, lf 6)	
Seq (various)	Seq (various)		Seq (various)	

Table 12.2   This is a compilation of similar construction methods for 5th, 7th and 11th order pan-diagonal magic squares.

In the table above we have put the construction descriptions next to each other so that they are easier to compare with each other. (We have also left room in the table above for the 9th order and 13th order construction methods.) Now, notice that the placement of the number 1 (shown by P1) is the same for each order. Likewise, the placement of the 2 with respect to the 1 (shown by P21) is also the same with each order. The characteristic that varies is the placement of the first number of the next set with respect to the last number of the previous set.

The 5th order places the 6 with respect to the 5 down two cells. The 7th order places the 8 with respect to the 7 down two and left two cells; and the 11th order places the 12 with respect to the 11 down two and left six cells. These three methods give us enough information to establish a pattern; and we can use that pattern to fill in the gaps in our table above.

Extending the pattern just mentioned to the 9th and the 13th orders, we would say that the P10 9 parameter for the 9th order would be down two and left four cells, and the P14 13 parameter for the 13th order would be down two and left eight cells.

5th Order	7th Order	9th Order	11th Order	13th Order
P1 (1, 1)	P1 (1, 1)	P1 (1, 1)	P1 (1, 1)	P1 (1, 1)
P21 (rt 2, dn 1)	P21 (rt 2, dn 1)	P21 (rt 2, dn 1)	P21 (rt 2, dn 1)	P21 (rt 2, dn 1)
P65 (dn 2, lf 0)	P87 (dn 2, lf 2)	P10 9 (dn 2, lf 4)	P12 11 (dn 2, lf 6)	P14 13 (dn 2, lf 8)
Seq (various)	Seq (various)	Seq (various)	Seq (various)	Seq (various)

Table 12.3   An extension of the previous table that also includes 9th and 13th order descriptions of construction methods. The grayed area shows the properties that are different between orders.

Using these descriptions to build magic squares for the 9th and 13th orders will not give us any type of magic squares with regard to the 9th order square, but it gives us pan-diagonal magic squares with regard to the 13th order. Therefore, the table above describes a construction method for pan-diagonal magic squares for the 5th, 7th, 11th and 13th orders of magic squares.

And there are still more odd-ordered pan-diagonal magic squares that we can build from scratch. Here is a description of another set of pan-diagonal construction methods that are related to each other:

5th Order	7th Order	9th Order	11th Order	13th Order
P1 (1, 1)	P1 (1, 1)	P1 (1, 1)	P1 (1, 1)	P1 (1, 1)
P21 (rt 2, dn 1)	P21 (rt 3, dn 2)	P21 (rt 4, dn 3)	P21 (rt 5, dn 4)	P21 (rt 6, dn 5)
P65 (rt 0, up 3)	P87 (rt 0, up 3)	P10 9 (rt 0, up 3)	P12 11 (rt 0, up 3)	P14 13 (rt 0, up 3)
Seq (various)	Seq (various)	Seq (various)	Seq (various)	Seq (various)

Table 12.4   Another table of construction methods for odd-ordered pan-diagonal magic squares from the 5th order through the 13th order. The grayed area shows the properties that are different between orders.

In the table above, the 5th order construction method is actually the same as that in the previous table, although they are described just a little differently from each other. That difference is with the P65 parameter. In the previous table it is shown as down two while in the table above it is shown as up 3. For a 5th order array either description will yield the same result. The reason why I changed the P65 parameter is so that the pattern becomes apparent when compared with the similar parameter for the other orders of magic squares in the table. In each case, that particular parameter is right zero and up three.

The differences between these pan-diagonal magic squares (in the table above), can be seen in the P21 parameter. The 5th order places the 2 with respect to the 1 such that the 2 is right two cells and down one. The 7th order places it right three, and down two cells, and so on. And as before, the 5th, 7th, 11th and 13th order descriptions are pan-diagonal while the 9th order will not create any kind of magic square at all.

A Mapping of Odd-Ordered Pan-Diagonal Construction Methods
Within a pan-diagonal magic square, the location of the 1 can actually be anywhere within the array. Thus, the P1 parameter is practically meaningless when we talk about a construction method for an odd-ordered pan-diagonal magic square. Likewise, if we predefine the sequence in an ascending order so that it is from 1 ... n (where n is the order of the magic square); then the only two parameters left are the P21 parameter, and the "P n+1, n" parameter (such as the P65 parameter for a 5th order magic square, and the P87 parameter for a 7th order magic square, etc.) Knowing this, we can create a table where we compile instances of the P21 parameter with the corresponding "P n+1, n" parameter used to create the pan-diagonal magic square.

The following table is a compilation of construction methods for 7th order pan-diagonal magic squares using the regular sequence of 1 2 3 4 5 6 7. Note that the numbers next to the "rt" and "r" refer to the number of cells to the right while the "up" and "u" refers to the number of cells up.

P21	rt 0	rt 1	rt 2	rt 3	rt 4	rt 5	rt 6
up 6			r0, u2 r0, u3 r1, u2 r3, u1 r5, u4	r1, u5 r4, u4	r1, u4 r3, u3 r5, u3 r6, u0	r0, u3 r2, u4	
up 5		r2, u2 r4, u0		r1, u6 r5, u1	r0, u6 r3, u1		
up 4		r0, u1 r1, u4	r4, u0 r4, u5				
up 3		r0, u6 r4, u1 r4, u4					
up 2		r4, u4					
up 1							
up 0							

Table 12.5   This is a compilation of construction methods for 7th order pan-diagonal magic squares using the sequence 1 2 3 4 5 6 7. (Note that this table is not by any means believed to be complete.)

The way the table is read is that the P21 relationship is described along the left-side and bottom of the table, while the P87 relationship is described in the array. For example, here are four of the 7th order pan-diagonal construction methods shown in the table above:

P1 (anywhere)	P1 (anywhere)	P1 (anywhere)	P1 (anywhere)
P21 r1, u2	P21 r2, u4	P21 r2, u6	P21 r5, u6
P87 r4, u4	P87 r4, u5	P87 r0, u3	P87 r2, u4
Seq 1234567	Seq 1234567	Seq 1234567	Seq 1234567

The table above lists twenty-seven 7th order pan-diagonal magic squares that I have found that use a "1 2 3 4 5 6 7" sequence; however, there could be many more. I compiled the table in order to look for patterns so that more pan-diagonal magic squares could be discovered, but I have been unable to see a pattern in the data presented by the table.

The following table on the next page is a compilation of construction methods for 11th order pan-diagonal magic squares using the sequence of 1 2 3 4 5 6 7 8 9 10 11. Again, note that the numbers next to the "rt" and "r" refer to the number of cells to the right while the "up" and "u" refers to the number of cells up.

Again, the P21 relationship is described along the left-side and bottom of the table, while the P12 11 relationship is described in the array.

P21	rt 0	rt 1	rt 2	rt 3	rt 4	rt 5	rt 6	rt 7	rt 8	rt 9	rt 10
up10			r5, u9 r7, u1 r10, u9	r7, u2 r9, u7	r3, u1 r3, u5 r8, u0 r9, u1	r0, u3	r0, u9 r1, u6 r10, u5	r2, u9 r9, u7	r5, u0 r10, u4	r8, u7	
up 9		r3, u3		r2, u5	r3, u1	r0, u6 r4, u6 r8, u8 r10, u0	r5, u4	r1, u0 r4, u4	r2, u7 r4, u4		
up 8		r0, u5 r5, u3 r9, u3	r0, u5 r8, u1 r10, u3		r3, u3	r1, u3 r6, u6 r7, u2		r2, u5 r4, u7			
up 7		r0, u10	r7, u9 r9, u2	r9, u6		r0, u3 r4, u1 r10, u0	r1, u0				
up 6		r0, u1 r2, u0 r7, u2 r9, u5	r4, u0	r0, u1 r6, u5	r3, u3						
up 5		r0, u10 r9, u6	r2, u0 r7, u6 r8, u2	r2, u0 r4, u10 r9, u9	r10, u4						
up 4			r1, u7 r6, u10								
up 3		r4, u2 r8, u9									
up 2											
up 1											
up 0											

Table 12.6　This is a compilation of construction methods for 11th order pan-diagonal magic squares using the sequence 1 2 3 4 5 6 7 8 9 10 11. (Again, the table is not meant by any means to be complete.)

You may have also noticed in these two tables that the values are located in only the upper left triangular section of the table. In other words, there is data located at position rt 1, up3, but nothing is located at the rt 3, up 1 position. This is because any description of data at the rt 3, up 1 position is a mirror-rotation of the data at rt 1, up 3. For example, the table shows:

P21	rt 1, up 3
P12 11	r 4, u 2

We could also create a pan-diagonal magic square with this construction method:

P21   rt 3, up 1
P12 11   r 2, u 4

And these two construction methods describe the same magic square where the second magic square is a mirror-rotation of the first.

An Infinite Number (Or at Least a Really Big Number) of Magic Squares
Of course, the possibilities of creating magic squares are virtually limitless. With the various methods and techniques that we've seen in this chapter and the preceding chapters, it is possible to create literally tens of thousands of distinct magic squares (many of which can be pan-diagonal). Then by interchanging pairs of rows that are equidistant (or symmetrical) from the center of the square, performing quadrant swaps, and by using other transformation methods mentioned within these pages, these tens of thousands of magic squares can turn into hundreds of thousands, millions, billions, and even more magic squares. It is literally true that there are so many possibilities, and so little time.

And as amazing as the methods are that can be used to create magic squares, there are still other aspects of these arrays of squares that are just as amazing. Within the remaining chapters of this book we will look at some of the many patterns and symmetries that can be found within these magnificently marvelous magically magical magic squares.

Chapter 13
Multi-Grades and Magic Squares

Multi-Grades
A multi-grade is a sum-of-squares relationship that isn't limited to just squares. We could say that it is the sum of a series of numbers raised to a power that is equal to the sum of another series of numbers raised to the same power; where in addition to this, the relation between the two series of sums is true for more than just a single power. Let's look at an example of a second order multi-grade:

$$1 + 6 + 8 + 9 \ \overset{2}{=}\ 3 + 4 + 6 + 11$$

The expression above means that the following is true:

$$1^1 + 6^1 + 8^1 + 9^1 = 3^1 + 4^1 + 6^1 + 11^1 \qquad (\text{ both sides} = \ 24)$$
$$1^2 + 6^2 + 8^2 + 9^2 = 3^2 + 4^2 + 6^2 + 11^2 \qquad (\text{ both sides} = \ 182)$$

This series is true when each of the numbers are all raised to the first power, and again when each of the numbers is raised to the second power. To denote a multi-grade, we write the power, or order of the multi-grade over the equal sign. In this chapter we will write the order of the multi-grade as a super-scripted number between two equal signs. In addition to this, as a kind of shorthand, we can also list the numbers on each side of the equality separated by commas instead of separated with the plus sign. Therefore, the second degree multi-grade shown above can also be represented like this:

$$1, 6, 8, 9 \ \overset{2}{=}\ 3, 4, 6, 11$$

There are many kinds of multi-grades, and they appear to be more common than you would think. And sometimes they can appear in unexpected places, as is the case with multi-grades and magic squares.

Unexpected Multi-grades
Let's take a look at the relation between magic squares and multi-grades. We'll start with a basic, third order magic square:

8 1 6
3 5 7
4 9 2

The sum of the numbers within each row, column and diagonal add up to 15. However, within this magic square we also have two sets of second order multi-grades. One multi-grade consists of the left column set equal to the right column:

$$3, 4, 8 \ =^2= \ 2, 6, 7$$

And the other consists of the top row set equal to the bottom row:

$$1, 6, 8 \ =^2= \ 2, 4, 9$$

In other words, when the left column is set equal to the right column, we have:

$$3^1 + 4^1 + 8^1 \ = \ 2^1 + 6^1 + 7^1 \qquad \text{(both sides equal 15)}$$
$$3^2 + 4^2 + 8^2 \ = \ 2^2 + 6^2 + 7^2 \qquad \text{(both sides equal 89)}$$

And when the top row is set equal to the bottom row, we have:

$$1^1 + 6^1 + 8^1 \ = \ 2^1 + 4^1 + 9^1 \qquad \text{(both sides equal 15)}$$
$$1^2 + 6^2 + 8^2 \ = \ 2^2 + 4^2 + 9^2 \qquad \text{(both sides equal 101)}$$

Interesting! Now, let's extend our search for multi-grades and look at a fourth order magic square. In this magic square, each row, column and diagonal adds up to 34:

$$\begin{array}{cccc} 1 & 15 & 14 & 4 \\ 12 & 6 & 7 & 9 \\ 8 & 10 & 11 & 5 \\ 13 & 3 & 2 & 16 \end{array}$$

Within this magic square we can find four sets of second order multi-grades. We create two sets of multi-grades from the columns, and two sets from the rows. In the case of the columns, one multi-grade is made from the sum of the two outside columns set equal to each other, and the second is made from the two inside columns:

$$1, 8, 12, 13 \ =^2= \ 4, 5, 9, 16$$
$$3, 6, 10, 15 \ =^2= \ 2, 7, 11, 14$$

To create the multi-grades from the rows, we set the top and bottom rows equal to each other for the first multi-grade, and set the two inside rows equal to each other for the second multi-grade:

$$1, 4, 14, 15 \ =^2= \ 2, 3, 13, 16$$
$$6, 7, 9, 12 \ =^2= \ 5, 8, 10, 11$$

Something that I find is an interesting aspect of these four multi-grades is that we are using the numbers one through 16 to create the first pair of multi-grades, and then to create the second pair of multi-grades (which is unique from the first pair), we again use the same numbers, but in different combinations. Not all magic squares have this kind of symmetry where they can create multi-grades in this manner; however, as we will soon see, magic squares do appear to order the numbers so that they are conducive to creating multi-grades.

More Multi-grades from Magic Squares

There is still more we can do with these magic squares. Let's look a little closer at another fourth order magic square. In the table below, we have our magic square along with two additional columns to the right of the magic square; and two additional rows beneath the magic square (these additional columns and rows are shaded). The first column to the right shows the sum of each row, and the next column to the right shows the sum of the squares of the numbers within each row. Similarly, the first row beneath the magic square shows the sum of each column, and the row beneath that shows the sum of the squares of the numbers within each column.

4 x 4 Magic Square				Sum of Row	Row Sum of Squares
1	12	14	7	34	390
15	6	4	9	34	358
8	13	11	2	34	358
10	3	5	16	34	390
Sum of Column	34	34	34	34	
Column Sum of Squares	390	358	358	390	

Table 13.1 A fourth order magic square along with the sums of the rows and columns and the sums of the squares of the rows and columns.

We see that the sums of the squares of the top and bottom rows are both equal to 390. We also see that the sums of the squares of the first and last columns are also both equal to 390. From what we've seen earlier in this chapter, we can generate the following second order multi-grades from this magic square:

$$1, 7, 12, 14 \; =\!\!\overset{2}{=} \; 3, 5, 10, 16 \qquad \text{(from the top and bottom rows)}$$
$$1, 8, 10, 15 \; =\!\!\overset{2}{=} \; 2, 7, 9, 16 \qquad \text{(from the first and last columns)}$$

And we can go even further with this. We can set the top row equal to the first column; and we can set the bottom row equal to the last column to give us these two multi-grades:

$$1, 7, 12, 14 \; =\!\!\overset{2}{=} \; 1, 8, 10, 15 \quad \text{(from the top row and first column)}$$
$$3, 5, 10, 16 \; =\!\!\overset{2}{=} \; 2, 7, 9, 16 \quad \text{(from the bottom row and last column)}$$

There is no rule that says that you can't have the same number on both the right and left sides of a multi-grade; however, it is redundant to do so, and if the same number is on both sides of a multi-grade, it can be eliminated. The first multi-grade above shares a 1 on the right and left side, and the second multi-grade shares a 16 on the right and left side; therefore, each of these multi-grades can be condensed so that there are now only three numbers on right and left sides of each instead of four:

$$7, 12, 14 \;\overset{2}{=}\; 8, 10, 15 \qquad \text{(from the top row and first column)}$$
$$3, 5, 10 \;\overset{2}{=}\; 2, 7, 9 \qquad \text{(from the bottom row and last column)}$$

Similarly we can set the top row equal to the last column and the bottom row equal to the first column to again create two other multi-grades that each share a common number. These multi-grades can then be reduced so that they have only three numbers on the right and left sides. Without going through all the steps, here are those resulting multi-grades:

$$1, 12, 14 \;\overset{2}{=}\; 2, 9, 16 \qquad \text{(from the top row and last column)}$$
$$1, 8, 15 \;\overset{2}{=}\; 3, 5, 16 \qquad \text{(from the bottom row and first column)}$$

And that is only half of the possibilities with this magic square. This particular fourth order magic square also has two rows (the second and third row) and two columns (the second and third column) whose sum-of-squares each add up to 358. We can do the same thing by setting these rows and columns equal to each other and thus creating even more second order multi-grades from this one magic square.

Life gets even more interesting when we look at 5th order magic squares. In the following table, we have a 5th order magic square. This table is similar to the previous table, except that we have also added a column to the right that shows the sum of the cubes for each row, and we've added an additional row at the bottom of the table that shows the sum of the cubes for each column.

5 x 5 Magic Square					Sum of Row	Row Sum of Squares	Row Sum of Cubes
1	24	17	15	8	65	1155	22625
14	7	5	23	16	65	1055	19475
22	20	13	6	4	65	1105	21125
10	3	21	19	12	65	1055	18875
18	11	9	2	25	65	1155	23525
Sum of Column	65	65	65	65	65			
Column Sum of Squares	1105	1155	1005	1155	1105			
Column Sum of Cubes	20225	23525	17225	22625	22025			

Table 13.2 A fifth order magic square along with the sums of the rows and columns, the sums of the squares of the rows and columns, and the sums of the cubes of the rows and columns.

In this table we can see that sums of the squares of the first and last rows are both equal to 1155; and the sum-of-the-squares of the second and fourth columns are also equal to

1155. This means that we can pair these rows and columns in various ways to create second order multi-grades just as we did in the previous example with the 4th order magic square. We can also do the same thing with the second and fourth rows and the first and last columns whose sums of squares are all equal to 1105.

Now, draw your attention to the sums of cubes of the columns and rows. Notice that the second column and the fifth row both have a sum of squares equal to 1155, and a sum of cubes equal to 23525. This means that we can create a 3rd order multi-grade from this column and row, which would be:

$$3, 7, 11, 20, 24 \; =^3= \; 2, 9, 11, 18, 25 \qquad \text{(from the second column and fifth row)}$$

And since there is an 11 on both sides of this multi-grade, we can remove it from both sides to get this third order multi-grade:

$$3, 7, 20, 24 \; =^3= \; 2, 9, 18, 25 \qquad \text{(from the second column and fifth row)}$$

And in a similar way we also notice that the sum of cubes of the first row and the fourth column are both equal to 22625. And since their sum-of-squares are also equal to each other, we can create this 3rd order multi-grade from that row and column:

$$2, 6, 15, 19, 23 \; =^3= \; 1, 8, 15, 17, 24 \quad \text{(from the fourth column and first row)}$$

And after removing the number that is common on both sides (which is the 15), we have this more condensed, 3rd order multi-grade:

$$2, 6, 19, 23 \; =^3= \; 1, 8, 17, 24 \qquad \text{(from the fourth column and first row)}$$

A closer examination of the rows and columns of this magic square will reveal that there are still several other rows and columns that can be combined to create various multi-grades. Finally, before we leave this topic, let's briefly look at two more magic squares, a 5th order magic square, and a 7th order magic square.

First, we'll look at a 5th order magic square (as shown in the next figure). This magic square is interesting in that each of the five sums of cubes for the rows corresponds to a sum of cubes from the columns. This means that in addition to the many 2nd order multi-grades that can be created from the magic square, we can also generate five unique 3rd order multi-grades from this magic square. Here are the 3rd order multi-grades:

$$1, 9, 12, 20, 23 \; =^3= \; 1, 8, 15, 17, 24 \qquad \text{(1^{st} row and 1^{st} column)}$$
$$4, 7, 15, 18, 21 \; =^3= \; 3, 10, 12, 19, 21 \qquad \text{(4^{th} row and 2^{nd} column)}$$
$$2, 10, 13, 16, 24 \; =^3= \; 4, 6, 13, 20, 22 \qquad \text{(3^{rd} row and 3^{rd} column)}$$
$$5, 8, 11, 19, 22 \; =^3= \; 5, 7, 14, 16, 23 \qquad \text{(2^{nd} row and 4^{th} column)}$$
$$3, 6, 14, 17, 25 \; =^3= \; 2, 9, 11, 18, 25 \qquad \text{(5^{th} row and 5^{th} column)}$$

And here is the magic square that they were taken from:

5 x 5 Magic Square					Sum of Row	Row Sum of Squares	Row Sum of Cubes
1	12	20	23	9	65	1155	22625
8	19	22	5	11	65	1055	19475
24	10	13	16	2	65	1105	21125
15	21	4	7	18	65	1055	18875
17	3	6	14	25	65	1155	23525
Sum of Column	65	65	65	65	65		
Column Sum of Squares	1155	1055	1105	1055	1155		
Column Sum of Cubes	22625	18875	21125	19475	23525		

Table 13.3 A second 5 x 5 magic square along with the sums of the rows and columns, the sums of the squares of rows and columns, and the sums of the cubes of rows and columns with five unique 3rd order multi-grades.

And since we are combining a row with a column in each of these multi-grades, it means that each multi-grade also has a number that is common between the left and right sides. Now, after removing the common number from each side of these multi-grades, we get the following 3rd order condensed multi-grades:

$$9, 12, 20, 23 \;=^3=\; 8, 15, 17, 24 \qquad (1^{st}\text{ row and } 1^{st}\text{ column})$$
$$4, 7, 15, 18 \;=^3=\; 3, 10, 12, 19 \qquad (4^{th}\text{ row and } 2^{nd}\text{ column})$$
$$2, 10, 16, 24 \;=^3=\; 4, 6, 20, 22 \qquad (3^{rd}\text{ row and } 3^{rd}\text{ column})$$
$$8, 11, 19, 22 \;=^3=\; 7, 14, 16, 23 \qquad (2^{nd}\text{ row and } 4^{th}\text{ column})$$
$$3, 6, 14, 17 \;=^3=\; 2, 9, 11, 18 \qquad (5^{th}\text{ row and } 5^{th}\text{ column})$$

The next figure shows the 7th order magic square that I promised. If you look closely at the sums of cubes for each row, and compare them with the sums of cubes for the columns, you will see that there are four numbers that correspond between the two lists of numbers. Since their sums of squares also correspond with each other, we follow the same process used in the preceding example and generate four 3rd order multi-grades from this magic square.

And as higher order magic squares are inspected in this way (by comparing the sums of squares of the rows and columns, and then also comparing the sums of cubes of the rows and columns), it is possible to see many more 2nd and 3rd order multi-grades within these magic squares.

7 x 7 Magic Square							Sum of Row	Row Sum of Squares	Row Sum of Cubes
1	27	46	16	42	12	31	175	5971	226723
39	9	35	5	24	43	20	175	5677	204379
28	47	17	36	13	32	2	175	5775	212317
10	29	6	25	44	21	40	175	5579	199675
48	18	37	14	33	3	22	175	5775	216433
30	7	26	45	15	41	11	175	5677	209671
19	38	8	34	4	23	49	175	5971	231427

Sum of Column	175	175	175	175	175	175	175
Column Sum of Squares	5971	5677	5775	5579	5775	5677	5971
Column Sum of Cubes	226723	209671	214081	199675	214699	204379	231427

Table 13.4 A 7 x 7 magic square along with the sums of the rows and columns, the sums of the squares of rows and columns, and the sums of the cubes of rows and columns with five unique 3rd order multi-grades.

Building Multi-grades from Multi-grades

Once you have a multi-grade, it is possible to build higher order multi-grades from an existing multi-grade. Therefore, we can take the multi-grades that we found within our magic squares, and we can use the method in this section to build higher order multi-grades. For example, let's start with a basic multi-grade that works through the second power:

$$2, 3, 7 \ =^2= \ 1, 5, 6$$

In order to create another multi-grade from this one, we take some number k, and add it to each term in the relation to create a second relation.

the left side has: 2+k, 3+k, 7+k   the right side has: 1+k, 5+k, 6+k

Now, we combine the first and second relations so that the left side has the original left-side numbers 2, 3 and 7 as well as the numbers from the right side with an added k. On the right side we also place the original right-side numbers, which are 1, 5 and 6 and also the three left-side numbers with an added k. With this, we now have a formula from a second order multi-grade that will create an unlimited supply of third order multi-grades:

$$2, 3, 7, (1+k), (5+k), (6+k) \ =^3= \ 1, 5, 6, (2+k), (3+k), (7+k)$$

In this example, k can equal any number. In addition to this, where we started with a second order multi-grade, we now have a formula for a third order multi-grade. Since the value of k can now be equal to any integer, by letting k equal 1, 2 and 3, we get these third order multi-grades:

$$k = 1 \qquad 2, 3, 7, 2, 6, 7 \;=^3=\; 1, 5, 6, 3, 4, 8$$
$$k = 2 \qquad 2, 3, 7, 3, 7, 8 \;=^3=\; 1, 5, 6, 4, 5, 9$$
$$k = 3 \qquad 2, 3, 7, 4, 8, 9 \;=^3=\; 1, 5, 6, 5, 6, 10$$

In the example in this section we took a second order multi-grade, and applied a method of building multi-grades that could actually create any number of third order multi-grades.

As we can start to see, it is possible to get long lists of numbers on both sides of the relationship; however, in some cases it is also possible to trim the lists of numbers a little. For example, in the example above where k = 1, we have this third order multi-grade:

$$k = 1 \qquad 2, 3, 7, 2, 6, 7 \;=^3=\; 1, 5, 6, 3, 4, 8$$

Notice how there is a 3 and a 6 on both sides of the relationship. Since the same numbers are on both sides, we can remove those numbers from each side to get a trimmed down multi-grade:

$$2, 7, 2, 7 \;=^3=\; 1, 5, 4, 8$$

We now have a third order multi-grade with only four numbers on each side of the relationship.

We can now take any of these third order multi-grades from the example above, apply this method again, and create fourth order multi-grades. And we can continue in the same way and create fifth order and higher order multi-grades. The possibilities are virtually limitless. And there is still much more we could discuss about multi-grades, but, what we have must suffice for now as we return our focus to the many patterns and properties of magic squares.

Chapter 14
Other Perspectives of the Magic Square

The Aa and the Zero-Centered Patterns
Magic squares have many other properties and patterns within them. These patterns
become apparent as we devise other ways to look at the magic square. In this chapter
we'll change our perspective just a little. We'll start by considering two of these alternate
representations, along with a few corresponding magic squares. (We briefly saw these
patterns in chapter 3.)

The first of these patterns we'll call the Aa pattern. To construct this pattern, we recall
that an nth order magic square is made up of n numbers taken n times, or in other words,
we have n sets of n numbers. For a 4th order magic square this means that we have four
sets of four numbers. To give each number in a 4th order magic square a unique
designation, we use the uppercase letters A thru D to denote the set, and we use the
lowercase letters a thru d to denote which number it is within the set. Therefore, the
numbers 1 thru 4 would all have an uppercase A, the numbers 5 thru 8 would have an
uppercase B, and so on for C and D. Likewise, the number 1 is the first number of its set,
as are the numbers 5, 9, and 13. These four numbers are represented by the lowercase a.
The numbers 2, 6, 10 and 14 are the second numbers of their sets, so they are represented
by the lowercase b. And so on for the letters c and d. Using this notation to list the
numbers 1 thru 16 in ascending order, we have the following sequence:

Aa, Ab, Ac, Ad, Ba, Bb, Bc, Bd, Ca, Cb, Cc, Cd, Da, Db, Dc, Dd

The second pattern we'll consider is what we'll call the zero-centered pattern. In our
study, this pattern will be created differently for even and odd-ordered magic squares.
For an odd-ordered magic square, we simply subtract the middle number (of the range)
from each element in each cell, thus giving us a zero-centered pattern (there will be the
same amount of numbers less than zero as there are greater than zero, with zero in the
middle). To create a zero-centered pattern for an even-ordered magic square, there are
several ways it could be done, but the way we will do it is to first double each number,
and then subtract the order squared plus one (or $n^2 + 1$) from each number.

By zero-centering an even-order magic square in this way, zero will be the middle
number (although it won't appear in the array), and it will still keep the individual
numbers evenly spaced (with respect to each other), causing half the numbers to be below
zero while the other half are above zero.

1	15	14	4
12	6	7	9
8	10	11	5
13	3	2	16

Aa	Dc	Db	Ad
Cd	Bb	Bc	Ca
Bd	Cb	Cc	Ba
Da	Ac	Ab	Dd

-15	13	11	-9
7	-5	-3	1
-1	3	5	-7
9	-11	-13	15

Figure 14.1   A magic square from chapter 4 along with the Aa pattern and the zero-centered pattern.

The listing above shows the first magic square that we created in chapter 4, along with its corresponding Aa pattern, and zero-centered pattern. Notice that within the Aa array pattern, the uppercase letters A and D are only on the first and fourth rows while the lowercase letters a thru d are on each row; however, when we look at the columns, the lowercase letters a and d are only in the first and last column while the uppercase letters A thru D are in each column. We can see the same type of distribution with the uppercase B and C letters as well as the lower case b and c letters.

We also see an interesting arrangement in the zero-centered pattern above. If we disregard the negative sign for a moment, notice how the numbers start in the upper left corner with the number 15 and proceed left to right and top to bottom in descending order until they reach 1 and then continue from 1 back up to 15. Now, if we consider the negative signs, we see that each column and each row has two numbers that are negative (and two that are positive), and the sum of each row, column, and main diagonals are all equal to zero. Within this square, we can also see a symmetry between each negative number and its corresponding positive number about the center of the magic square.

Now let's look at the 4th order diabolical magic square that we saw in chapter 4. In the listing below, we see that the Aa pattern has a uppercase A thru D on every row and in every column. Similarly, there is a lowercase a thru d on every row and in every column. There are also some interesting relationships when we look at the main diagonals and the broken diagonals.

14	11	5	4
1	8	10	15
12	13	3	6
7	2	16	9

Db	Cc	Ba	Ad
Aa	Bd	Cb	Dc
Cd	Da	Ac	Bb
Bc	Ab	Dd	Ca

11	5	-7	-9
-15	-1	3	13
7	9	-11	-5
-3	-13	15	1

Figure 14.2   Diabolic magic square from chapter 4 with the Aa and the zero-centered patterns.

The zero-centered patterns show another type of symmetry. Every positive number is diagonally two cells away for its corresponding negative number; and again, the sum of every row, column and diagonal is equal to zero. It is an interesting balance to a balanced magic square.

Now let's consider a couple of 5th order magic squares. The 5th order Aa pattern is created in the same way that the 4th order was, except that with a 5th order magic square we have five sets of five numbers. For this reason we use the uppercase letters A thru E to represent the sets and the lowercase letters a thru e to represent each number within the set. The following 5th order array is just a regular magic square. In the Aa pattern, we can easily see that the uppercase letters A thru E as well as the lowercase letters a thru e are used once in each row and each column.

1	20	9	23	12
19	8	22	11	5
7	21	15	4	18
25	14	3	17	6
13	2	16	10	24

Aa	De	Bd	Ec	Cb
Dd	Bc	Eb	Ca	Ae
Bb	Ea	Ce	Ad	Dc
Ee	Cd	Ac	Db	Ba
Cc	Ab	Da	Be	Ed

-12	7	-4	10	-1
6	-5	9	-2	-8
-6	8	2	-9	5
12	1	-10	4	-7
0	-11	3	-3	11

Figure 14.3   Magic square from chapter 5 with the Aa and the zero-centered patterns.

We also see an interesting structure in the two main diagonals. The main diagonal going from the upper-left to lower-right uses the uppercase letters A thru E, and in order. The diagonal going from the lower-left to upper-right only uses the uppercase C letter.

For an odd-order magic square, such as this 5th order square, we create the zero-centered pattern differently than we did for the fourth order square. A 5th order magic square uses the numbers 1 thru 25 with the middle number being 13. Therefore, we simply subtract 13 from each number in the array to create the zero-centered pattern.

The zero-centered pattern above has an interesting arrangement. The first column and last row has one type of symmetry while the remaining 4 x 4 array has another type of symmetry. In the first column the 6's are one-above-the-other and surrounded by the 12's. Similarly, on the bottom row the 3's are next to each other surrounded by the 11's. The remaining 4 x 4 array has each number diagonally opposite its opposite-signed counterpart. As an exercise, shift the rows and columns so that the 13 is in the center cell. Now construct the Aa pattern and the zero-centered pattern and note how the symmetry has changed.

Now let's look at the 5th order diabolical magic square from chapter 5 with its Aa and zero-centered patterns. In this example, the 13 is in the center cell.

1	15	24	8	17
23	7	16	5	14
20	4	13	22	6
12	21	10	19	3
9	18	2	11	25

Aa	Ce	Ed	Bc	Db
Ec	Bb	Da	Ae	Cd
De	Ad	Cc	Eb	Ba
Cb	Ea	Be	Dd	Ac
Bd	Dc	Ab	Ca	Ee

-12	2	11	-5	4
10	-6	3	-8	1
7	-9	0	9	-7
-1	8	-3	6	-10
-4	5	-11	-2	12

Figure 14.4   Diabolic magic square from chapter 5 with the Aa and the zero-centered patterns.

The Aa pattern has the uppercase letters A thru E and the lowercase letters a thru e in every row and in every column. What is interesting about this magic square is that the letter combinations Aa, Bb, Cc, Dd, and Ee are lined up on the main diagonal that runs from the upper left to the lower right. These numbers along that main diagonal are in their original positions (we'll look at other patterns of magic squares with numbers in their original positions a little later in this chapter).

The zero-centered pattern appears very balanced. Every number is diagonally opposite its opposite signed counterpart. In other words, the 12 is diagonally opposite the -12, as is the 11 and the -11, and so on all the way down to the 1 and the -1.

Every magic square is different, and these representations provide an alternate way to see patterns within the magic square. In many cases, by changing the way we look at these squares we find a balance and symmetry that wasn't apparent before. However, keep in mind that there are exceptions that go both ways. For example, it is possible to use some of the odd-order methods to create a 9th order square that isn't a magic square, yet it still has the balance and symmetry of other squares that are magic. Going the other way, we can create a magic square that doesn't appear to have a balance or symmetry, and yet it is a magic square. Therefore, these other methods of looking at magic squares help bring out the patterns, but they don't define the magic square itself.

As magic squares get larger, it becomes more difficult to pull out patterns from the Aa and zero-centered representations of the magic square, so we won't spend too much more time with these patterns; however, we'll look at a few 7th order magic squares before we move on.

Within the next two magic squares, the numbers along the main diagonal running from the upper-left to lower-right are all in their original positions. Therefore, we see the letters Aa, Bb, Cc, etc., running along the main diagonal from the upper-left to lower-right. In addition to this, if we also consider the broken diagonals that also run from the upper-left to lower right, we can still find the uppercase letters A thru G, and in order (although each broken diagonal starts with a different letter, such as E, B, F, etc.). Here is our 7th order magic square:

1	34	11	37	21	47	24
32	9	42	19	45	22	6
14	40	17	43	27	4	30
38	15	48	25	2	35	12
20	46	23	7	33	10	36
44	28	5	31	8	41	18
26	3	29	13	39	16	49

Aa	Ef	Bd	Fb	Cg	Ge	Dc
Ed	Bb	Fg	Ce	Gc	Da	Af
Bg	Fe	Cc	Ga	Df	Ad	Eb
Fc	Ca	Gf	Dd	Ab	Eg	Be
Cf	Gd	Db	Ag	Ee	Bc	Fa
Gb	Dg	Ae	Ec	Ba	Ff	Cd
De	Ac	Ea	Bf	Fd	Cb	Gg

-24	9	-14	12	-4	22	-1
7	-16	17	-6	20	-3	-19
-11	15	-8	18	2	-21	5
13	-10	23	0	-23	1	-13
-5	21	-2	-18	8	-15	11
19	3	-20	6	-17	16	-7
1	-22	4	-12	14	-9	24

Figure 14.5   A 7th order magic square on the upper-left along with the Aa pattern on the lower-left (with the lowercase a's shaded) and the zero-centered pattern on the lower-right.

Now, if we look at the main and broken diagonals running in the opposite direction (from the lower-left to the upper right), we find that each diagonal consists of the same uppercase letter within its diagonal. For example, the main diagonal has the uppercase D. Moving to the right, the next diagonal (a broken diagonal) has the uppercase A, followed by the letter E, and so on.

The lowercase letters follow the knights move (as in knights from a chess game). Start with any lower case letter, and the same lower case letter will be found again two cells to the right and up one.

This next magic square also has the numbers along the upper-left to lower-right main diagonal in their original positions. Again, this means that the Aa, Bb, Cc, etc. letters are also along this main diagonal. However, the rolls of the uppercase and lowercase have reversed. Now, if we look at the diagonals running in the opposite direction, we find that the lowercase letters run along the same diagonal while the upper case letters are a knights move away from each other (over two and up one).

1	40	23	13	45	35	18
26	9	48	31	21	4	36
44	34	17	7	39	22	12
20	3	42	25	8	47	30
38	28	11	43	33	16	6
14	46	29	19	2	41	24
32	15	5	37	27	10	49

Aa	Fe	Db	Bf	Gc	Eg	Cd
De	Bb	Gf	Ec	Cg	Ad	Fa
Gb	Ef	Cc	Ag	Fd	Da	Be
Cf	Ac	Fg	Dd	Ba	Ge	Eb
Fc	Dg	Bd	Ga	Ee	Cb	Af
Bg	Gd	Ea	Ce	Ab	Ff	Dc
Ed	Ca	Ae	Fb	Df	Bc	Gg

-24	15	-2	-12	20	10	-7
1	-16	23	6	-4	-21	11
19	9	-8	-18	14	-3	-13
-5	-22	17	0	-17	22	5
13	3	-14	18	8	-9	-19
-11	21	4	-6	-23	16	-1
7	-10	-20	12	2	-15	24

Figure 14.6   Another 7th order magic square on the upper-left along with the Aa pattern on the lower-left (with the uppercase A's shaded) and the zero-centered pattern on the lower-right.

In both of the examples above, the zero-centered patterns are symmetrical through the center of the magic square. So, if we take any negative number, go through the zero at the center of the magic square to the opposite side of the magic square, we'll find the positive number.

The following magic square has the numbers in the opposite main diagonal (the diagonal running from the lower-left to the upper-right) in their original positions. This causes the lowercase letters a thru g to run in order from the lower left to upper right, while the uppercase letters A thru G run in order from the upper right to the lower left.

In this example the uppercase letters are each in their own diagonal running from upper-left to lower-right. The lowercase letters are a knight's move away from each other, moving right one and down two cells.

24	48	16	40	8	32	7
5	22	46	21	38	13	30
35	3	27	44	19	36	11
9	33	1	25	49	17	41
39	14	31	6	23	47	15
20	37	12	29	4	28	45
43	18	42	10	34	2	26

Dc	Gf	Cb	Fe	Ba	Ed	Ag
Ae	Da	Gd	Cg	Fc	Bf	Eb
Eg	Ac	Df	Gb	Ce	Fa	Bd
Bb	Ee	Aa	Dd	Gg	Cc	Ff
Fd	Bg	Ec	Af	Db	Ge	Ca
Cf	Fb	Be	Ea	Ad	Dg	Gc
Ga	Cd	Fg	Bc	Ef	Ab	De

-1	23	-9	15	-17	7	-18
-20	-3	21	-4	13	-12	5
10	-22	2	19	-6	11	-14
-16	8	-24	0	24	-8	16
14	-11	6	-19	-2	22	-10
-5	12	-13	4	-21	3	20
18	-7	17	-15	9	-23	1

Figure 14.7   A third 7th order magic square on the upper-left with the Aa pattern on the lower-left and zero-centered pattern on the lower-right.

The following magic square is a pan-diagonal 7th order magic square. It also has the numbers along the lower-left to upper right main diagonal in their original position. The property that is unique about this magic square is that there is one uppercase letter from the set [A, B, C, D, E, F, G] in every row and column, as well as in each diagonal running both directions. The same is true for the lowercase letters. There is one lowercase letter from the set [a, b, c, d, e, f, g] in each row, column, and diagonal in both directions.

9	18	27	29	38	47	7
24	33	42	44	4	13	15
39	48	1	10	19	28	30
5	14	16	25	34	36	45
20	22	31	40	49	2	11
35	37	46	6	8	17	26
43	3	12	21	23	32	41

Bb	Cd	Df	Ea	Fc	Ge	Ag
Dc	Ee	Fg	Gb	Ad	Bf	Ca
Fd	Gf	Aa	Bc	Ce	Dg	Eb
Ae	Bg	Cb	Dd	Ef	Fa	Gc
Cf	Da	Ec	Fe	Gg	Ab	Bd
Eg	Fb	Gd	Af	Ba	Cc	De
Ga	Ac	Be	Cg	Db	Ed	Ff

-16	-7	2	4	13	22	-18
-1	8	17	19	-21	-12	-10
14	23	-24	-15	-6	3	5
-20	-11	-9	0	9	11	20
-5	-3	6	15	24	-23	-14
10	12	21	-19	-17	-8	1
18	-22	-13	-4	-2	7	16

Figure 14.8   A 7th order pan-diagonal magic square on the upper-left with the Aa pattern and zero-centered pattern on the lower left and right.

Again, the zero-centered patterns in both of the examples above are symmetrical through the center of the zero-centered array. Of course, we could go on and on, but there are still other patterns to consider as we will see in this next section.

Numbers in Their Original Position

The next set of patterns are made up of numbers in their original positions. We have to carefully define what we mean by "original position." After all, in several of the previous chapters we looked at a great variety of initial squares, and each of those would be considered the original position for that magic square. However, for now, we will consider the original position to be where the 1 is in the upper left corner, and numbers increase from left to right and from top to bottom.

The following two groups of 5th order magic squares have numbers along one main diagonal or the other that remained in their original positions. There are three magic squares of each type shown here. In each type two are regular magic squares, and the third is pan-diagonal. Even with this similarity, each magic square is still unique to the others. (The shaded numbers show the numbers that remained in their original position.)

1	24	17	15	8		1	18	10	22	14
14	7	5	23	16		20	7	24	11	3
22	20	13	6	4		9	21	13	5	17
10	3	21	19	12		23	15	2	19	6
18	11	9	2	25		12	4	16	8	25

The next magic square is pan-diagonal:

1	15	24	8	17
23	7	16	5	14
20	4	13	22	6
12	21	10	19	3
9	18	2	11	25

This next set shows three magic squares where numbers in the other main diagonal remained in their original positions.

8	11	19	22	5		14	23	7	16	5
20	23	1	9	12		2	11	25	9	18
2	10	13	16	24		20	4	13	22	6
14	17	25	3	6		8	17	1	15	24
21	4	7	15	18		21	10	19	3	12

The next magic square is pan-diagonal:

7	14	16	23	5
18	25	2	9	11
4	6	13	20	22
15	17	24	1	8
21	3	10	12	19

Figure 14.9 Six 5th order magic squares where numbers in one main diagonal or the other remained in their same position.

In earlier chapters you may have noticed that I highlighted numbers that remained in their original positions instead of the ones that moved (or changed positions). This is because the numbers that remain in their original positions often form patterns that could then be compared with similar patterns in other magic squares as well as other orders of magic squares. I think it is especially intriguing when we do find a similar pattern as we compare orders of magic squares. In this next figure, we have a similar diagonal pattern that can be seen repeating in these 4th, 5th, 7th, 8th, and 10th order magic squares.

1	12	14	7
15	6	4	9
8	13	11	2
10	3	5	16

1	14	22	10	18
24	7	20	3	11
17	5	13	21	9
15	23	6	19	2
8	16	4	12	25

1	34	11	37	21	47	24
32	9	42	19	45	22	6
14	40	17	43	27	4	30
38	15	48	25	2	35	12
20	46	23	7	33	10	36
44	28	5	31	8	41	18
26	3	29	13	39	16	49

1	9	40	48	31	23	58	50
2	10	39	47	32	24	57	49
54	62	19	27	44	36	13	5
53	61	20	28	43	35	14	6
59	51	30	22	37	45	4	12
60	52	29	21	38	46	3	11
16	8	41	33	18	26	55	63
15	7	42	34	17	25	56	64

1	20	21	70	50	60	40	71	81	91
2	12	29	39	59	49	69	72	82	92
8	83	23	68	58	48	38	73	13	93
97	87	77	34	44	54	64	27	14	7
96	86	76	35	45	55	65	26	16	5
95	85	75	36	46	56	66	25	15	6
94	84	74	37	47	57	67	24	17	4
3	18	28	63	53	43	33	78	88	98
99	19	22	62	52	42	32	79	89	9
10	11	80	61	51	41	31	30	90	100

Figure 14.10 These 4th, 5th, 7th, 8th, and 10th order magic squares have numbers along a diagonal that remained in their original positions.

A similar pattern is the "X" pattern where numbers along both diagonals have remained in their original positions. In this next example we have a series of 4th, 6th, and 8th order magic squares with numbers along both diagonals in their original positions.

1	15	14	4
12	6	7	9
8	10	11	5
13	3	2	16

1	32	4	33	35	6
30	8	28	27	11	7
24	17	15	16	20	19
13	23	21	22	14	18
12	26	9	10	29	25
31	5	34	3	2	36

1	9	52	61	60	53	16	8
2	10	54	59	62	51	15	7
31	47	19	38	35	22	42	26
40	24	45	28	29	44	17	33
32	48	21	36	37	20	41	25
39	23	43	30	27	46	18	34
58	50	14	3	6	11	55	63
57	49	12	5	4	13	56	64

Figure 14.11 Numbers along both diagonals remained in their original positions.

The next two examples show an "XO" patterns where an "X" is overlaid by an "O". Shown below are an 8th order and a 10th order example of these magic squares.

1	63	62	4	5	59	58	8
56	10	54	13	12	51	15	49
48	47	19	21	20	22	42	41
25	34	35	28	29	38	39	32
33	26	27	36	37	30	31	40
24	23	43	45	44	46	18	17
16	50	14	53	52	11	55	9
57	7	6	60	61	3	2	64

1	99	98	4	6	95	7	93	92	10
90	12	88	87	15	16	84	83	19	11
80	79	23	77	26	75	74	28	22	21
31	69	68	34	65	66	37	63	32	40
51	42	53	47	45	46	44	58	59	60
50	52	48	57	55	56	54	43	49	41
61	39	38	64	35	36	67	33	62	70
30	29	73	27	76	25	24	78	72	71
20	82	13	14	86	85	17	18	89	81
91	2	3	94	96	5	97	8	9	100

Figure 14.12 Magic Squares where an "X" pattern is overlaid by an "O" pattern.

Of course these are very basic patterns, and the patterns that we can find are limitless. In the following set of patterns, we have an 8th, 12th and 16th order magic square where half

The Methods, Marvels, and Madness of Magic Squares

the numbers in each square have remained in their original positions. The numbers that didn't move form a checkerboard pattern. Here are the magic squares:

1	63	3	61	60	6	58	8
56	10	54	12	13	51	15	49
17	47	19	45	44	22	42	24
40	26	38	28	29	35	31	33
32	34	30	36	37	27	39	25
41	23	43	21	20	46	18	48
16	50	14	52	53	11	55	9
57	7	59	5	4	62	2	64

1	143	3	141	5	139	138	8	136	10	134	12
132	14	130	16	128	18	19	125	21	123	23	121
25	119	27	117	29	115	114	32	112	34	110	36
108	38	106	40	104	42	43	101	45	99	47	97
49	95	51	93	53	91	90	56	88	58	86	60
84	62	82	64	80	66	67	77	69	75	71	73
72	74	70	76	68	78	79	65	81	63	83	61
85	59	87	57	89	55	54	92	52	94	50	96
48	98	46	100	44	102	103	41	105	39	107	37
109	35	111	33	113	31	30	116	28	118	26	120
24	122	22	124	20	126	127	17	129	15	131	13
133	11	135	9	137	7	6	140	4	142	2	144

1	255	3	253	5	251	7	249	248	10	246	12	244	14	242	16
240	18	238	20	236	22	234	24	25	231	27	229	29	227	31	225
33	223	35	221	37	219	39	217	216	42	214	44	212	46	210	48
208	50	206	52	204	54	202	56	57	199	59	197	61	195	63	193
65	191	67	189	69	187	71	185	184	74	182	76	180	78	178	80
176	82	174	84	172	86	170	88	89	167	91	165	93	163	95	161
97	159	99	157	101	155	103	153	152	106	150	108	148	110	146	112
144	114	142	116	140	118	138	120	121	135	123	133	125	131	127	129
128	130	126	132	124	134	122	136	137	119	139	117	141	115	143	113
145	111	147	109	149	107	151	105	104	154	102	156	100	158	98	160
96	162	94	164	92	166	90	168	169	87	171	85	173	83	175	81
177	79	179	77	181	75	183	73	72	186	70	188	68	190	66	192
64	194	62	196	60	198	58	200	201	55	203	53	205	51	207	49
209	47	211	45	213	43	215	41	40	218	38	220	36	222	34	224
32	226	30	228	28	230	26	232	233	23	235	21	237	19	239	17
241	15	243	13	245	11	247	9	8	250	6	252	4	254	2	256

Figure 14.13 These 8th, 12th, and 16th order magic squares have a checkerboard pattern of numbers that remained in their original positions.

The following 8th and 16th order magic squares have two interesting patterns. The first pattern is visible at its original orientation, and the second can be seen when we rotate the magic square 180 degrees. (The square on the left is the magic square in its original orientation while the square on the right is the same square after it has been rotated 180 degrees.) Here is the first 8th order magic square:

1	9	53	61	60	52	16	8
2	10	54	59	62	51	15	7
39	47	19	30	27	22	42	34
40	24	44	28	29	45	17	33
32	48	20	36	37	21	41	25
31	23	43	38	35	46	18	26
58	50	14	3	6	11	55	63
57	49	13	5	4	12	56	64

64	56	12	4	5	13	49	57
63	55	11	6	3	14	50	58
26	18	46	35	38	43	23	31
25	41	21	37	36	20	48	32
33	17	45	29	28	44	24	40
34	42	22	27	30	19	47	39
7	15	51	62	59	54	10	2
8	16	52	60	61	53	9	1

And here is the second 8th order magic square with the same property:

1	27	39	61	60	34	30	8
20	10	54	48	41	51	15	21
53	47	19	9	16	22	42	52
40	62	2	28	29	7	59	33
32	6	58	36	37	63	3	25
13	23	43	49	56	46	18	12
44	50	14	24	17	11	55	45
57	35	31	5	4	26	38	64

64	38	26	4	5	31	35	57
45	55	11	17	24	14	50	44
12	18	46	56	49	43	23	13
25	3	63	37	36	58	6	32
33	59	7	29	28	2	62	40
52	42	22	16	9	19	47	53
21	15	51	41	48	54	10	20
8	30	34	60	61	39	27	1

Figure 14.14 Two 8th order magic squares (top and bottom) where one orientation shows an X pattern (on the left), and after rotated 180 degrees shows an O pattern (on the right).

Next, we have a 16th order magic square with this same property.

1	17	33	49	201	217	233	249	248	232	216	200	64	48	32	16
2	18	34	50	202	218	234	250	247	231	215	199	63	47	31	15
3	19	35	51	203	219	235	251	246	230	214	198	62	46	30	14
4	20	36	52	204	220	236	252	245	229	213	197	61	45	29	13
141	157	173	189	69	85	101	117	124	108	92	76	180	164	148	132
142	158	174	190	70	86	102	118	123	107	91	75	179	163	147	131
143	159	175	191	71	87	103	119	122	106	90	74	178	162	146	130
144	160	176	192	72	88	104	120	121	105	89	73	177	161	145	129
128	112	96	80	184	168	152	136	137	153	169	185	65	81	97	113
127	111	95	79	183	167	151	135	138	154	170	186	66	82	98	114
126	110	94	78	182	166	150	134	139	155	171	187	67	83	99	115
125	109	93	77	181	165	149	133	140	156	172	188	68	84	100	116
244	228	212	196	60	44	28	12	5	21	37	53	205	221	237	253
243	227	211	195	59	43	27	11	6	22	38	54	206	222	238	254
242	226	210	194	58	42	26	10	7	23	39	55	207	223	239	255
241	225	209	193	57	41	25	9	8	24	40	56	208	224	240	256

Figure 14.15a   A 16th order magic square where one orientation shows an X pattern (above), and after rotated 180 degrees shows an O pattern (see the next figure).

And now when we rotate the above magic square 180 degrees, we still have the same magic square, but now the numbers in their original position form an O pattern:

256	240	224	208	56	40	24	8	9	25	41	57	193	209	225	241
255	239	223	207	55	39	23	7	10	26	42	58	194	210	226	242
254	238	222	206	54	38	22	6	11	27	43	59	195	211	227	243
253	237	221	205	53	37	21	5	12	28	44	60	196	212	228	244
116	100	84	68	188	172	156	140	133	149	165	181	77	93	109	125
115	99	83	67	187	171	155	139	134	150	166	182	78	94	110	126
114	98	82	66	186	170	154	138	135	151	167	183	79	95	111	127
113	97	81	65	185	169	153	137	136	152	168	184	80	96	112	128
129	145	161	177	73	89	105	121	120	104	88	72	192	176	160	144
130	146	162	178	74	90	106	122	119	103	87	71	191	175	159	143
131	147	163	179	75	91	107	123	118	102	86	70	190	174	158	142
132	148	164	180	76	92	108	124	117	101	85	69	189	173	157	141
13	29	45	61	197	213	229	245	252	236	220	204	52	36	20	4
14	30	46	62	198	214	230	246	251	235	219	203	51	35	19	3
15	31	47	63	199	215	231	247	250	234	218	202	50	34	18	2
16	32	48	64	200	216	232	248	249	233	217	201	49	33	17	1

Figure 14.15b   The same 16th order magic square as from the previous figure where this orientation shows an O pattern.

The following magic square is similar to the one above. The first orientation shows an "X" pattern that is two cells wide. Then when the same magic square is rotated 180 degrees, the new orientation shows an "O" pattern that is two cells wide:

1	2	126	125	133	134	250	249	248	247	139	140	116	115	15	16
17	18	110	109	149	150	234	233	232	231	155	156	100	99	31	32
176	175	35	36	220	219	87	88	89	90	214	213	45	46	162	161
192	191	51	52	204	203	71	72	73	74	198	197	61	62	178	177
193	194	190	189	69	70	58	57	56	55	75	76	180	179	207	208
209	210	174	173	85	86	42	41	40	39	91	92	164	163	223	224
160	159	19	20	236	235	103	104	105	106	230	229	29	30	146	145
144	143	3	4	252	251	119	120	121	122	246	245	13	14	130	129
128	127	243	244	12	11	135	136	137	138	6	5	253	254	114	113
112	111	227	228	28	27	151	152	153	154	22	21	237	238	98	97
33	34	94	93	165	166	218	217	216	215	171	172	84	83	47	48
49	50	78	77	181	182	202	201	200	199	187	188	68	67	63	64
80	79	195	196	60	59	183	184	185	186	54	53	205	206	66	65
96	95	211	212	44	43	167	168	169	170	38	37	221	222	82	81
225	226	158	157	101	102	26	25	24	23	107	108	148	147	239	240
241	242	142	141	117	118	10	9	8	7	123	124	132	131	255	256

Figure 14.16a   A 16th order magic square where one orientation shows an X pattern that is two cells wide (above), and after rotated 180 degrees shows an O pattern that is two cells wide (see the figure below).

256	255	131	132	124	123	7	8	9	10	118	117	141	142	242	241
240	239	147	148	108	107	23	24	25	26	102	101	157	158	226	225
81	82	222	221	37	38	170	169	168	167	43	44	212	211	95	96
65	66	206	205	53	54	186	185	184	183	59	60	196	195	79	80
64	63	67	68	188	187	199	200	201	202	182	181	77	78	50	49
48	47	83	84	172	171	215	216	217	218	166	165	93	94	34	33
97	98	238	237	21	22	154	153	152	151	27	28	228	227	111	112
113	114	254	253	5	6	138	137	136	135	11	12	244	243	127	128
129	130	14	13	245	246	122	121	120	119	251	252	4	3	143	144
145	146	30	29	229	230	106	105	104	103	235	236	20	19	159	160
224	223	163	164	92	91	39	40	41	42	86	85	173	174	210	209
208	207	179	180	76	75	55	56	57	58	70	69	189	190	194	193
177	178	62	61	197	198	74	73	72	71	203	204	52	51	191	192
161	162	46	45	213	214	90	89	88	87	219	220	36	35	175	176
32	31	99	100	156	155	231	232	233	234	150	149	109	110	18	17
16	15	115	116	140	139	247	248	249	250	134	133	125	126	2	1

Figure 14.16b   The same 16th order magic square as from the previous figure where this orientation shows an O pattern two cells wide.

Of course we could go on and on with other patterns of numbers in their original positions. Instead, we will move on and look at another type of pattern.

Positions of Even or Odd Numbers
Other interesting patterns can be seen when we look at the placement of just the even numbers or just the odd numbers within the magic square.

4	29	12	37	20	45	28
35	11	36	19	44	27	3
10	42	18	43	26	2	34
41	17	49	25	1	33	9
16	48	24	7	32	8	40
47	23	6	31	14	39	15
22	5	30	13	38	21	46

5	46	15	56	25	66	35	76	45
54	14	55	24	65	34	75	44	4
13	63	23	64	33	74	43	3	53
62	22	72	32	73	42	2	52	12
21	71	31	81	41	1	51	11	61
70	30	80	40	9	50	10	60	20
29	79	39	8	49	18	59	19	69
78	38	7	48	17	58	27	68	28
37	6	47	16	57	26	67	36	77

Figure 14.17 A 7th and 9th order magic square with shaded cells showing the position of even numbers.

Again, it is very interesting when we find similar patterns between orders of magic squares. The next example shows a 5th order and a 7th order magic square whose even numbers share a similar pattern.

1	8	24	15	17
12	19	10	21	3
20	22	13	4	6
23	5	16	7	14
9	11	2	18	25

1	12	42	16	46	27	31
30	41	15	45	26	7	11
48	3	33	14	37	18	22
10	21	44	25	6	29	40
28	32	13	36	17	47	2
39	43	24	5	35	9	20
19	23	4	34	8	38	49

Figure 14.18 A 5th order and 7th order magic square showing a similar placement pattern of the even numbers.

The following example shows two patterns of odd numbers that mirror each other (the odd numbers are shaded). Even though the magic squares have mirroring shaded patterns, the actual values show that they are completely different from each other.

1	18	40	35	23	45	13
24	41	14	2	46	19	29
34	44	17	12	7	22	39
47	8	30	25	20	42	3
11	28	43	38	33	6	16
21	31	4	48	36	9	26
37	5	27	15	10	32	49

1	21	11	47	34	24	37
45	9	6	42	22	19	32
23	36	33	20	7	46	10
35	48	38	25	12	2	15
40	4	43	30	17	14	27
18	31	28	8	44	41	5
13	26	16	3	39	29	49

Figure 14.19 Two unique 7th order magic squares with mirroring odd number patterns.

The next two magic squares are related to each other. We can start with either magic square, and then by swapping symmetrical rows and columns arrive at the other. Here are the magic squares with the pattern showing highlighted odd numbers:

23	15	55	72	31	80	7	47	39
13	5	54	62	21	70	78	37	29
63	46	14	22	71	30	38	6	79
64	56	24	32	81	40	48	16	8
33	25	65	73	41	9	17	57	49
74	66	34	42	1	50	58	26	18
3	76	44	52	11	60	68	36	19
53	45	4	12	61	20	28	77	69
43	35	75	2	51	10	27	67	59

14	6	79	22	71	30	63	46	38
4	77	69	12	61	20	53	45	28
75	67	59	2	51	10	43	35	27
24	16	8	32	81	40	64	56	48
65	57	49	73	41	9	33	25	17
34	26	18	42	1	50	74	66	58
55	47	39	72	31	80	23	15	7
54	37	29	62	21	70	13	5	78
44	36	19	52	11	60	3	76	68

Figure 14.20 Two 9th order magic squares with complimentary odd number pattern.

And here are a 9th and 11th order magic squares with a more complex similar shaded pattern created by the odd numbers within the square.

35	2	27	43	51	59	75	10	67
56	32	48	64	81	8	24	40	16
76	52	68	3	11	19	44	60	36
5	62	78	13	21	29	54	70	37
25	73	17	33	41	49	65	9	57
45	12	28	53	61	69	4	20	77
46	22	38	63	71	79	14	30	6
66	42	58	74	1	18	34	50	26
15	72	7	23	31	39	55	80	47

83	33	12	43	93	73	53	103	2	113	63
23	94	84	104	44	13	114	54	74	64	3
22	82	72	92	32	1	102	42	62	52	112
35	106	96	116	45	25	5	66	86	76	15
95	34	24	55	105	85	65	115	14	4	75
71	21	11	31	81	61	41	91	111	101	51
47	118	108	7	57	37	17	67	98	88	27
107	46	36	56	117	97	77	6	26	16	87
10	70	60	80	20	121	90	30	50	40	100
119	58	48	68	8	109	78	18	38	28	99
59	9	120	19	69	49	29	79	110	89	39

Figure 14.21 A 9th and 11th order magic square with similar odd number patterns.

In the 9th order pattern above, notice how all the numbers ending with 1's are in the center column. In the column to the right are the odd numbers ending with 9's and the column to the left has odd numbers ending with 3's. Then we have a column of 7's on the left followed by a column of 5's on the left outside column; while on the right we have a column of 5's followed by a column of 7's on the right outside column.

The 11th order magic square has its numbers ordered in rows such that the center row has numbers ending with 1's. The odd numbers on the two rows above the center row end with 5's, and the odd numbers in the two rows below the center row end with 7's. The odd numbers in the top two rows end with 3's while the bottom two rows end with 9's.

These patterns are all very interesting, and there are even more types of patterns that we could consider. For example, we could look for the position of the first half of the numbers in the square; or we could look for the position of the multiples of n (where n is equal to the order); or we could look at numbers that have swapped position through the center of the square. As you can see, there are many patterns that we could find within these magic squares, and the possibilities go on and on. However, I believe there is an even more interesting set of patterns that emerges when we start connecting "the dots" as we shall see in the next chapter.

Chapter 15
Connecting Dots within the Magic Square

Drawing Lines to Numbers Sequentially
As we will see in this chapter, there are several ways that we can draw lines within these magic squares. Probably the most basic method is to draw lines sequentially where we connect the 1 to the 2 and then to the 3, etc. But even within this method there are several variations, and we will look at a few of those. However, let's not get ahead of ourselves. In the figures below, the magic square is on the left, and the square on the right is the same magic square except that the numbers have been replaced with lines that connect where the numbers were. In these first two figures, lines are drawn from the 1 to the 2, to the 3, and so on.

7	10	64	49	33	48	26	23
34	47	25	24	8	9	63	50
59	51	4	12	29	21	38	46
30	22	37	45	60	52	3	11
54	62	13	5	20	28	43	35
19	27	44	36	53	61	14	6
15	2	56	57	41	40	18	31
42	39	17	32	16	1	55	58

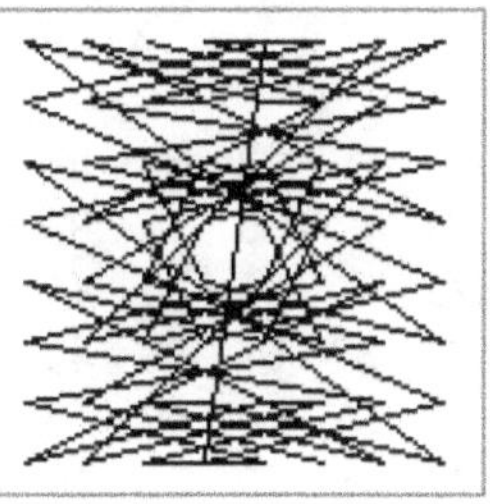

Figure 15.1   An 8th order magic square with the square on the right showing lines that connect the numbers sequentially.

3	58	14	55	50	11	63	6
60	1	53	16	9	52	8	61
30	39	19	42	47	22	34	27
37	32	44	17	24	45	25	36
29	40	20	41	48	21	33	28
38	31	43	18	23	46	26	35
4	57	13	56	49	12	64	5
59	2	54	15	10	51	7	62

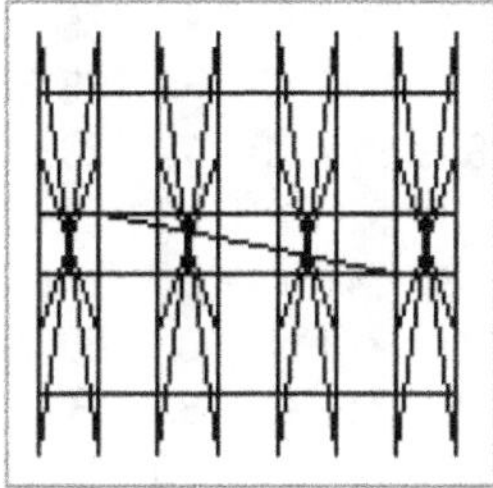

Figure 15.2   Another 8th order magic square with the square on the right showing lines that connect the numbers sequentially.

Sequence Pairs

A variation of the previous method is to only connect pairs of numbers. In the figure below, lines are drawn between the 1 and 2, and then between the 3 and 4, but not between the 2 and 3. I call this a sequence pair method of drawing lines:

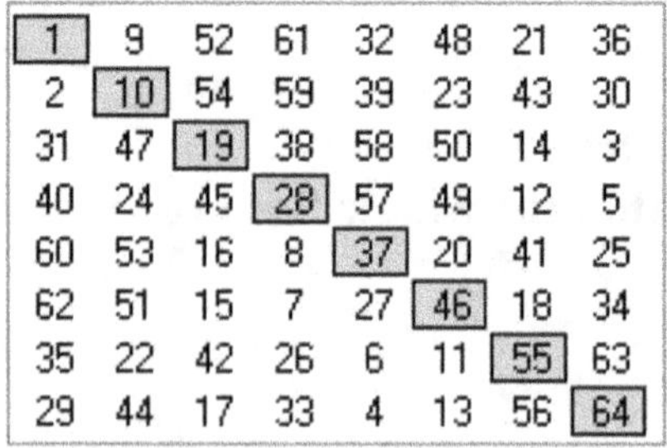 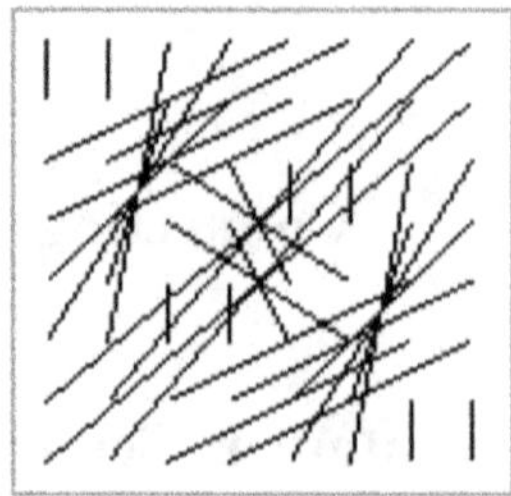

Figure 15.3   An 8th order magic square with the lines on the right showing numbers connected in sequential pairs.

By connecting numbers in this way we eliminate practically half the lines, and thus simplify the drawings. This also allows us find some very interesting patterns within various magic squares. This next set of figures show two 10th order magic squares, and three 12th order magic squares. In some cases these diagrams can also give an insight to how the magic square was made.

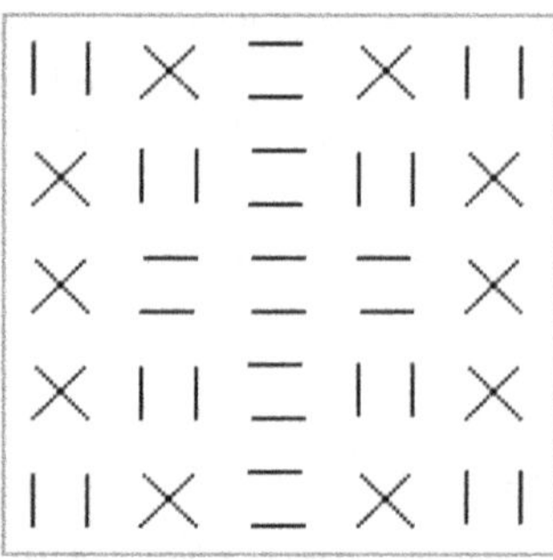

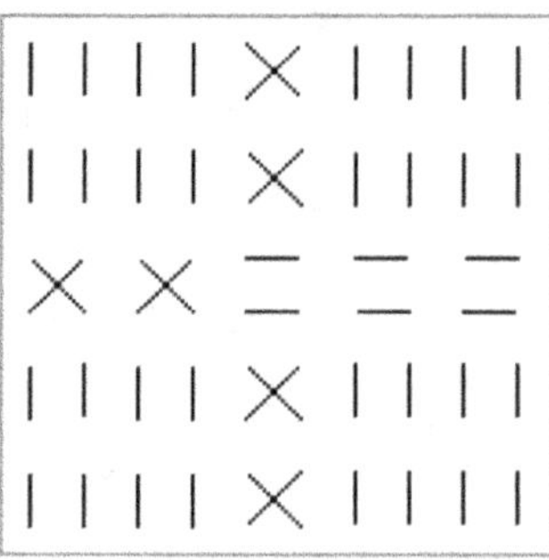

Figure 15.4   Two 10th order magic squares with the square on the right showing numbers connected in sequential pairs.

These three figures show 12th order magic squares. The square on the left has shaded boxes around the numbers that are in their original positions while the square on the right connects the numbers in sequential pairs:

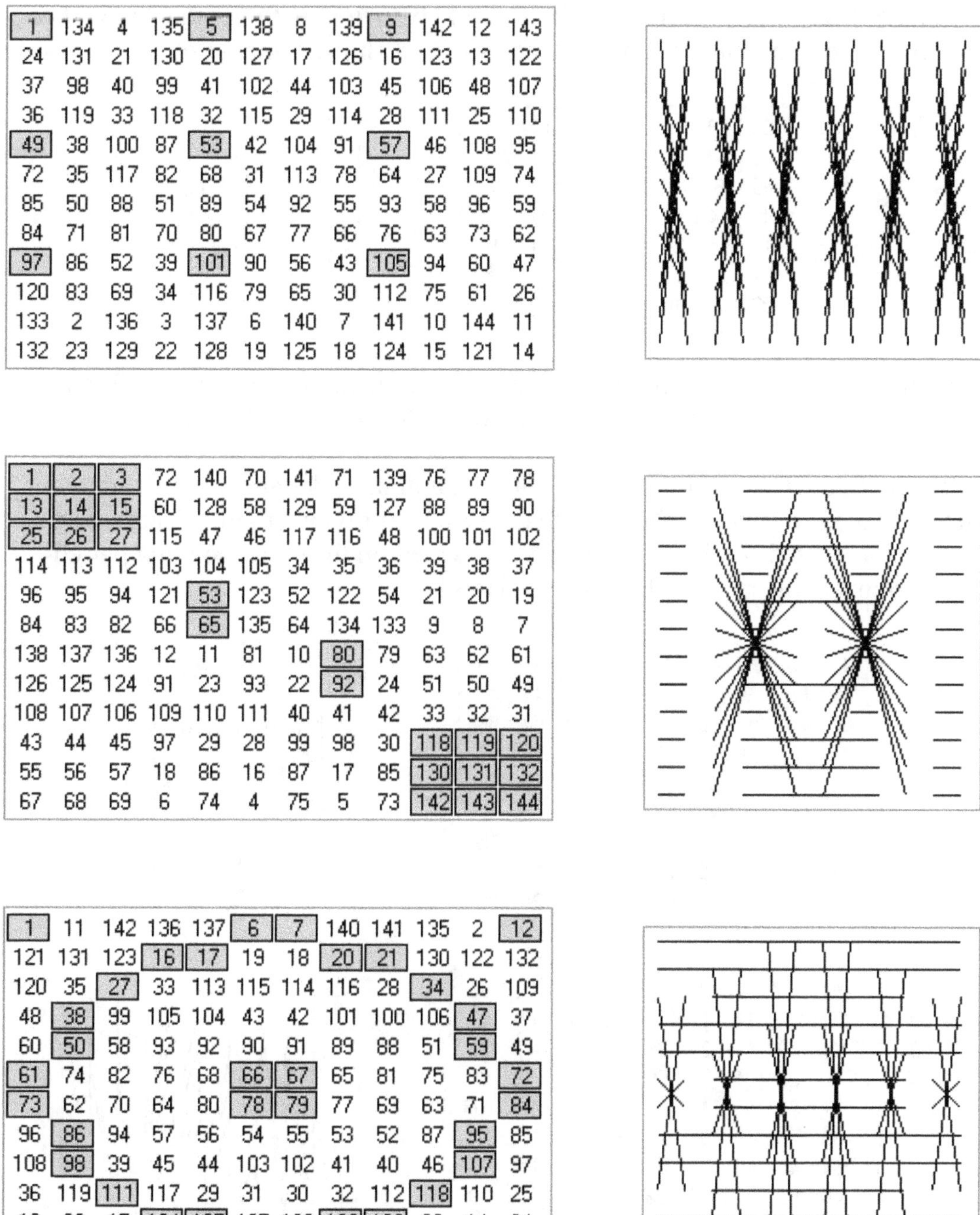

Figure 15.5   Three 12th order magic squares with the square on the right showing numbers connected in sequential pairs.

When working with odd-ordered magic squares, there are other ways that sequence pairs can be created. One method is to match the first number with the last, the second number with the next-to-last number, and so on. However, a method that I like is to match the first number with the number just following the middle number; the second number with the next number after that, etc. For example, with a seventh order magic square, the middle number is 25, which means that the 1 would be paired with 26, the 2 would be paired with 27, etc. The following figure shows a seventh order magic square with such a pairing.

30	39	48	1	10	19	28
38	47	7	9	18	27	29
46	6	8	17	26	35	37
5	14	16	25	34	36	45
13	15	24	33	42	44	4
21	23	32	41	43	3	12
22	31	40	49	2	11	20

Figure 15.6   A 7th order magic square where the middle number is 25; and therefore the 1 is paired with 26, the 2 with 27, etc.

The following figure has two more magic squares, a ninth order and an eleventh order, constructed in the same way as the seventh order in the previous figure. These magic squares are constructed similarly, which is very evident in the corresponding paired-number diagrams to the right of the magic squares.

47	58	69	80	1	12	23	34	45
57	68	79	9	11	22	33	44	46
67	78	8	10	21	32	43	54	56
77	7	18	20	31	42	53	55	66
6	17	19	30	41	52	63	65	76
16	27	29	40	51	62	64	75	5
26	28	39	50	61	72	74	4	15
36	38	49	60	71	73	3	14	25
37	48	59	70	81	2	13	24	35

68	81	94	107	120	1	14	27	40	53	66
80	93	106	119	11	13	26	39	52	65	67
92	105	118	10	12	25	38	51	64	77	79
104	117	9	22	24	37	50	63	76	78	91
116	8	21	23	36	49	62	75	88	90	103
7	20	33	35	48	61	74	87	89	102	115
19	32	34	47	60	73	86	99	101	114	6
31	44	46	59	72	85	98	100	113	5	18
43	45	58	71	84	97	110	112	4	17	30
55	57	70	83	96	109	111	3	16	29	42
56	69	82	95	108	121	2	15	28	41	54

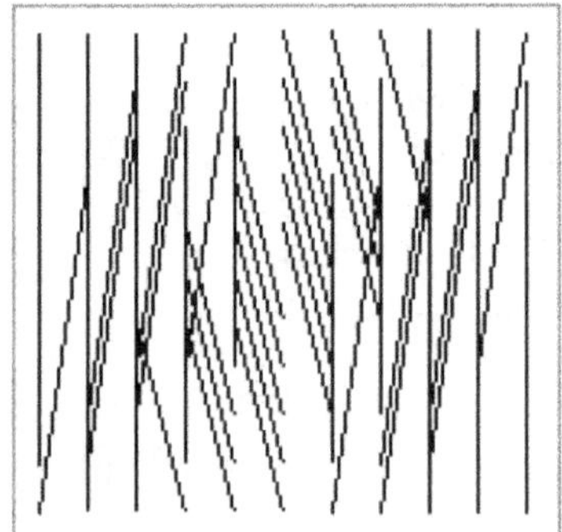

Figure 15.7   A 9th and an 11th order magic square where the 1 is paired with the number just after the middle number, the 2 is paired with the number just after that, etc.

Odd and Even Sequence Patterns

Another variation of the sequence drawing method is to sequentially connect only the odd numbers or only the even numbers. From there we can overlay the odd and even patterns on top of each other, or look at them individually. In many cases the individual odd pattern or even pattern doesn't have much symmetry by itself, but when the two patterns are overlain on top of each other, the symmetry becomes very apparent. For example, here is an 8th order magic square. The square on the right connects the even numbers in sequential order. The lower square connects the odd numbers sequentially. Neither pattern seems to have very much symmetry.

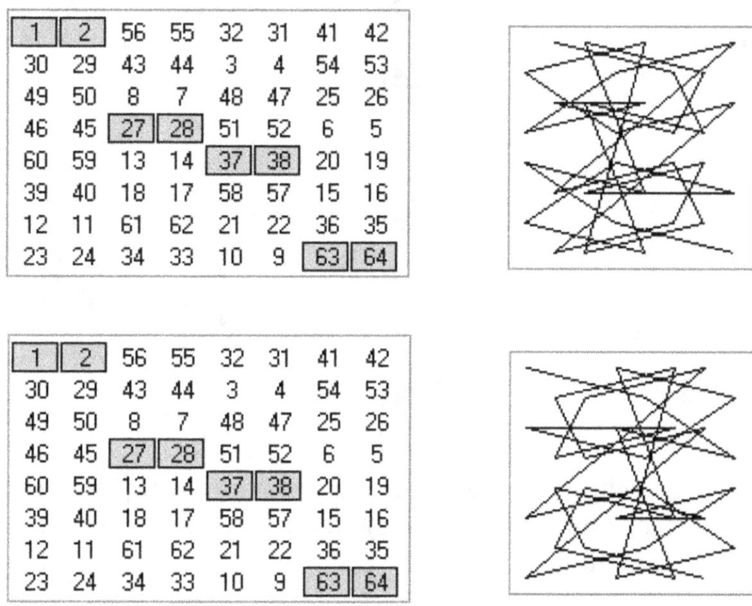

Figure 15.8a   An 8th order magic square with the top figure on the right connecting even sequential numbers while the figure on the bottom connects odd sequential numbers.

Now, we'll take the same magic square and overlay the odd pattern over the even pattern, and this is what we get:

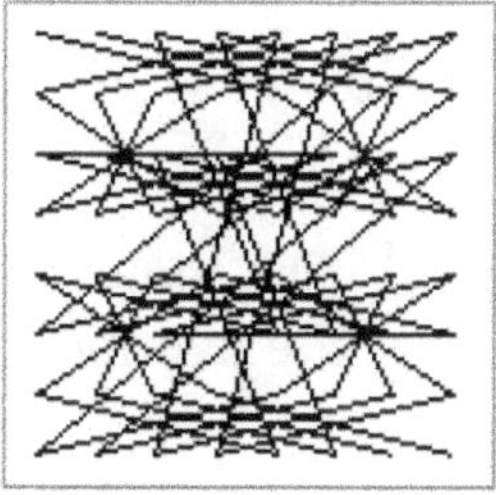

Figure 15.8b   The same 8th order magic square with the square on the right showing the pattern when the even and odd sequences are overlain on top of each other.

The illustration with the odd number pattern laid over the even number pattern appears symmetrical. The following magic square is another example (again using an 8th order magic square). First, the pattern on top connects the even numbers sequentially and the pattern on the bottom connects the odd numbers sequentially:

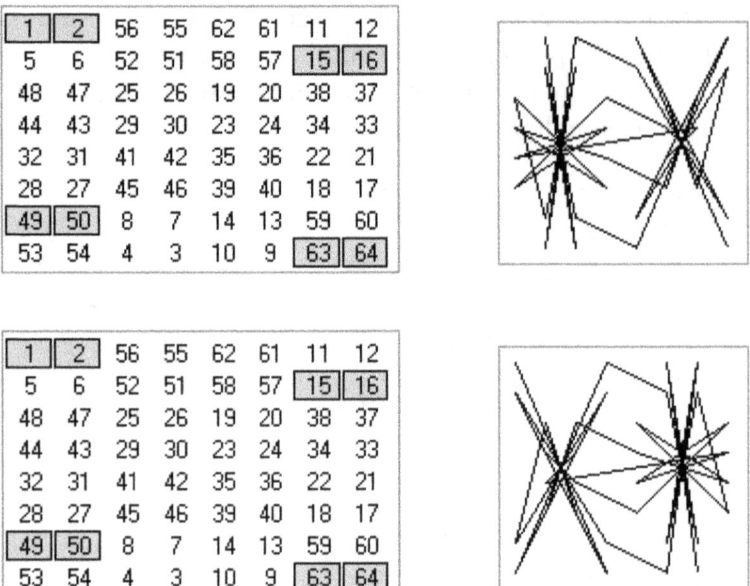

Figure 15.9a   Another 8th order magic square with the top figure connecting even sequential numbers while the figure on the bottom connects odd sequential numbers.

And finally, the odd number pattern is overlain on top of the even number pattern:

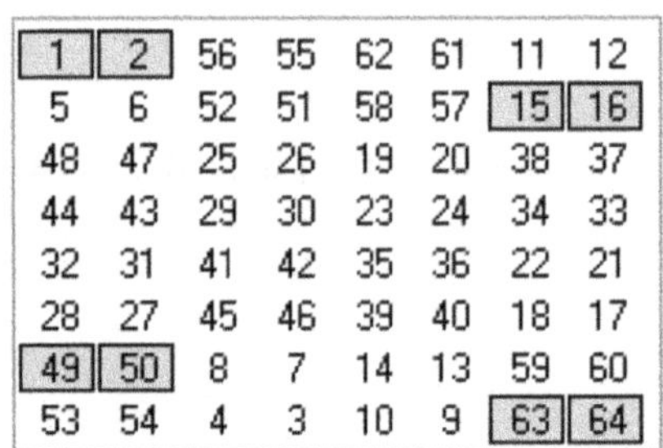
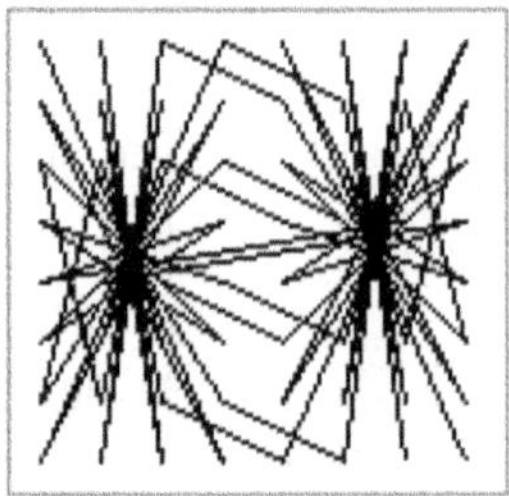

Figure 15.9b   The same 8th order magic square with the even and odd sequences are overlain on top of each other in the right figure.

Individually they have no apparent symmetry, but overlain together the pattern is symmetrical. Here is another interesting set of odd and even patterns within a 16th order magic square.

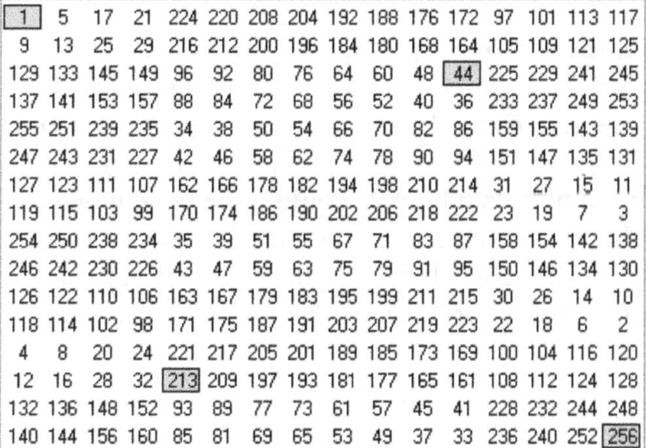

1	5	17	21	224	220	208	204	192	188	176	172	97	101	113	117
9	13	25	29	216	212	200	196	184	180	168	164	105	109	121	125
129	133	145	149	96	92	80	76	64	60	48	44	225	229	241	245
137	141	153	157	88	84	72	68	56	52	40	36	233	237	249	253
255	251	239	235	34	38	50	54	66	70	82	86	159	155	143	139
247	243	231	227	42	46	58	62	74	78	90	94	151	147	135	131
127	123	111	107	162	166	178	182	194	198	210	214	31	27	15	11
119	115	103	99	170	174	186	190	202	206	218	222	23	19	7	3
254	250	238	234	35	39	51	55	67	71	83	87	158	154	142	138
246	242	230	226	43	47	59	63	75	79	91	95	150	146	134	130
126	122	110	106	163	167	179	183	195	199	211	215	30	26	14	10
118	114	102	98	171	175	187	191	203	207	219	223	22	18	6	2
4	8	20	24	221	217	205	201	189	185	173	169	100	104	116	120
12	16	28	32	213	209	197	193	181	177	165	161	108	112	124	128
132	136	148	152	93	89	77	73	61	57	45	41	228	232	244	248
140	144	156	160	85	81	69	65	53	49	37	33	236	240	252	256

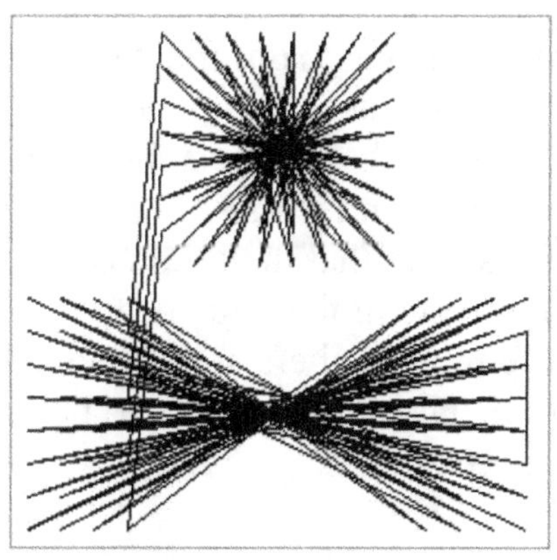

Even Sequence

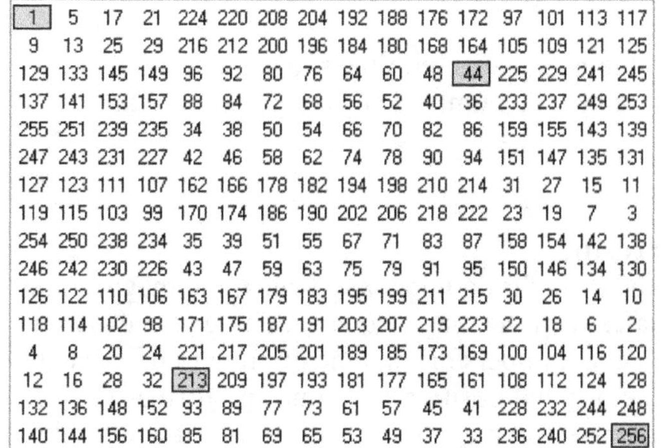

1	5	17	21	224	220	208	204	192	188	176	172	97	101	113	117
9	13	25	29	216	212	200	196	184	180	168	164	105	109	121	125
129	133	145	149	96	92	80	76	64	60	48	44	225	229	241	245
137	141	153	157	88	84	72	68	56	52	40	36	233	237	249	253
255	251	239	235	34	38	50	54	66	70	82	86	159	155	143	139
247	243	231	227	42	46	58	62	74	78	90	94	151	147	135	131
127	123	111	107	162	166	178	182	194	198	210	214	31	27	15	11
119	115	103	99	170	174	186	190	202	206	218	222	23	19	7	3
254	250	238	234	35	39	51	55	67	71	83	87	158	154	142	138
246	242	230	226	43	47	59	63	75	79	91	95	150	146	134	130
126	122	110	106	163	167	179	183	195	199	211	215	30	26	14	10
118	114	102	98	171	175	187	191	203	207	219	223	22	18	6	2
4	8	20	24	221	217	205	201	189	185	173	169	100	104	116	120
12	16	28	32	213	209	197	193	181	177	165	161	108	112	124	128
132	136	148	152	93	89	77	73	61	57	45	41	228	232	244	248
140	144	156	160	85	81	69	65	53	49	37	33	236	240	252	256

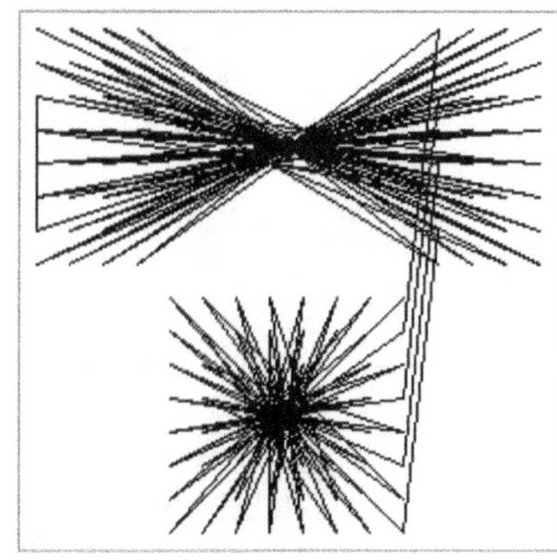

Odd Sequence

1	5	17	21	224	220	208	204	192	188	176	172	97	101	113	117
9	13	25	29	216	212	200	196	184	180	168	164	105	109	121	125
129	133	145	149	96	92	80	76	64	60	48	44	225	229	241	245
137	141	153	157	88	84	72	68	56	52	40	36	233	237	249	253
255	251	239	235	34	38	50	54	66	70	82	86	159	155	143	139
247	243	231	227	42	46	58	62	74	78	90	94	151	147	135	131
127	123	111	107	162	166	178	182	194	198	210	214	31	27	15	11
119	115	103	99	170	174	186	190	202	206	218	222	23	19	7	3
254	250	238	234	35	39	51	55	67	71	83	87	158	154	142	138
246	242	230	226	43	47	59	63	75	79	91	95	150	146	134	130
126	122	110	106	163	167	179	183	195	199	211	215	30	26	14	10
118	114	102	98	171	175	187	191	203	207	219	223	22	18	6	2
4	8	20	24	221	217	205	201	189	185	173	169	100	104	116	120
12	16	28	32	213	209	197	193	181	177	165	161	108	112	124	128
132	136	148	152	93	89	77	73	61	57	45	41	228	232	244	248
140	144	156	160	85	81	69	65	53	49	37	33	236	240	252	256

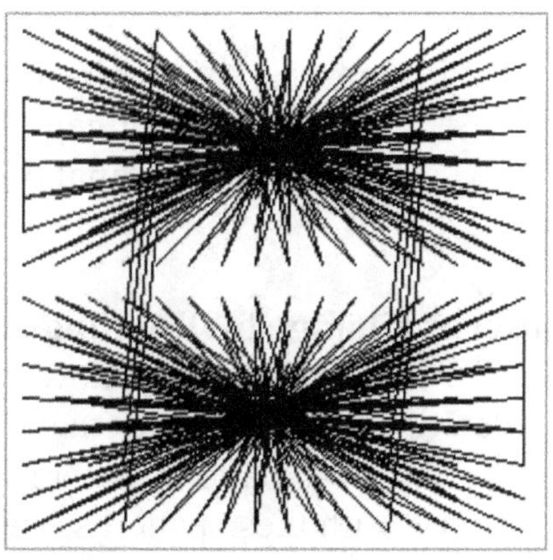

Odd/Even Overlay

Figure 15.10   A set of 16th order magic squares with the squares on the right connecting even and then odd sequential numbers. The bottom image shows the odd and even patterns overlain on top of each other.

Drawing Lines from the Number's Original to Current Position
In the previous chapter, one of the methods we discussed was highlighting numbers that remained in their original position. Now, we will carry that concept one step further and draw lines connecting a number's original position with its new position. In other words, we'll assume that the original position of a number is where it would have been had all the numbers in the array been listed with the lowest number in the upper left corner and succeeding numbers are placed in ascending order from left-to-right, and from top-to-bottom. The current (or new) position for these numbers is their location in the magic square itself.

8	7	62	61	33	34	27	28
16	15	54	53	41	42	19	20
52	51	10	9	21	22	47	48
60	59	2	1	29	30	39	40
25	26	35	36	64	63	6	5
17	18	43	44	56	55	14	13
45	46	23	24	12	11	50	49
37	38	31	32	4	3	58	57

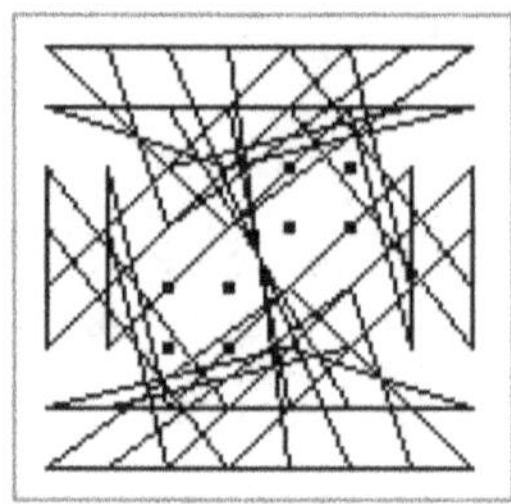

8	7	53	54	60	59	9	10
16	15	61	62	52	51	1	2
17	18	36	35	45	46	32	31
25	26	44	43	37	38	24	23
42	41	27	28	22	21	39	40
34	33	19	20	30	29	47	48
63	64	14	13	3	4	50	49
55	56	6	5	11	12	58	57

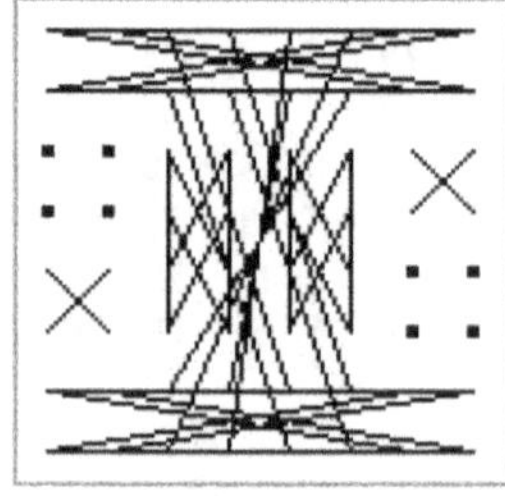

Figure 15.11   A pair of 8th order magic squares with the squares on the right showing lines that connect a numbers current position with its original position.

Patterns Change by Rotating and Mirroring
Most of the line patterns that we've looked at so far do not change when the magic square is rotated; instead, the pattern simply rotates with the magic square. Things are different when we look at patterns where we connect a numbers original position with its current position. With these patterns, we keep the original position the same and then as the magic square is rotated or mirrored the numbers change their position with regard to their original position. For example, let's start with the following 8th order magic square (on the following page). We'll call this the 0 degree orientation. With this orientation for this magic square, we have all of the numbers in the main diagonal that runs from the upper left to the lower right in their original positions.

In the following illustrations, the magic square is on the left; while the square on the right shows lines drawn from the number's "original" position to its new (or current) position. The boxed numbers on the left show numbers that did not move and that are still in their original position. These same numbers are represented by dots in the illustration on the right. Here is our first magic square:

1	9	52	61	32	48	21	36
2	**10**	54	59	39	23	43	30
31	47	**19**	38	58	50	14	3
40	24	45	**28**	57	49	12	5
60	53	16	8	**37**	20	41	25
62	51	15	7	27	**46**	18	34
35	22	42	26	6	11	**55**	63
29	44	17	33	4	13	56	**64**

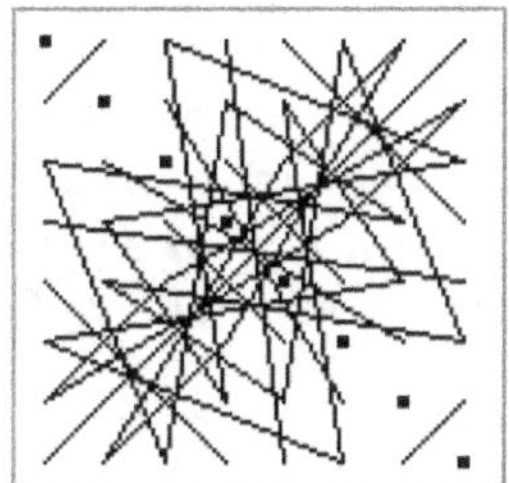

Figure 15.12 A typical 8th order magic square at a 0 degree orientation.

Now we'll take the same magic square, and rotate it clockwise 90 degrees.

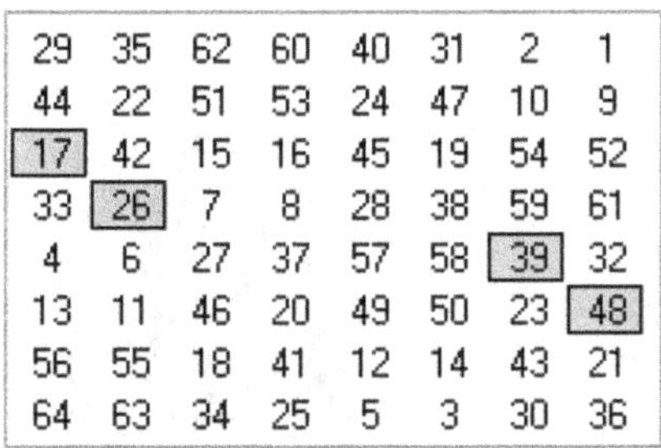

29	35	62	60	40	31	2	1
44	22	51	53	24	47	10	9
17	42	15	16	45	19	54	52
33	**26**	7	8	28	38	59	61
4	6	27	37	57	58	**39**	32
13	11	46	20	49	50	23	**48**
56	55	18	41	12	14	43	21
64	63	34	25	5	3	30	36

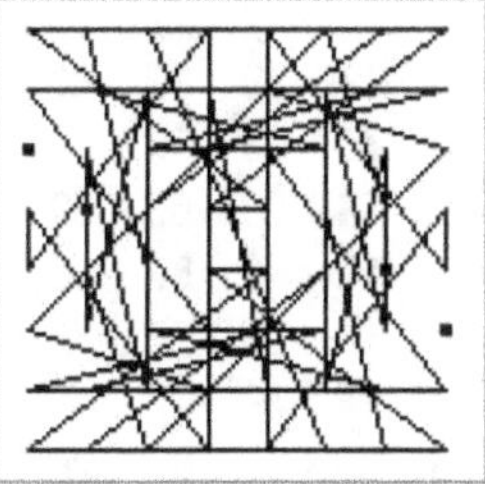

Figure 15.13 The same 8th order magic square at a 90 degree orientation.

Even though this magic square is identical to the previous one (except that it was rotated 90 degrees), we find that there are now different numbers in their original position (highlighted on the left); and the illustration on the right is completely different. Next, we'll rotate this magic square another 90 degrees clockwise:

64	56	13	**4**	33	17	44	29
63	55	**11**	6	26	42	22	35
34	**18**	46	27	7	15	51	62
25	41	20	37	8	16	53	60
5	12	49	57	28	45	24	**40**
3	14	50	58	38	19	**47**	31
30	43	23	39	59	**54**	10	2
36	21	48	32	**61**	52	9	1

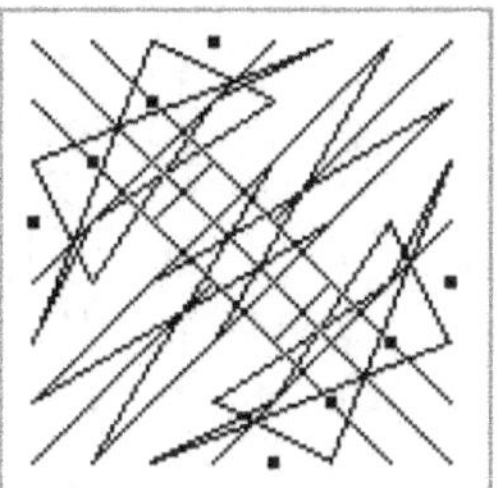

Figure 15.14 The same 8th order magic square at a 180 degree orientation.

The magic square above has now been rotated 180 degrees from the original orientation. Notice that at this orientation we again have yet other numbers that are now in their original location (as shown by the boxed numbers on the left, and the dots on the right). Finally, we'll rotate this magic square one more time so that it is now 270 degrees from its original position.

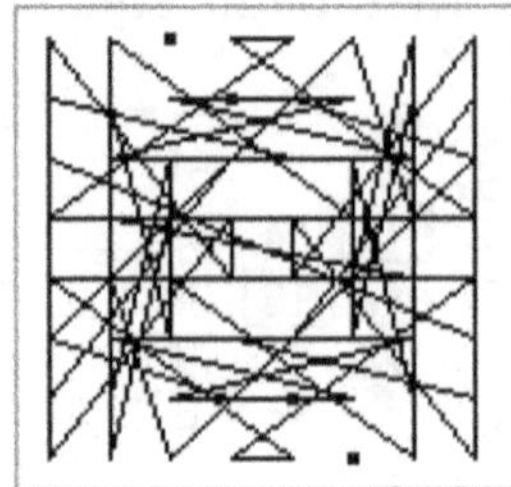

Figure 15.15 The same 8th order magic square at a 270 degree orientation.

The line drawings of the 90 degree and 270 degree orientations appear to be 90 degree rotations of each other while the 0 degree and 180 degree orientations are unique to each other; yet all the while it is the same magic square. Now let's look at the mirrorings of this same magic square.   This next example is the mirror of the 0 degree orientation

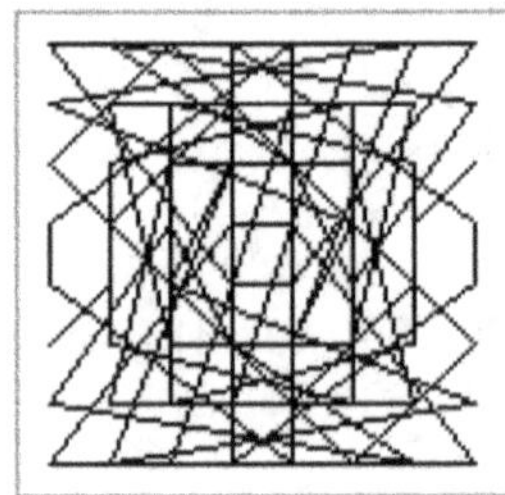

Figure 15.16 The same magic square, with a mirrored 0 degree orientation.

The next illustration is a mirror of the 90 degree rotation (from earlier), and is a 90 degree counter-clockwise rotation of the magic square from the previous figure.

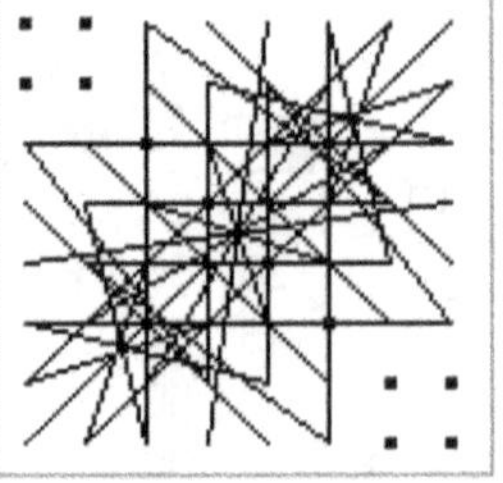

Figure 15.17 The same magic square, with a mirrored 90 degree orientation.

Now we have the mirror of the 180 degree rotation (which again is a 90 degree counter-clockwise rotation of the magic square in the previous figure).

192

29	44	17	33	4	13	56	64
35	22	42	26	6	11	55	63
62	51	15	7	27	46	18	34
60	53	16	8	37	20	41	25
40	24	45	28	57	49	12	5
31	47	19	38	58	50	14	3
2	10	54	59	39	23	43	30
1	9	52	61	32	48	21	36

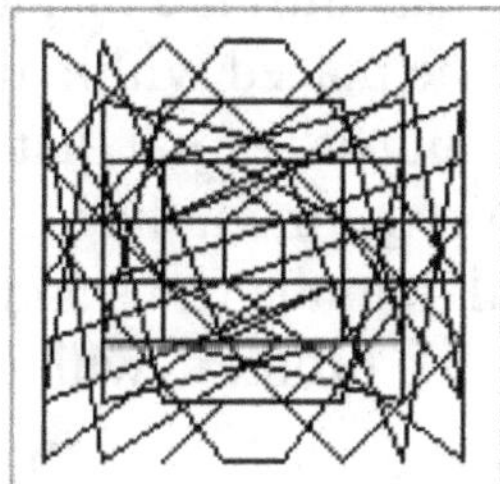

Figure 15.18 The same magic square, with a mirrored 180 degree orientation.

And finally in the following figure we have the mirror of the 270 degree rotation.

64	63	34	25	[5]	3	30	36
56	55	18	41	12	[14]	43	21
13	11	46	[20]	49	50	[23]	48
4	6	[27]	37	57	58	39	[32]
[33]	26	7	8	28	[38]	59	61
17	[42]	15	16	[45]	19	54	52
44	22	[51]	53	24	47	10	9
29	35	62	[60]	40	31	2	1

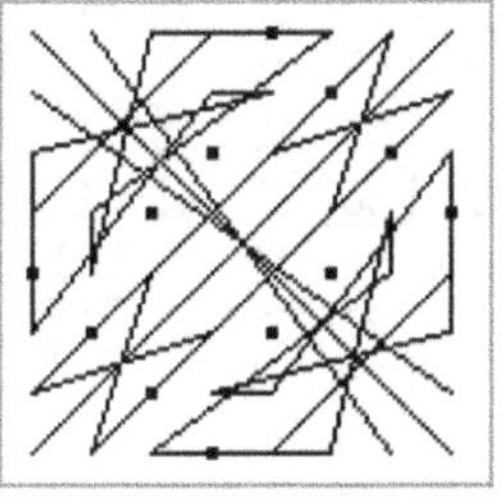

Figure 15.19 The same magic square, with a mirrored 270 degree orientation.

All eight of the previous illustrations used the same magic square, except that in each case the magic square was mirrored and/or rotated from its original orientation. This only illustrates that even though we usually don't count a rotation or a mirroring of a magic square as a unique magic square, in some ways (such as this particular way) they are still unique to each other.

Symmetry While the Center Number Is in the Center
Drawing lines from a number's original position to its new position also helps us easily see the symmetry (or lack of symmetry) in a magic square. Notice the right-side pattern in the pan-diagonal 7th order magic square below. There doesn't appear to be anything that might be considered a regular pattern, with any kind of symmetry.

38	44	1	14	20	26	32
13	19	25	31	37	43	7
30	36	49	6	12	18	24
5	11	17	23	29	42	48
22	35	41	47	4	10	16
46	3	9	15	28	34	40
21	27	33	39	45	2	8

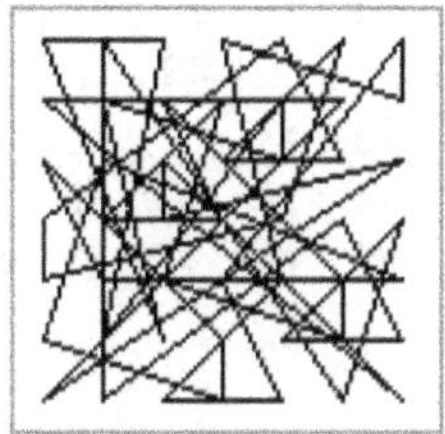

Figure 15.20 A 7th order pan-diagonal magic square with the lines on the right drawn from a numbers original position to its current position.

In some cases we can shift the magic square so that the center number of an odd-ordered magic square can be placed within the middle cell of the square. Shifting the rows or columns won't change the total that the rows and columns add up to; however, it changes the numbers within the main diagonals, sometimes causing the main diagonals to add up to a different total than the rows and columns; thus making the magic square semi-magic. However, this isn't a problem with pan-diagonal magic squares.

For a 7th order magic square, the numbers used in this square range from 1 to 49; which means that 25 is the middle number. Now, we take the pan-diagonal magic square from the previous figure, and shift its members until the 25 is in the center cell. We still have a magic square, and this time the line drawing on the right has a symmetrical pattern. Here it is:

40	46	3	9	15	28	34
8	21	27	33	39	45	2
32	38	44	1	14	20	26
7	13	19	25	31	37	43
24	30	36	49	6	12	18
48	5	11	17	23	29	42
16	22	35	41	47	4	10

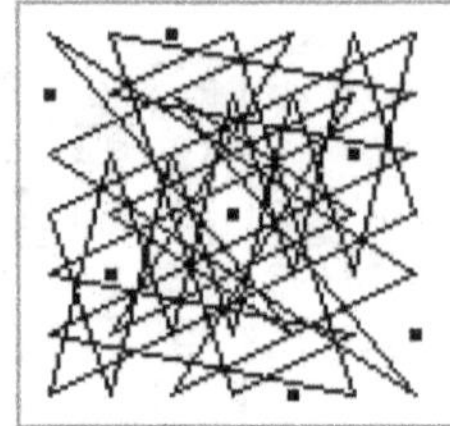

Figure 15.21 The 7th order pan-diagonal magic square with 25 shifted to the center cell. Lines are drawn from the original to current position.

This appears to be true with all odd-ordered magic squares. When the center number is placed in the center of the array, the resulting magic square has a symmetrical pattern when drawn. We can see the same thing again in this standard 7th order magic square:

48	40	16	8	32	7	24
17	9	41	33	1	25	49
42	34	10	2	26	43	18
11	3	35	27	44	19	36
5	46	22	21	38	13	30
29	28	4	45	20	37	12
23	15	47	39	14	31	6

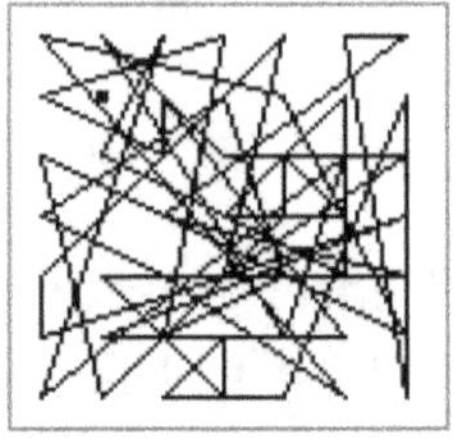

4	45	20	37	12	29	28
47	39	14	31	6	23	15
16	8	32	7	24	48	40
41	33	1	25	49	17	9
10	2	26	43	18	42	34
35	27	44	19	36	11	3
22	21	38	13	30	5	46

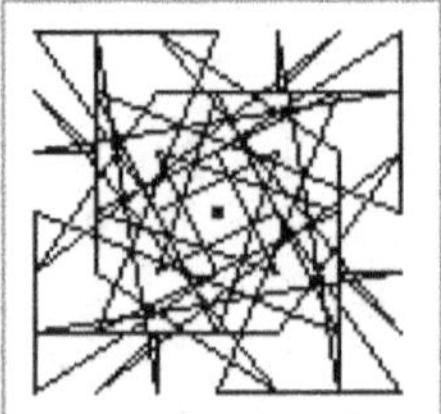

Figure 15.22 A second example using a regular 7th order magic square. The top pattern shows no symmetry. Then when the 25 is shifted to the center cell a symmetrical pattern appears. In both views the lines on the right are drawn from the numbers original position to its current position.

194

In the figure on the previous page, both squares represent the same magic square, with the difference being that in the lower magic square, the rows and columns have been shifted so that the 25 is now in the center position. (However, also keep in mind that since neither of the squares is a rotation or mirror of the other, these two magic squares are actually considered to be unique to each other.)

Similar Patterns between Orders of Magic Squares
Drawing lines from a number's original position to its new position also helps us to see similar patterns between differing orders of magic squares. Here is an interesting set of patterns that appear to be very similar between the 5th order, 7th order, 11th order and 13th order. Notice that in each of these magic squares, the numbers in the upper-left-to-lower-right diagonal are in their original locations.

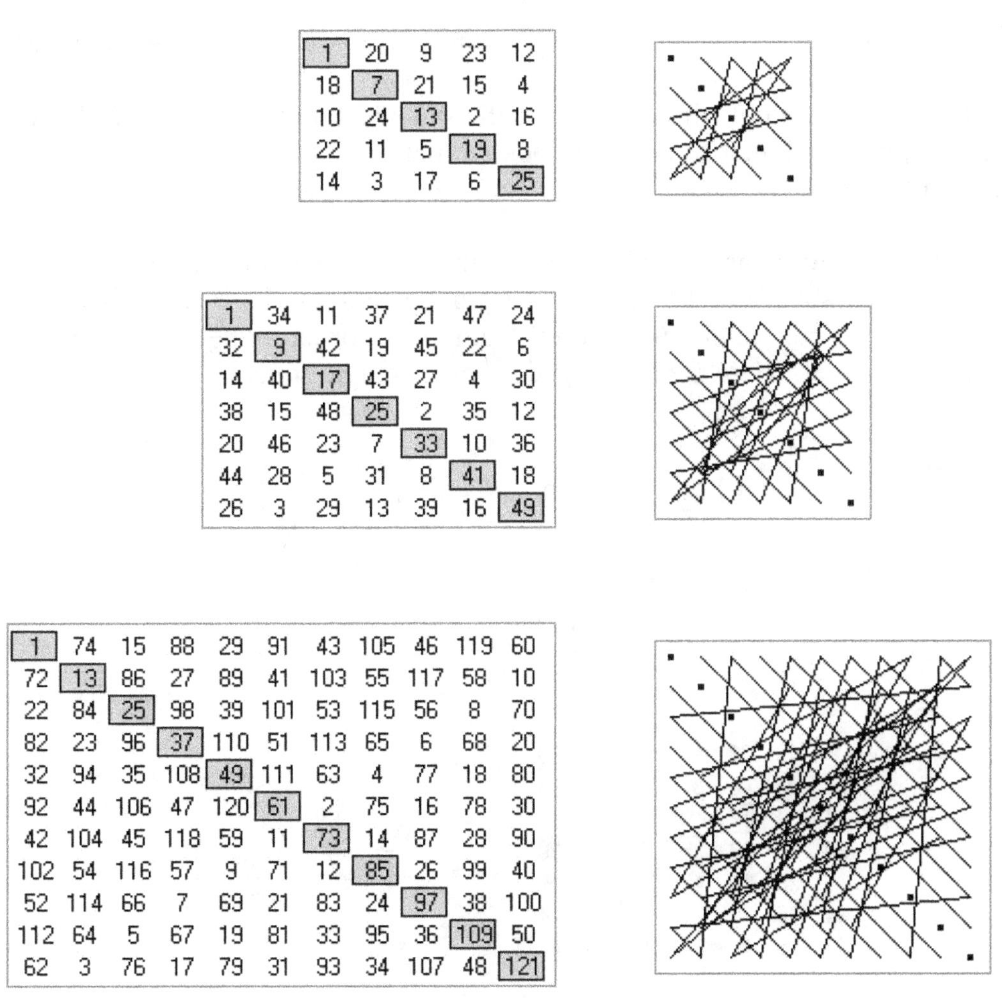

Figure 15.23a A 5th, 7th, and 11th order magic square with similar patterns within both the left and right squares.

And here is the 13th order magic square that is similar to the previous magic squares:

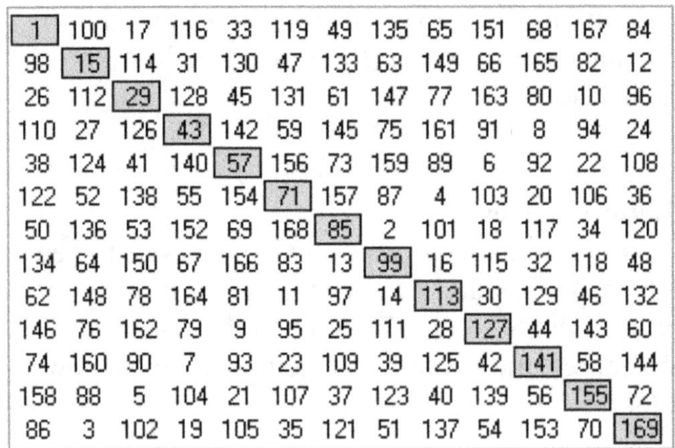

1	100	17	116	33	119	49	135	65	151	68	167	84
98	15	114	31	130	47	133	63	149	66	165	82	12
26	112	29	128	45	131	61	147	77	163	80	10	96
110	27	126	43	142	59	145	75	161	91	8	94	24
38	124	41	140	57	156	73	159	89	6	92	22	108
122	52	138	55	154	71	157	87	4	103	20	106	36
50	136	53	152	69	168	85	2	101	18	117	34	120
134	64	150	67	166	83	13	99	16	115	32	118	48
62	148	78	164	81	11	97	14	113	30	129	46	132
146	76	162	79	9	95	25	111	28	127	44	143	60
74	160	90	7	93	23	109	39	125	42	141	58	144
158	88	5	104	21	107	37	123	40	139	56	155	72
86	3	102	19	105	35	121	51	137	54	153	70	169

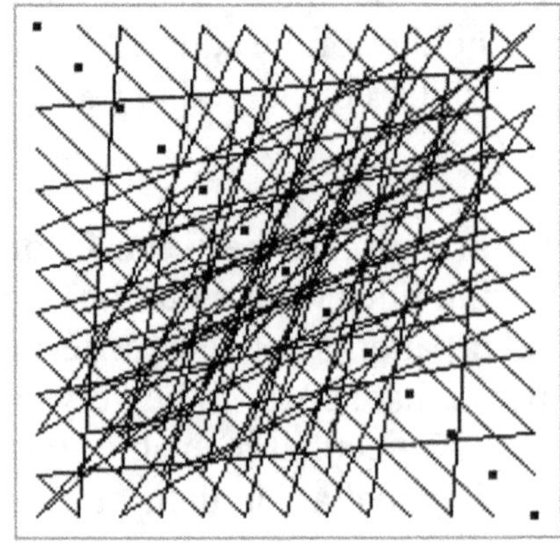

Figure 15.23b   A 13th order magic square with a similar right-side pattern.

It is also interesting to note that we can use the same technique to create a 9th order and a 15th order square, and their figures will look similar to these magic squares, but the actual arrays won't be magic squares.

The next four magic squares also keep their same numbers in their original locations. Again lines are drawn from the number's original position to its current position. And again we see a similar pattern as we move from the 5th order to the 7th, and then 11th and finally the 13th order. Here are the 5th and 7th order magic squares:

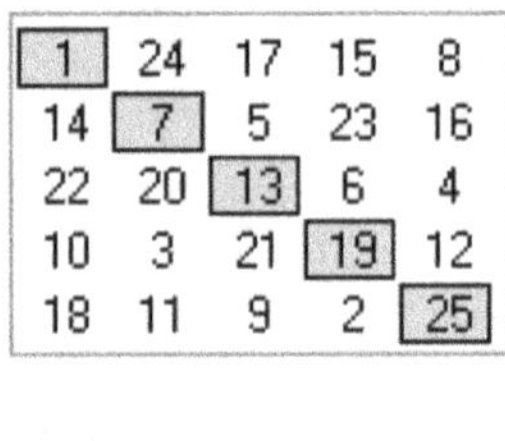

1	24	17	15	8
14	7	5	23	16
22	20	13	6	4
10	3	21	19	12
18	11	9	2	25

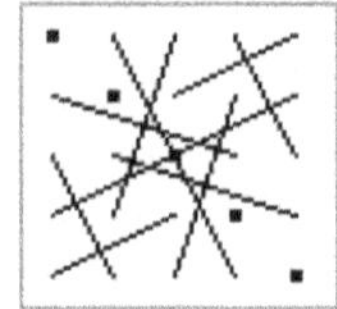

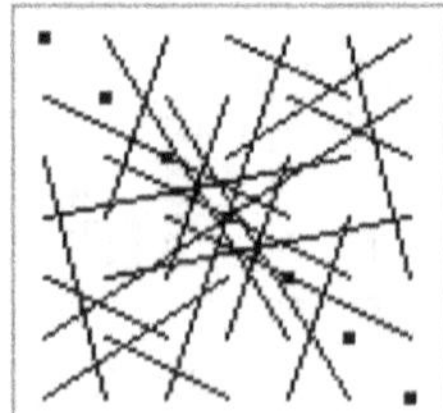

1	40	23	13	45	35	18
26	9	48	31	21	4	36
44	34	17	7	39	22	12
20	3	42	25	8	47	30
38	28	11	43	33	16	6
14	46	29	19	2	41	24
32	15	5	37	27	10	49

Figure 15.24a A 5th and 7th order magic squares with a similar pattern within the
left and right-side squares.

And here are the 11th and 13th order magic squares that share similar patterns within the left and right-side squares:

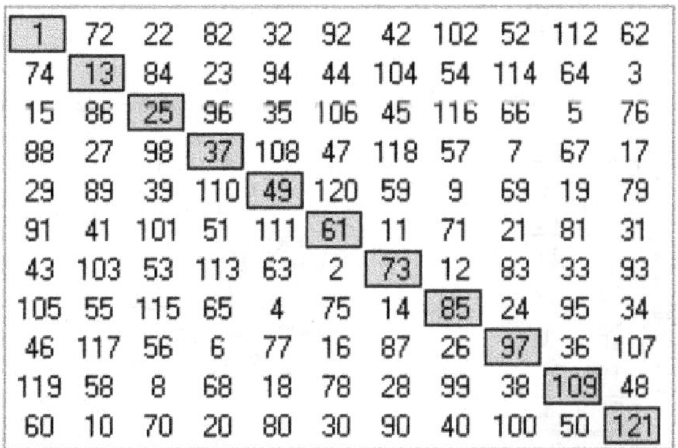

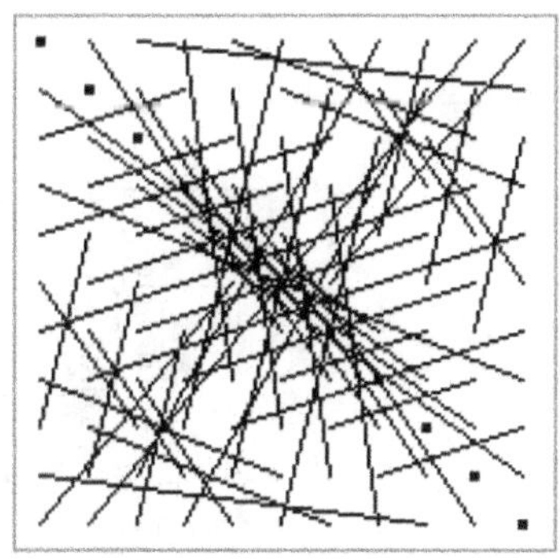

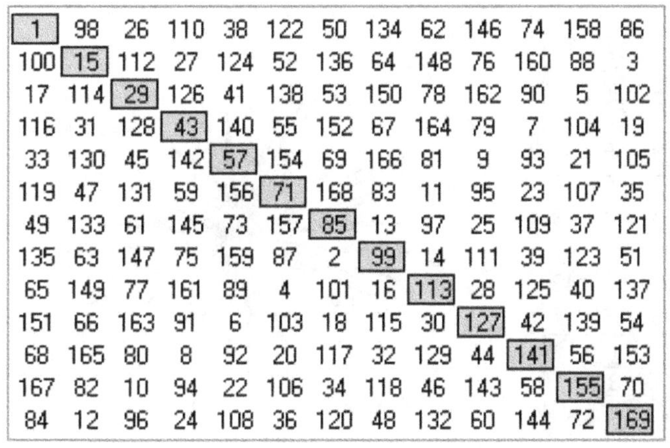

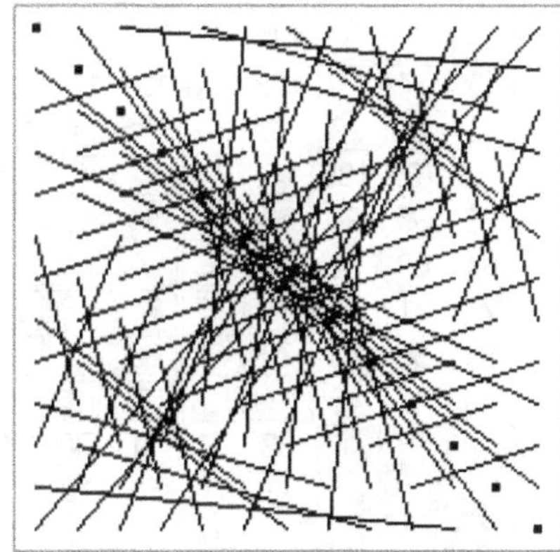

Figure 15.24b An 11th and 13th order magic squares with a similar pattern within the left and right-side squares.

There are still many other magic squares that share similar patterns as we move from a lower order to a higher order. In this next set of examples, we will compare a 4th order magic square with an 8th order, a 12th order and a 16th order magic square. In each of these magic squares, half the numbers in that array remain in their original position, while the other half trade places diagonally through the center of the square with their corresponding counterpart on the opposite side of the square.

By drawing lines from the number's original position to its current location, we can see exactly how these magic squares were made. Actually, the illustration itself acts as a better description of the method used to create these magic squares than anything else could. Here are the double even magic squares that share this similar pattern.

1	15	14	4
12	6	7	9
8	10	11	5
13	3	2	16

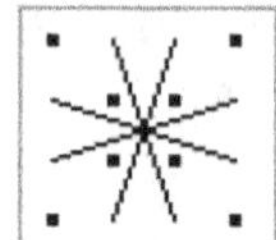

1	2	62	61	60	59	7	8
9	10	54	53	52	51	15	16
48	47	19	20	21	22	42	41
40	39	27	28	29	30	34	33
32	31	35	36	37	38	26	25
24	23	43	44	45	46	18	17
49	50	14	13	12	11	55	56
57	58	6	5	4	3	63	64

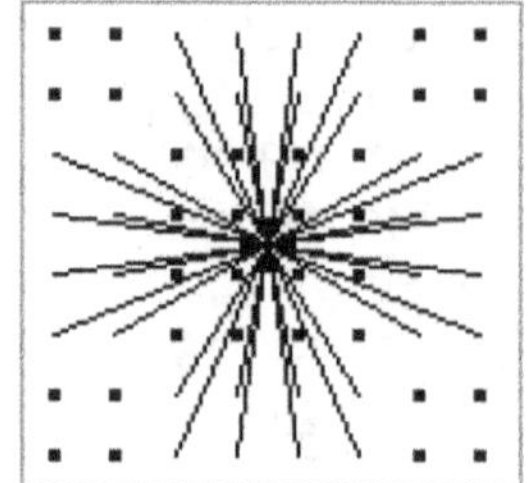

1	2	3	141	140	139	138	137	136	10	11	12
13	14	15	129	128	127	126	125	124	22	23	24
25	26	27	117	116	115	114	113	112	34	35	36
108	107	106	40	41	42	43	44	45	99	98	97
96	95	94	52	53	54	55	56	57	87	86	85
84	83	82	64	65	66	67	68	69	75	74	73
72	71	70	76	77	78	79	80	81	63	62	61
60	59	58	88	89	90	91	92	93	51	50	49
48	47	46	100	101	102	103	104	105	39	38	37
109	110	111	33	32	31	30	29	28	118	119	120
121	122	123	21	20	19	18	17	16	130	131	132
133	134	135	9	8	7	6	5	4	142	143	144

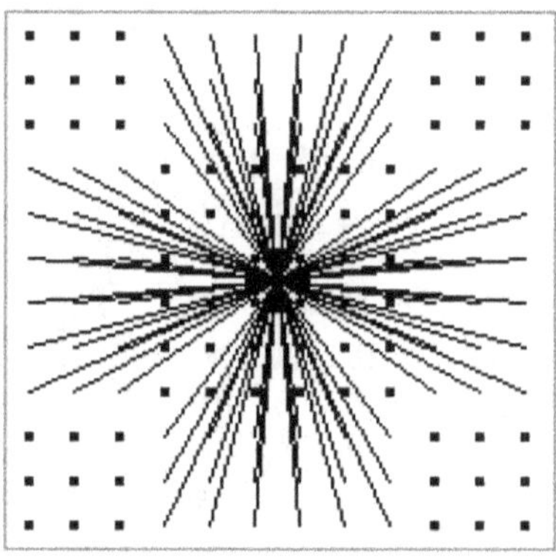

1	2	3	4	252	251	250	249	248	247	246	245	13	14	15	16
17	18	19	20	236	235	234	233	232	231	230	229	29	30	31	32
33	34	35	36	220	219	218	217	216	215	214	213	45	46	47	48
49	50	51	52	204	203	202	201	200	199	198	197	61	62	63	64
192	191	190	189	69	70	71	72	73	74	75	76	180	179	178	177
176	175	174	173	85	86	87	88	89	90	91	92	164	163	162	161
160	159	158	157	101	102	103	104	105	106	107	108	148	147	146	145
144	143	142	141	117	118	119	120	121	122	123	124	132	131	130	129
128	127	126	125	133	134	135	136	137	138	139	140	116	115	114	113
112	111	110	109	149	150	151	152	153	154	155	156	100	99	98	97
96	95	94	93	165	166	167	168	169	170	171	172	84	83	82	81
80	79	78	77	181	182	183	184	185	186	187	188	68	67	66	65
193	194	195	196	60	59	58	57	56	55	54	53	205	206	207	208
209	210	211	212	44	43	42	41	40	39	38	37	221	222	223	224
225	226	227	228	28	27	26	25	24	23	22	21	237	238	239	240
241	242	243	244	12	11	10	9	8	7	6	5	253	254	255	256

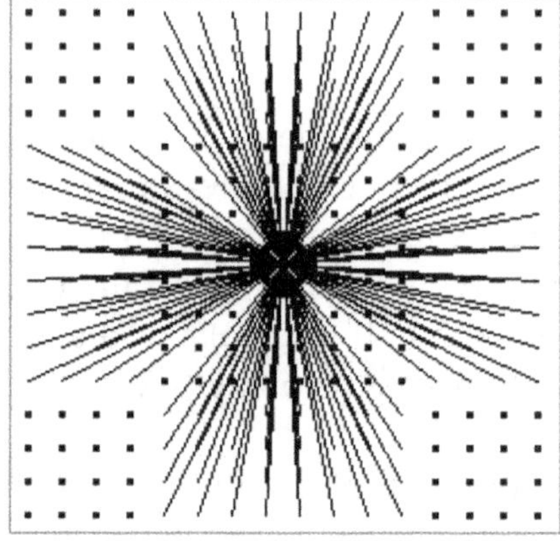

Figure 15.25   The 4th, 8th, 12th, and 16th order magic squares that share a similar pattern.

Template and Nested Magic Squares

As we saw at the beginning of this chapter the most obvious and straight-forward method of drawing line patterns is to draw a line from the 1 to the 2, to the 3, etc. Many template and nested magic squares have interesting patterns when lines are drawn in this manner. We'll look at some of these interesting patterns in this section. In this next example we'll take the pattern created from a 5th order magic square, and then we'll see the pattern repeated nine times in a 15th order template magic square where the 15 x 15 array is divided into nine 5 x 5 arrays. First, here is the 5th order magic square:

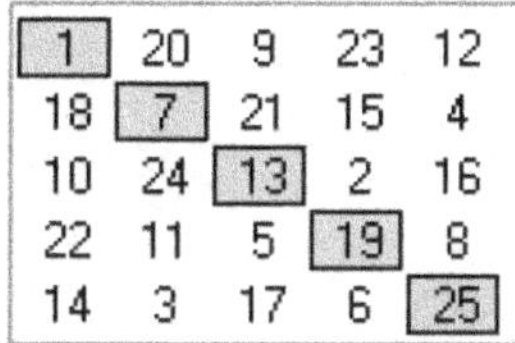
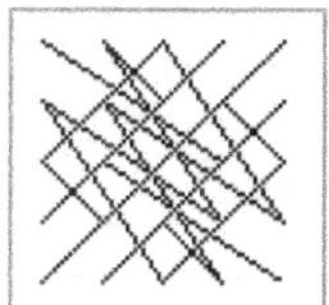

Figure 15.26   A 5th order magic square that will be used as a template within a 15th order magic square.

The following is a 15th order template magic square. Notice the nine identical 5th order magic squares nested within the 15 x 15 array.

176	195	184	198	187	1	20	9	23	12	126	145	134	148	137
193	182	196	190	179	18	7	21	15	4	143	132	146	140	129
185	199	188	177	191	10	24	13	2	16	135	149	138	127	141
197	186	180	194	183	22	11	5	19	8	147	136	130	144	133
189	178	192	181	200	14	3	17	6	25	139	128	142	131	150
51	70	59	73	62	101	120	109	123	112	151	170	159	173	162
68	57	71	65	54	118	107	121	115	104	168	157	171	165	154
60	74	63	52	66	110	124	113	102	116	160	174	163	152	166
72	61	55	69	58	122	111	105	119	108	172	161	155	169	158
64	53	67	56	75	114	103	117	106	125	164	153	167	156	175
76	95	84	98	87	201	220	209	223	212	26	45	34	48	37
93	82	96	90	79	218	207	221	215	204	43	32	46	40	29
85	99	88	77	91	210	224	213	202	216	35	49	38	27	41
97	86	80	94	83	222	211	205	219	208	47	36	30	44	33
89	78	92	81	100	214	203	217	206	225	39	28	42	31	50

Figure 15.27   The 15th order magic square created from nine 5th order magic squares.

Next, we'll do the same thing with a 4th order magic square, and then a 16th order template magic square made up of sixteen nested 4 x 4 arrays:

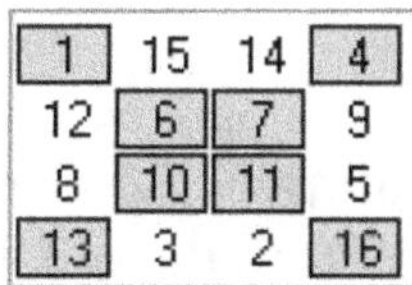
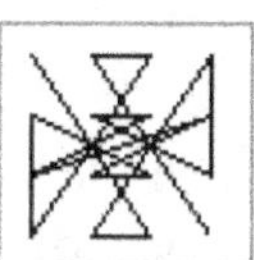

Figure 15.28   A 4th order magic square that will be used as a template within a 16th order magic square.

In the 16th order template magic square below we can easily see that there are sixteen 4th order magic squares nested within the 16 x 16 array.

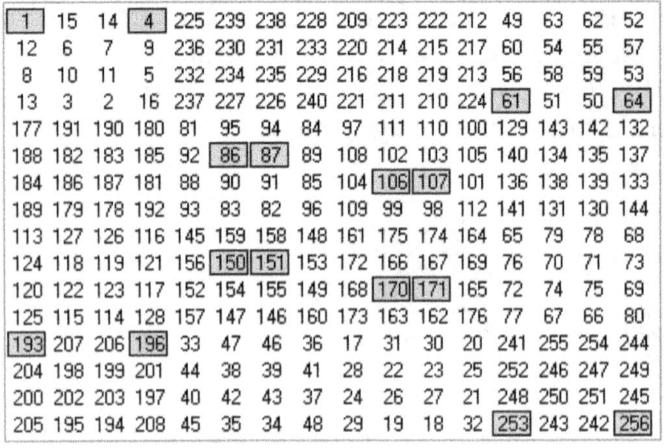

1	15	14	4	225	239	238	228	209	223	222	212	49	63	62	52
12	6	7	9	236	230	231	233	220	214	215	217	60	54	55	57
8	10	11	5	232	234	235	229	216	218	219	213	56	58	59	53
13	3	2	16	237	227	226	240	221	211	210	224	61	51	50	64
177	191	190	180	81	95	94	84	97	111	110	100	129	143	142	132
188	182	183	185	92	86	87	89	108	102	103	105	140	134	135	137
184	186	187	181	88	90	91	85	104	106	107	101	136	138	139	133
189	179	178	192	93	83	82	96	109	99	98	112	141	131	130	144
113	127	126	116	145	159	158	148	161	175	174	164	65	79	78	68
124	118	119	121	156	150	151	153	172	166	167	169	76	70	71	73
120	122	123	117	152	154	155	149	168	170	171	165	72	74	75	69
125	115	114	128	157	147	146	160	173	163	162	176	77	67	66	80
193	207	206	196	33	47	46	36	17	31	30	20	241	255	254	244
204	198	199	201	44	38	39	41	28	22	23	25	252	246	247	249
200	202	203	197	40	42	43	37	24	26	27	21	248	250	251	245
205	195	194	208	45	35	34	48	29	19	18	32	253	243	242	256

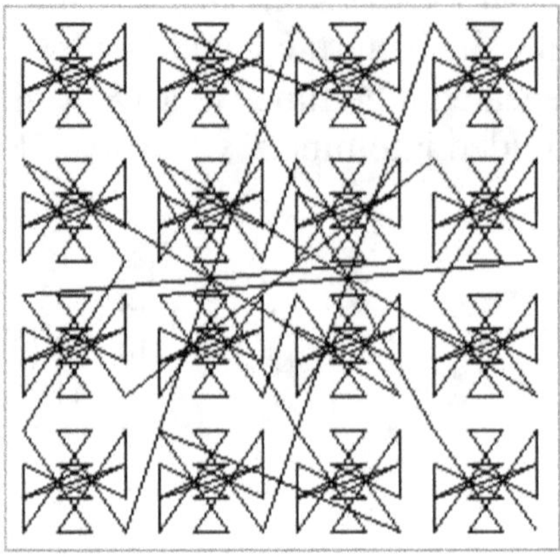

Figure 15.29   The 16th order magic square created from sixteen 4th order magic squares.

Of course there is more than one way to do this. The following magic square uses the 4th order template in the four corners, and then uses a variation of this template in the middle section of the four sides as well as the center section of the magic square itself:

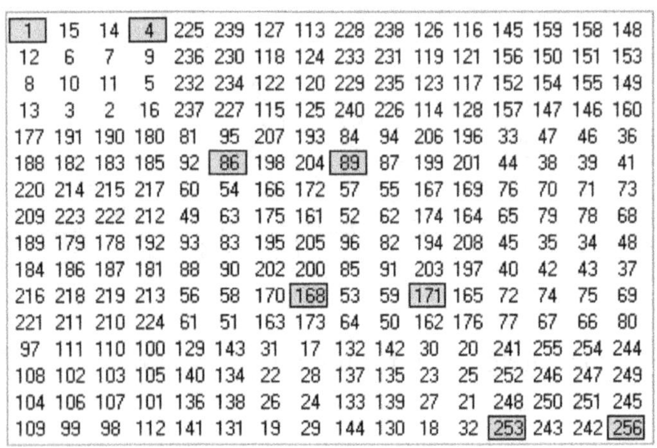

1	15	14	4	225	239	127	113	228	238	126	116	145	159	158	148
12	6	7	9	236	230	118	124	233	231	119	121	156	150	151	153
8	10	11	5	232	234	122	120	229	235	123	117	152	154	155	149
13	3	2	16	237	227	115	125	240	226	114	128	157	147	146	160
177	191	190	180	81	95	207	193	84	94	206	196	33	47	46	36
188	182	183	185	92	86	198	204	89	87	199	201	44	38	39	41
220	214	215	217	60	54	166	172	57	55	167	169	76	70	71	73
209	223	222	212	49	63	175	161	52	62	174	164	65	79	78	68
189	179	178	192	93	83	195	205	96	82	194	208	45	35	34	48
184	186	187	181	88	90	202	200	85	91	203	197	40	42	43	37
216	218	219	213	56	58	170	168	53	59	171	165	72	74	75	69
221	211	210	224	61	51	163	173	64	50	162	176	77	67	66	80
97	111	110	100	129	143	31	17	132	142	30	20	241	255	254	244
108	102	103	105	140	134	22	28	137	135	23	25	252	246	247	249
104	106	107	101	136	138	26	24	133	139	27	21	248	250	251	245
109	99	98	112	141	131	19	29	144	130	18	32	253	243	242	256

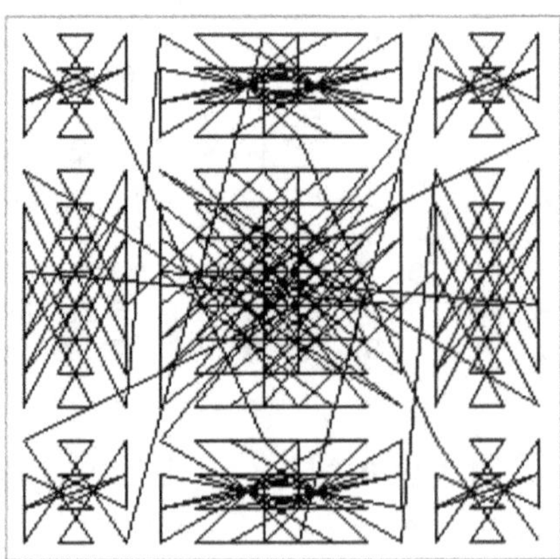

Figure 15.30   A variation of the 16th order magic square created from sixteen 4th order magic squares.

In an earlier chapter we saw a method of creating odd-ordered, magic ringed, magic squares. That method built part of the square from the outside moving in, then when the center 3 x 3 array was filled, the process completed the square by filling the remaining regions from the inside moving out. When we draw lines connecting the numbers sequentially, we can easily see the pattern grow as we move from a smaller ordered magic square to a larger ordered magic square. The following figure shows a 9th ordered, 11th ordered and 13th ordered set of magic squares that were all built using the same construction method.

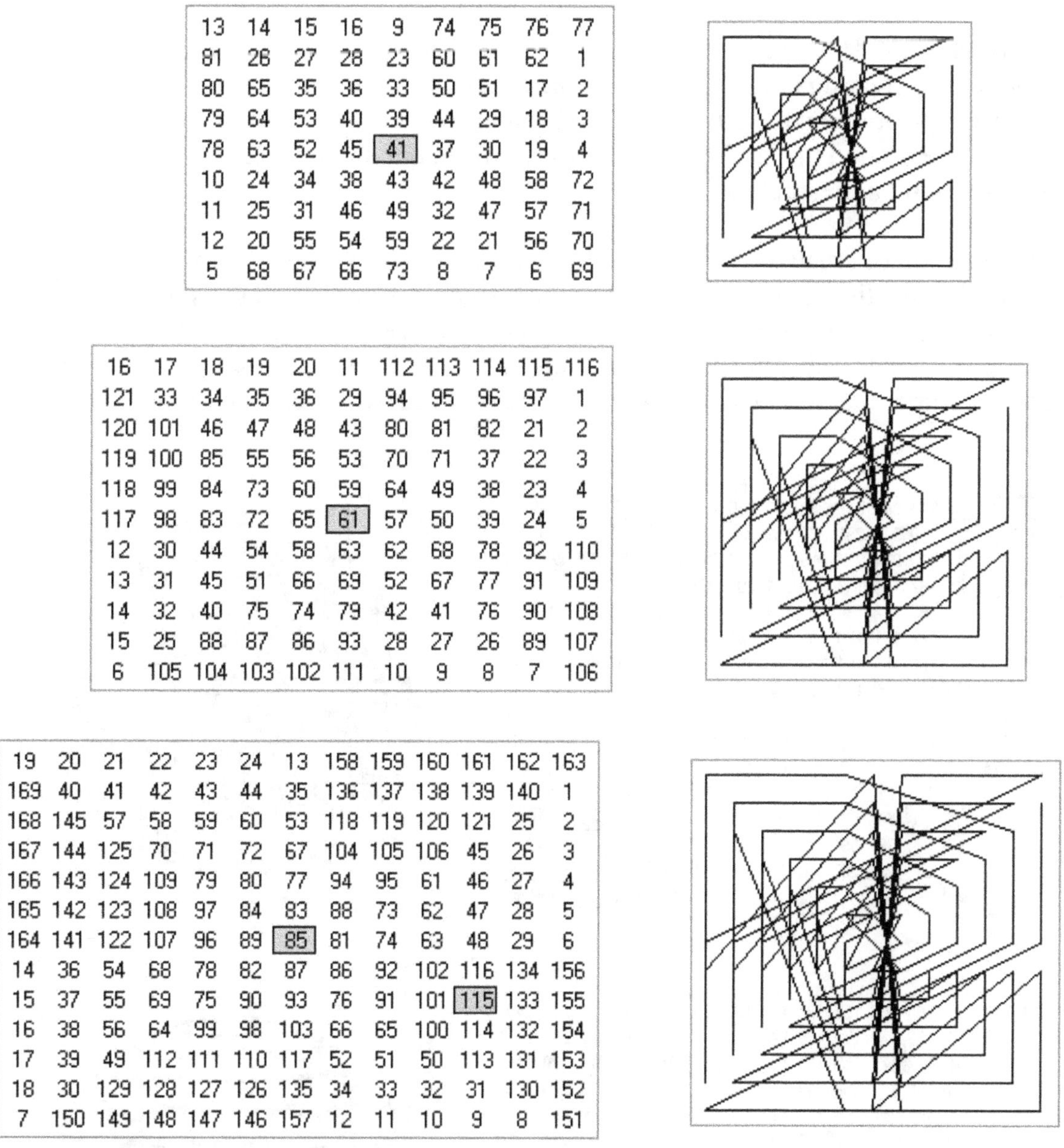

13	14	15	16	9	74	75	76	77
81	26	27	28	23	60	61	62	1
80	65	35	36	33	50	51	17	2
79	64	53	40	39	44	29	18	3
78	63	52	45	41	37	30	19	4
10	24	34	38	43	42	48	58	72
11	25	31	46	49	32	47	57	71
12	20	55	54	59	22	21	56	70
5	68	67	66	73	8	7	6	69

16	17	18	19	20	11	112	113	114	115	116
121	33	34	35	36	29	94	95	96	97	1
120	101	46	47	48	43	80	81	82	21	2
119	100	85	55	56	53	70	71	37	22	3
118	99	84	73	60	59	64	49	38	23	4
117	98	83	72	65	61	57	50	39	24	5
12	30	44	54	58	63	62	68	78	92	110
13	31	45	51	66	69	52	67	77	91	109
14	32	40	75	74	79	42	41	76	90	108
15	25	88	87	86	93	28	27	26	89	107
6	105	104	103	102	111	10	9	8	7	106

19	20	21	22	23	24	13	158	159	160	161	162	163
169	40	41	42	43	44	35	136	137	138	139	140	1
168	145	57	58	59	60	53	118	119	120	121	25	2
167	144	125	70	71	72	67	104	105	106	45	26	3
166	143	124	109	79	80	77	94	95	61	46	27	4
165	142	123	108	97	84	83	88	73	62	47	28	5
164	141	122	107	96	89	85	81	74	63	48	29	6
14	36	54	68	78	82	87	86	92	102	116	134	156
15	37	55	69	75	90	93	76	91	101	115	133	155
16	38	56	64	99	98	103	66	65	100	114	132	154
17	39	49	112	111	110	117	52	51	50	113	131	153
18	30	129	128	127	126	135	34	33	32	31	130	152
7	150	149	148	147	146	157	12	11	10	9	8	151

Figure 15.31   Similar construction methods can be seen between this 9th ordered, 11th ordered, and 13th ordered set of magic squares. (This method builds magic ringed, magic squares, and was discussed in the chapter about nested magic squares.)

And there are still many other ways to connect the dots and look at patterns within these magic squares. As another example, we could take the magic square, and replace each number with its corresponding Aa representation (see chapter 14). Then we could draw lines connecting all of the lowercase a's together (keeping the uppercase letters in order), then draw lines connecting the lowercase b's, and so on. And we could go on with still other methods; however, before closing this chapter we will consider a unique way of graphing magic squares using what I simply call circular plots.

Circular Plots

One method of creating circular plots is to draw lines from the number's original position to its current position on the circle. For a 16th order magic square, list the numbers 1 thru 256 clockwise around the circle. Now draw lines from the numbers original position to its current position on the circle.

Another method of creating circular plots is to draw lines in sequential order around the circle. To do this, the numbers are listed as they are read from the magic square from left to right and from top to bottom. Now, start at the 1 (wherever it is on the circle), and sequentially connect the numbers on the circle. The following figure shows a sampling of several 16th order magic squares drawn with one of these two circular plot methods.

1	2	3	4	245	246	247	248	249	250	251	252	13	14	15	16
17	18	19	20	229	230	231	232	233	234	235	236	29	30	31	32
33	34	35	36	213	214	215	216	217	218	219	220	45	46	47	48
49	50	51	52	197	198	199	200	201	202	203	204	61	62	63	64
80	79	78	77	188	187	186	185	184	183	182	181	68	67	66	65
96	95	94	93	172	171	170	169	168	167	166	165	84	83	82	81
112	111	110	109	156	155	154	153	152	151	150	149	100	99	98	97
128	127	126	125	140	139	138	137	136	135	134	133	116	115	114	113
144	143	142	141	124	123	122	121	120	119	118	117	132	131	130	129
160	159	158	157	108	107	106	105	104	103	102	101	148	147	146	145
176	175	174	173	92	91	90	89	88	87	86	85	164	163	162	161
192	191	190	189	76	75	74	73	72	71	70	69	180	179	178	177
193	194	195	196	53	54	55	56	57	58	59	60	205	206	207	208
209	210	211	212	37	38	39	40	41	42	43	44	221	222	223	224
225	226	227	228	21	22	23	24	25	26	27	28	237	238	239	240
241	242	243	244	5	6	7	8	9	10	11	12	253	254	255	256

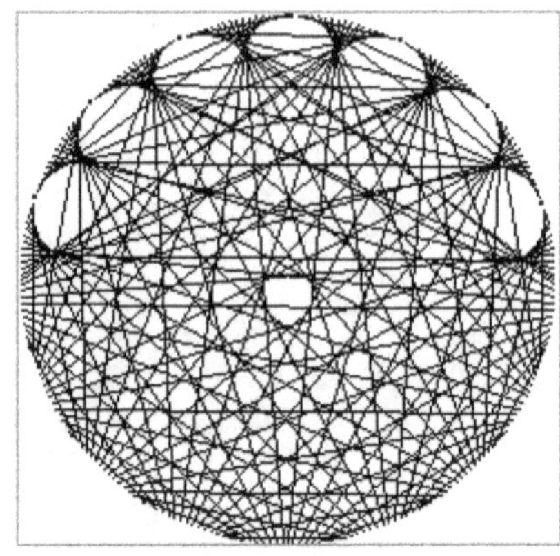

Circular Orig to Curr

8	24	40	56	185	169	153	137	121	105	89	73	200	216	232	248
7	23	39	55	186	170	154	138	122	106	90	74	199	215	231	247
6	22	38	54	187	171	155	139	123	107	91	75	198	214	230	246
5	21	37	53	188	172	156	140	124	108	92	76	197	213	229	245
253	237	221	205	68	84	100	116	132	148	164	180	61	45	29	13
254	238	222	206	67	83	99	115	131	147	163	179	62	46	30	14
255	239	223	207	66	82	98	114	130	146	162	178	63	47	31	15
256	240	224	208	65	81	97	113	129	145	161	177	64	48	32	16
241	225	209	193	80	96	112	128	144	160	176	192	49	33	17	1
242	226	210	194	79	95	111	127	143	159	175	191	50	34	18	2
243	227	211	195	78	94	110	126	142	158	174	190	51	35	19	3
244	228	212	196	77	93	109	125	141	157	173	189	52	36	20	4
12	28	44	60	181	165	149	133	117	101	85	69	204	220	236	252
11	27	43	59	182	166	150	134	118	102	86	70	203	219	235	251
10	26	42	58	183	167	151	135	119	103	87	71	202	218	234	250
9	25	41	57	184	168	152	136	120	104	88	72	201	217	233	249

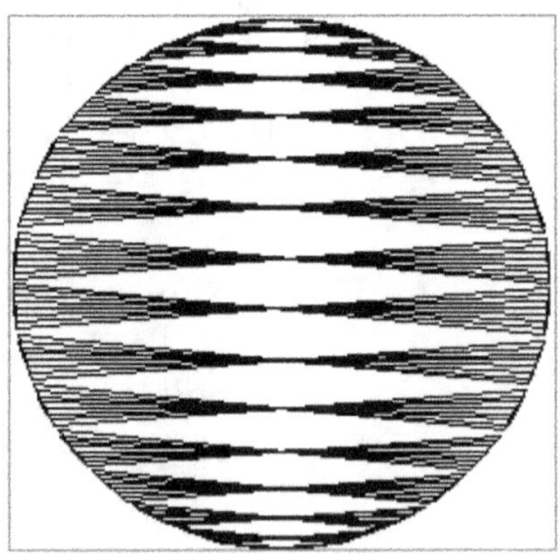

Circular Orig to Curr

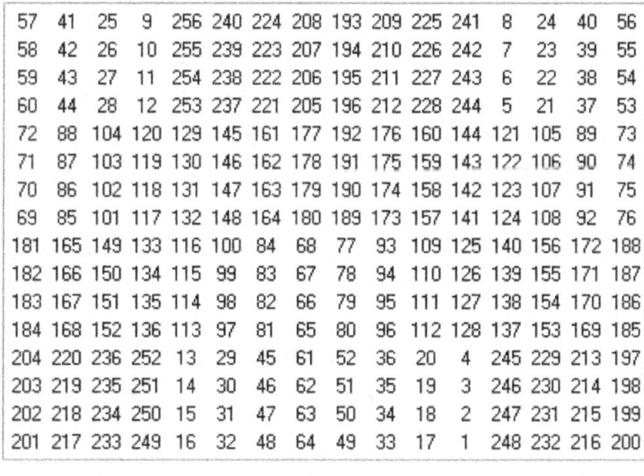

57	41	25	9	256	240	224	208	193	209	225	241	8	24	40	56
58	42	26	10	255	239	223	207	194	210	226	242	7	23	39	55
59	43	27	11	254	238	222	206	195	211	227	243	6	22	38	54
60	44	28	12	253	237	221	205	196	212	228	244	5	21	37	53
72	88	104	120	129	145	161	177	192	176	160	144	121	105	89	73
71	87	103	119	130	146	162	178	191	175	159	143	122	106	90	74
70	86	102	118	131	147	163	179	190	174	158	142	123	107	91	75
69	85	101	117	132	148	164	180	189	173	157	141	124	108	92	76
181	165	149	133	116	100	84	68	77	93	109	125	140	156	172	188
182	166	150	134	115	99	83	67	78	94	110	126	139	155	171	187
183	167	151	135	114	98	82	66	79	95	111	127	138	154	170	186
184	168	152	136	113	97	81	65	80	96	112	128	137	153	169	185
204	220	236	252	13	29	45	61	52	36	20	4	245	229	213	197
203	219	235	251	14	30	46	62	51	35	19	3	246	230	214	198
202	218	234	250	15	31	47	63	50	34	18	2	247	231	215	199
201	217	233	249	16	32	48	64	49	33	17	1	248	232	216	200

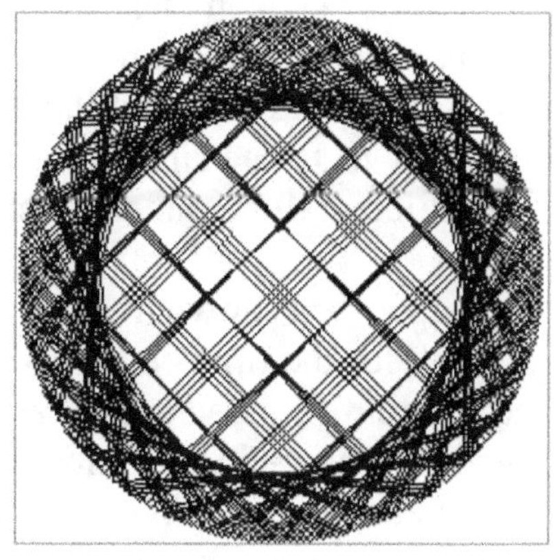

Circular Orig to Curr

1	65	2	66	254	190	253	189	252	188	251	187	7	71	8	72
129	193	130	194	126	62	125	61	124	60	123	59	135	199	136	200
9	73	10	74	246	182	245	181	244	180	243	179	15	79	16	80
137	201	138	202	118	54	117	53	116	52	115	51	143	207	144	208
240	176	239	175	19	83	20	84	21	85	22	86	234	170	233	169
112	48	111	47	147	211	148	212	149	213	150	214	106	42	105	41
232	168	231	167	27	91	28	92	29	93	30	94	226	162	225	161
104	40	103	39	155	219	156	220	157	221	158	222	98	34	97	33
224	160	223	159	35	99	36	100	37	101	38	102	218	154	217	153
96	32	95	31	163	227	164	228	165	229	166	230	90	26	89	25
216	152	215	151	43	107	44	108	45	109	46	110	210	146	209	145
88	24	87	23	171	235	172	236	173	237	174	238	82	18	81	17
49	113	50	114	206	142	205	141	204	140	203	139	55	119	56	120
177	241	178	242	78	14	77	13	76	12	75	11	183	247	184	248
57	121	58	122	198	134	197	133	196	132	195	131	63	127	64	128
185	249	186	250	70	6	69	5	68	4	67	3	191	255	192	256

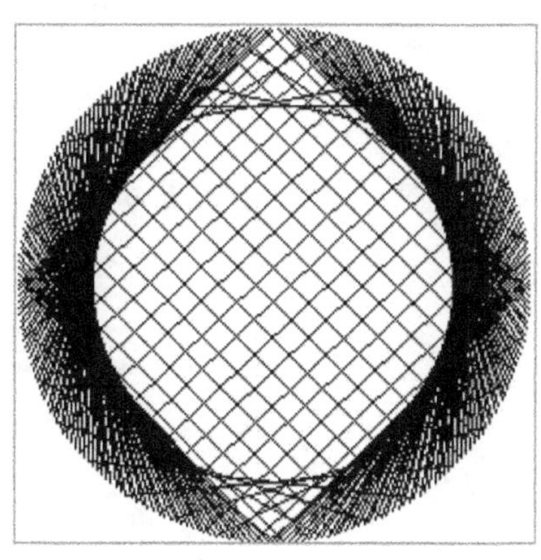

Circular Std Sequence

2	1	224	223	208	207	18	17	98	97	192	191	176	175	114	113
130	129	96	95	80	79	146	145	226	225	64	63	48	47	242	241
123	124	165	166	181	182	107	108	27	28	197	198	213	214	11	12
251	252	37	38	53	54	235	236	155	156	69	70	85	86	139	140
121	122	167	168	183	184	105	106	25	26	199	200	215	216	9	10
249	250	39	40	55	56	233	234	153	154	71	72	87	88	137	138
4	3	222	221	206	205	20	19	100	99	190	189	174	173	116	115
132	131	94	93	78	77	148	147	228	227	62	61	46	45	244	243
14	13	212	211	196	195	30	29	110	109	180	179	164	163	126	125
142	141	84	83	68	67	158	157	238	237	52	51	36	35	254	253
119	120	169	170	185	186	103	104	23	24	201	202	217	218	7	8
247	248	41	42	57	58	231	232	151	152	73	74	89	90	135	136
117	118	171	172	187	188	101	102	21	22	203	204	219	220	5	6
245	246	43	44	59	60	229	230	149	150	75	76	91	92	133	134
16	15	210	209	194	193	32	31	112	111	178	177	162	161	128	127
144	143	82	81	66	65	160	159	240	239	50	49	34	33	256	255

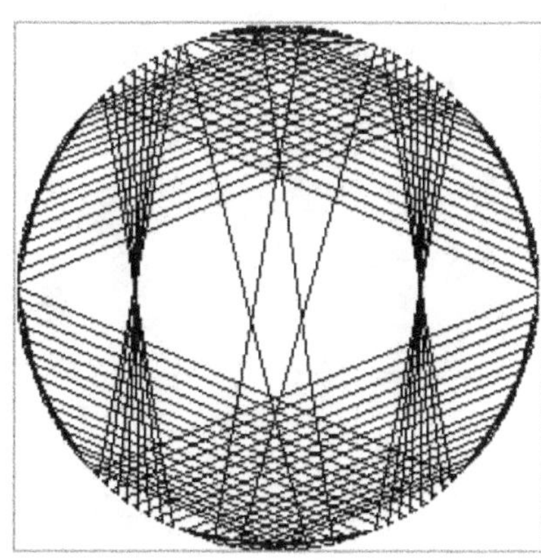

Circular Std Sequence

Figure 15.32   A sampling of five 16th order magic squares using one of two different types of circular plots.

In earlier chapters we looked at ways of defining magic squares so that we could easily see similarities between these arrays of numbers. These graphical methods of looking magic squares also make it much easier to place a magic square within family groups where the plots tend to be similar to each other. These methods of categorizing magic squares allow us to see similarities even though it is difficult to see these similar patterns by just looking at the numbers themselves in the array.

As this chapter has shown, the possibilities here are incredible. And while I've said more than enough for the time being about these magic squares, there is still so much more to say and see (but we need to face reality, and understand that as mere mortals there just isn't enough time to do everything that we would like to do). However, do not despair because I believe a good book on magic squares should not only show the methods that can be used to create a million billion magic squares, but it should also show scores and scores (and even more) magic squares themselves; therefore, in addition to those found within the preceding pages, the following pages contain an appendix where there are many more examples of magic squares grouped together by order and in some cases by pattern type.

Epilog

It was late in the night when Antonio finally finished discussing magic squares, but the crew was nowhere ready to stop their Mathematical pursuits. Many still feverishly punched away on their calculators, while others scribbled on the three white boards that had been brought up from the hold below.

Rubio and Seymour* each with a laptop, had already constructed scores of 8th order magic squares on a spreadsheet.

The captain, who had dozed off somewhere during Antonio's first sentence, was now decidedly awake and distraughtly aware of what he was seeing. He thought to himself, "Antonio could have easily taken over the ship while I slumbered;" and even though Antonio didn't have those ill-advised ambitions; still, in a way, that is exactly what he had done.

The captain sprang to his feet, ready to draw his saber; however, Antonio and his fellow shipmates didn't exactly follow the pirate's code of pillaging and plundering; instead, they seemed to be more of a refined and sophisticated sort. It was then that Antonio looked up, and read the fear in the captain's eyes and quickly realized that he needed to reassure the captain of his innocent intentions. He simply told the captain that mutiny had never entered his mind, and all they had discussed were magic squares. He even explained further that he was saving his discussion on the Pythagorean Theorem for another night. Finally, he assured the captain that all was well, and in fact, he even advised the captain to briefly breeze through the appendix for a few additional insights of information on magic squares, and there he would be beyond amazed at what he beheld.

Surprisingly, that is exactly what the captain did …, and surprisingly again, that is exactly what happened.

The End

* See the Introduction for our introduction to Antonio, Rubio, Seymour and the captain.

<h1 style="text-align:center">Appendices</h1>

The following pages contain over 300 selected magic squares, ranging from the 4th order through the 16th order. All the magic squares in these appendices are shown as a pair of squares, with the array on the left displaying numerical values while the right shows a graphical representation of the same square. The magic squares within the following pages are presented to show some of the various graphical differences and similarities within these magic square patterns.

The values in the left square are highlighted in one of two ways. (Highlighted numbers have a shaded box surrounding the number.) The scheme used for highlighting a number includes:
> Numbers that remain in their original position
> All odd or even numbers within the square

The graphical representation of the square is on the right, and shows the lines drawn between the numbers in one of several ways. The schemes that are used for drawing these lines include:
> Lines are drawn from a number's original position to its current position
> Lines are drawn in standard sequential order (from 1 to 2 to 3 to 4, etc.)
> Lines are drawn in odd, even, or odd-even overlaid sequential order
> Lines are drawn in sequential pairs (1 to 2, 3 to 4, 5 to 6, 7 to 8, etc.)
> Lines are drawn in a circular plot in a sequential or original-to-current position

Even as a Jedi Master must build his (or her) own light saber, so must a sincere follower of the mysteries of Magic Squares write his (or her) own program to build and manipulate higher orders of magic squares. The software used to create the magic squares within this book, as well as the corresponding graphics was developed by the author. The author's software manages a database of over 5000 magic squares (from the 3rd order through the 17th order), and this program allows the combination of any of the numerical highlighting schemes listed above to be combined with any of the afore-mentioned line drawing schemes, and then apply them to any magic square.

While it is usually fairly obvious which highlighting scheme is used, an effort has been made to group together and label the highlighting schemes and line-drawing methods within each section of the appendix. The magic squares in these appendices are either regular (standard) magic squares or pan-diagonal magic squares. I have also made a determined effort to label these two types as they are shown. There are no semi-magic squares within these appendices (although many of those are also very interesting). Beginning with the 8th order magic squares in the appendix, selected circular plots are also included. Keep in mind that these magic squares are just a sampling of the millions and billions of patterns that are characteristic of the magic square.

Appendix A – Selected 4th Order Magic Squares

The following are 4th order magic squares. The shaded numbers in the left-side squares are in their original positions (not all squares have numbers in their original positions). The right-side graphic uses lines to show the movement of the numbers from their original to their current position. The following are regular magic squares:

1	12	15	6
8	13	10	3
14	7	4	9
11	2	5	16

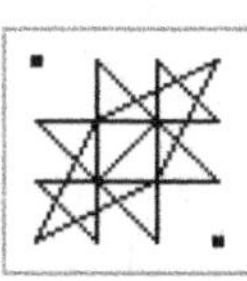

2	14	15	3
7	11	10	6
12	8	5	9
13	1	4	16

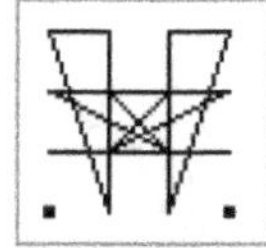

2	11	13	8
16	5	3	10
7	14	12	1
9	4	6	15

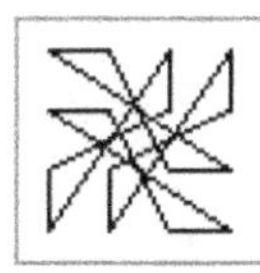

2	13	11	8
7	12	14	1
16	3	5	10
9	6	4	15

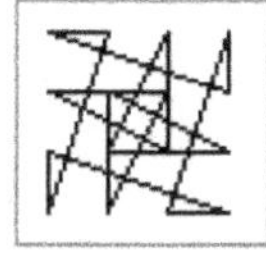

3	16	13	2
10	5	8	11
6	9	12	7
15	4	1	14

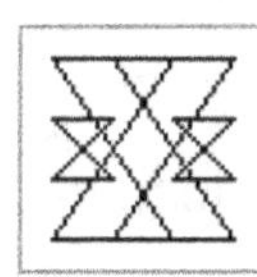

4	5	15	10
14	11	1	8
9	16	6	3
7	2	12	13

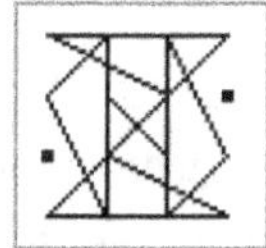

4	9	14	7
5	16	11	2
15	6	1	12
10	3	8	13

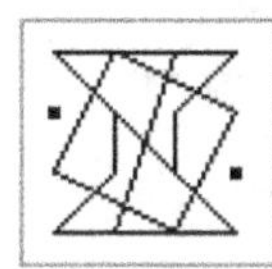

4	14	9	7
5	11	16	2
15	1	6	12
10	8	3	13

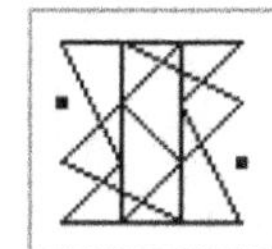

4	15	9	6
14	1	7	12
5	10	16	3
11	8	2	13

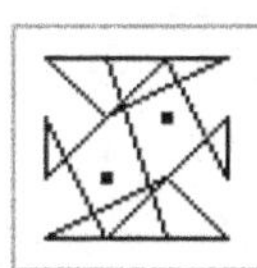

5	12	9	8
15	2	3	14
4	13	16	1
10	7	6	11

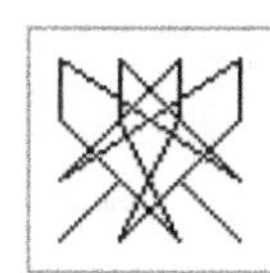

The next magic squares are regular. The left side shows numbers in their original positions while the right side uses lines to connect sequence pairs.

1	8	12	13
15	10	6	3
14	11	7	2
4	5	9	16

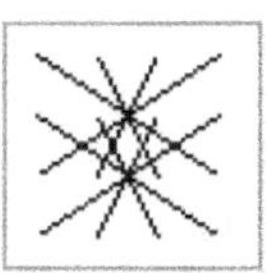

1	8	15	10
12	13	6	3
14	11	4	5
7	2	9	16

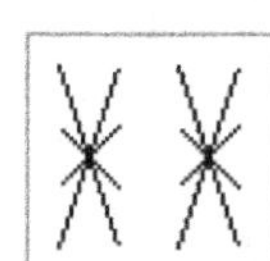

1	10	16	7
15	8	2	9
6	13	11	4
12	3	5	14

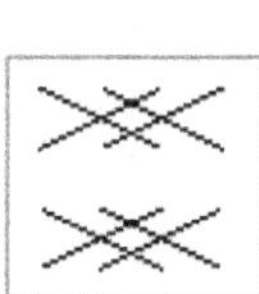

1	15	14	4
12	6	7	9
8	10	11	5
13	3	2	16

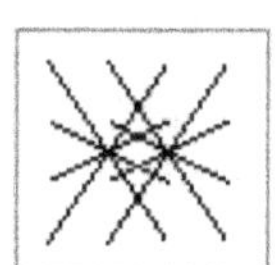

The next magic squares are also regular magic squares. The right-side graphic overlays the even sequential numbers on top of the odd sequential numbers.

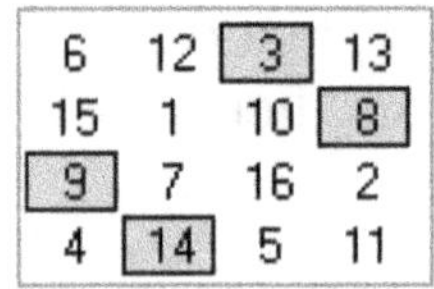

6	12	[3]	13
15	1	10	[8]
[9]	7	16	2
4	[14]	5	11

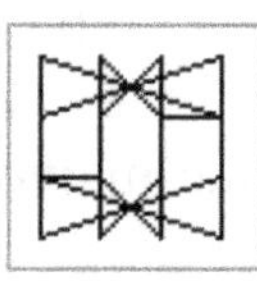

6	15	[3]	10
12	1	13	[8]
[9]	4	16	5
7	[14]	2	11

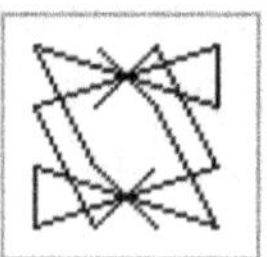

6	15	9	[4]
12	1	[7]	14
3	[10]	16	5
[13]	8	2	11

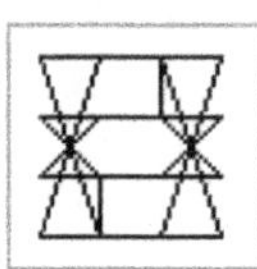

[1]	12	14	7
15	[6]	4	9
8	13	[11]	2
10	3	5	[16]

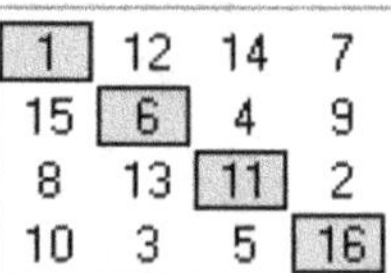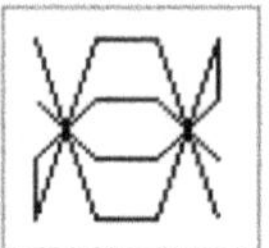

7	14	2	11
12	1	13	[8]
[9]	4	16	5
6	15	3	10

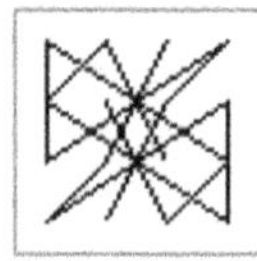

10	8	[3]	13
[5]	11	16	2
15	1	6	[12]
4	[14]	9	7

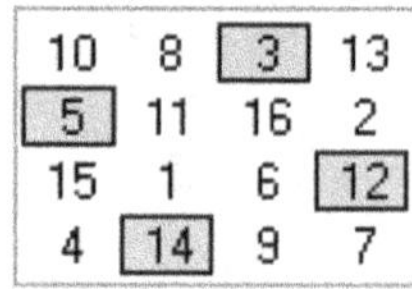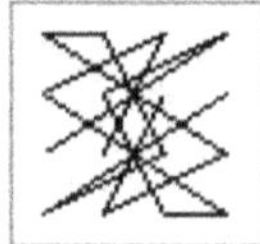

The remaining magic squares in this section are pan-diagonal, highlighting odd numbers on the left side, and connecting numbers sequentially on the right.

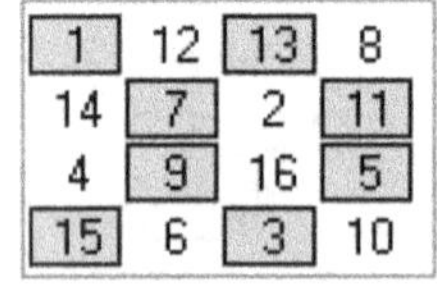

[1]	12	[13]	8
14	[7]	2	[11]
4	[9]	16	[5]
[15]	6	[3]	10

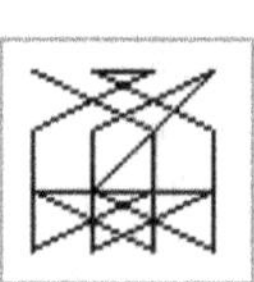

[1]	14	4	[15]
8	[11]	[5]	10
[13]	2	16	[3]
12	[7]	[9]	6

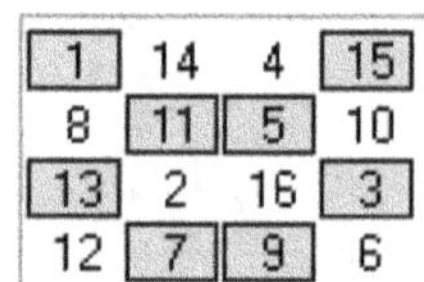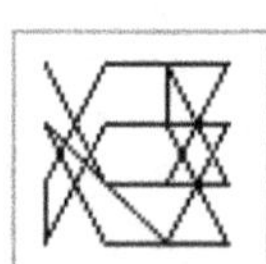

2	[11]	14	[7]
16	[5]	4	[9]
[3]	10	[15]	6
[13]	8	[1]	12

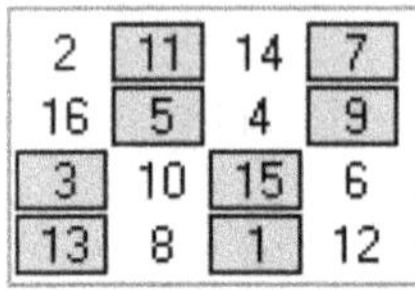

4	[9]	[7]	14
[15]	6	12	[1]
10	[3]	[13]	8
[5]	16	2	[11]

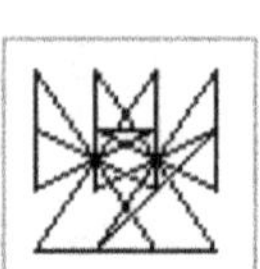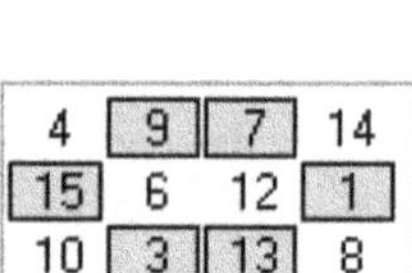

2	[7]	12	[13]
16	[9]	6	[3]
[5]	4	[15]	10
[11]	14	[1]	8

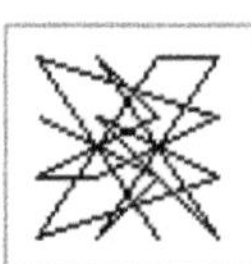

4	14	[7]	[9]
[15]	[1]	12	6
10	8	[13]	[3]
[5]	[11]	2	16

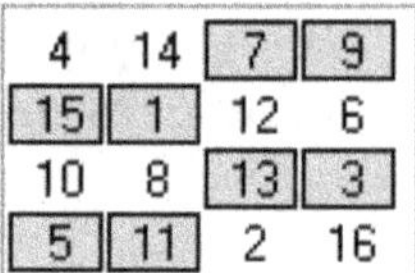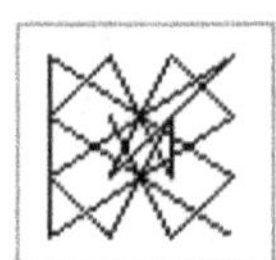

6	[15]	[1]	12
[3]	10	8	[13]
16	[5]	[11]	2
[9]	4	14	[7]

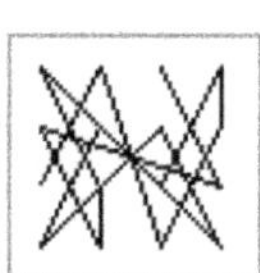

[7]	[9]	4	14
12	6	[15]	[1]
[13]	[3]	10	8
2	16	[5]	[11]

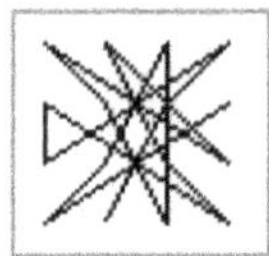

Appendix B – Selected 5th Order Magic Squares

The following are 5th order magic squares. The shaded numbers in the left-side squares are in their original positions. The right-side graphic squares show the movement of the numbers from their original to their current position.

The following four are regular magic squares.

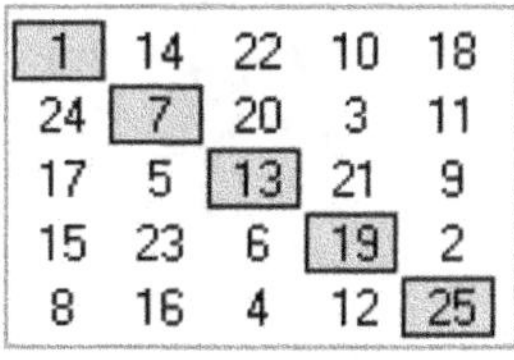 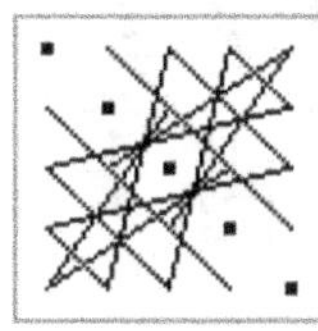 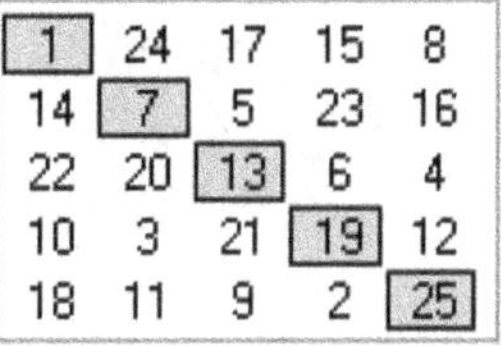 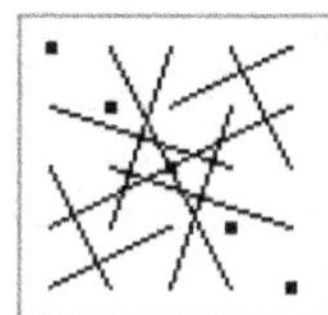

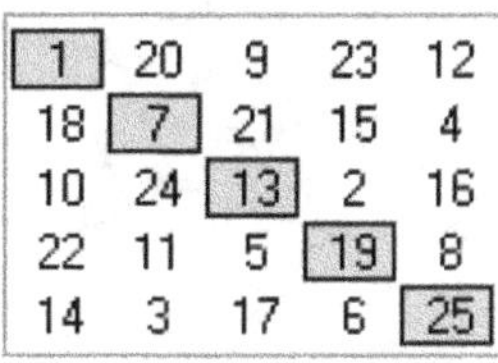 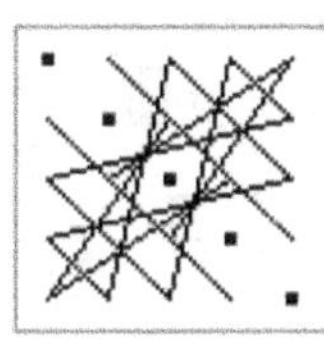 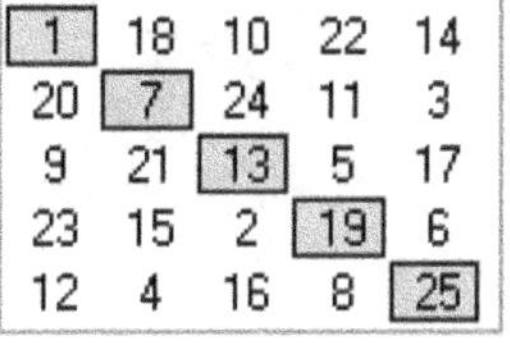 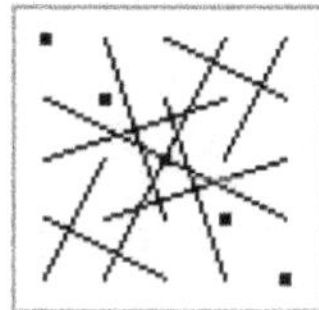

The next four are regular magic squares.

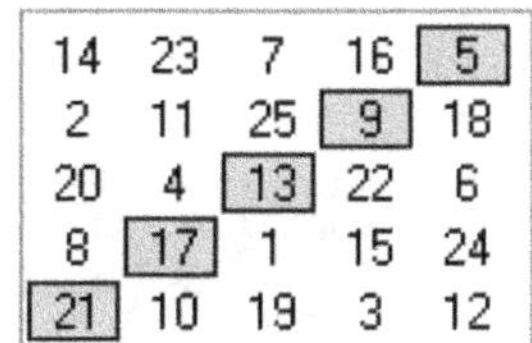 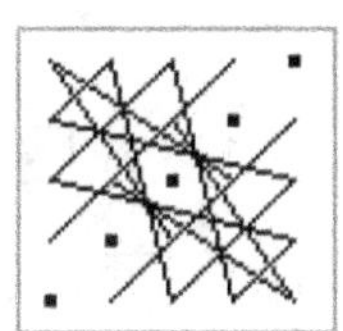 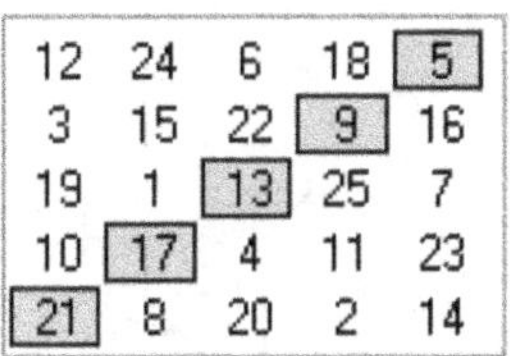 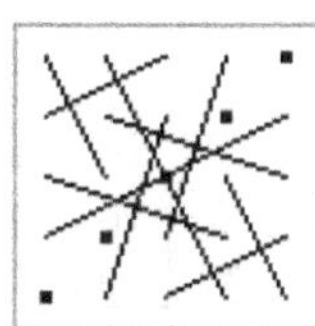

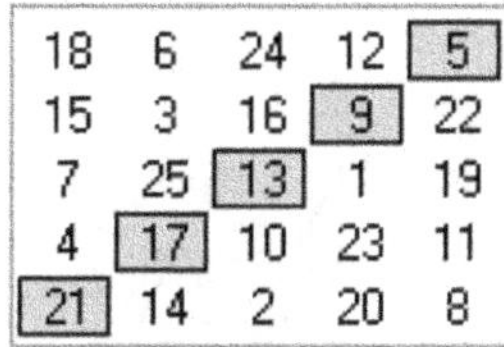 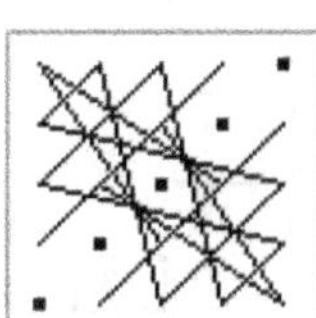 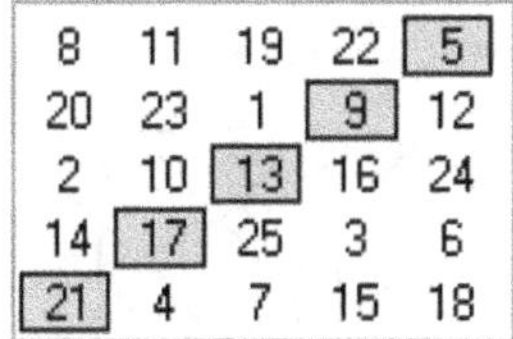 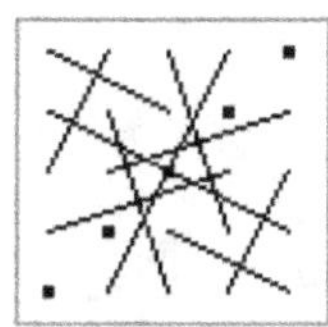

The next two are pan-diagonal magic squares.

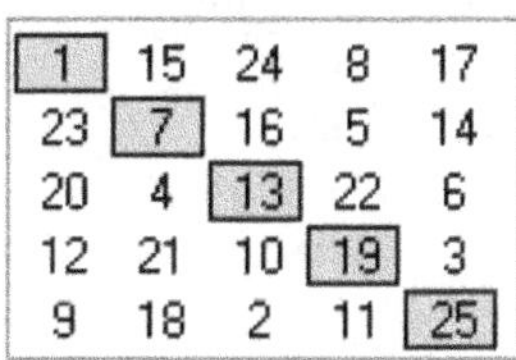 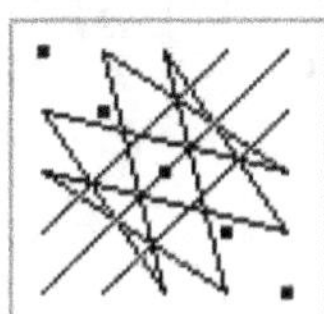 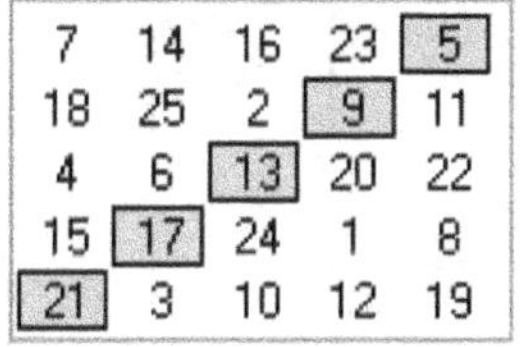 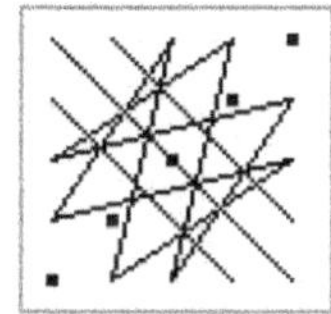

The next magic squares are regular magic squares.

8	**2**	21	20	14
4	23	17	11	**10**
25	19	**13**	7	1
16	15	9	3	22
12	6	5	**24**	18

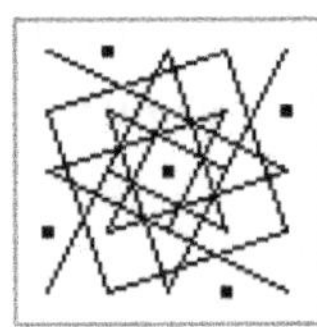

9	**2**	25	18	11
3	21	19	12	**10**
22	20	**13**	6	4
16	14	7	5	23
15	8	1	**24**	17

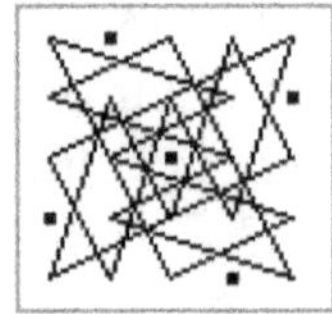

15	**2**	19	6	23
22	14	1	18	**10**
9	21	**13**	5	17
16	8	25	12	4
3	20	7	**24**	11

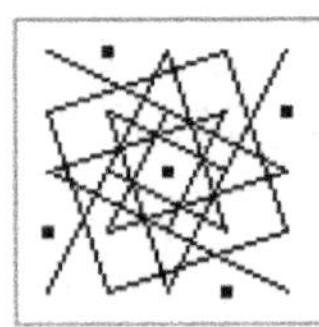

23	10	17	**4**	11
6	18	5	12	24
19	1	**13**	25	7
2	14	21	8	**20**
15	**22**	9	16	3

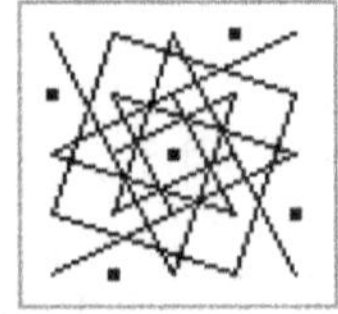

12	16	25	**4**	8
6	15	19	23	2
5	9	**13**	17	21
24	3	7	11	**20**
18	**22**	1	10	14

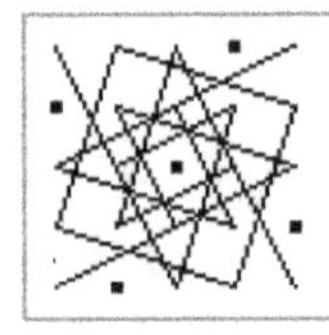

15	18	21	**4**	7
6	14	17	25	3
2	10	**13**	16	24
23	1	9	12	**20**
19	**22**	5	8	11

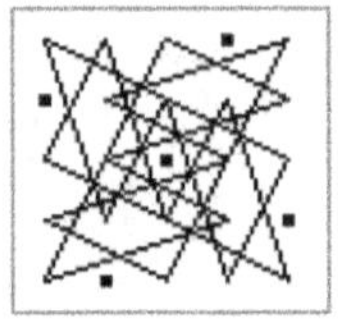

18	15	7	**4**	21
6	3	25	17	14
24	16	**13**	10	2
12	9	1	23	**20**
5	**22**	19	11	8

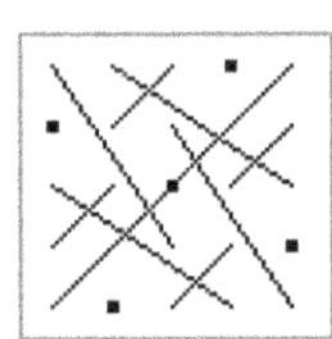

25	8	16	**4**	12
6	19	2	15	23
17	5	**13**	21	9
3	11	24	7	**20**
14	**22**	10	18	1

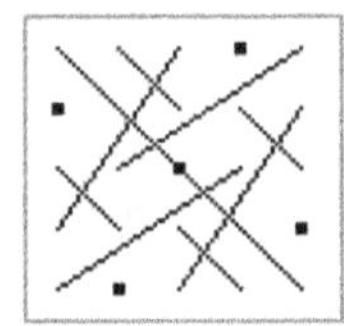

7	3	24	20	11
4	25	16	12	8
21	17	**13**	9	5
18	14	10	1	22
15	6	2	23	19

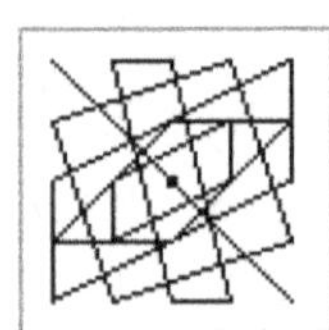

7	18	21	**4**	15
3	14	17	25	6
24	10	**13**	16	2
20	1	9	12	23
11	**22**	5	8	19

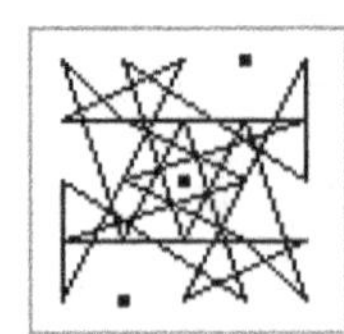

8	20	21	2	14
16	3	9	15	22
25	7	**13**	19	1
4	11	17	23	10
12	24	5	6	18

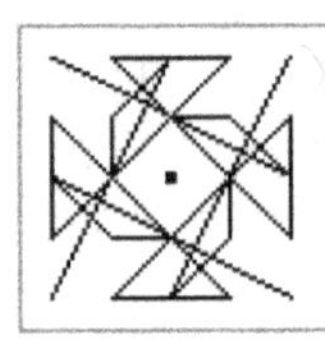

11	**2**	25	18	9
23	14	7	5	16
4	20	**13**	6	22
10	21	19	12	3
17	8	1	**24**	15

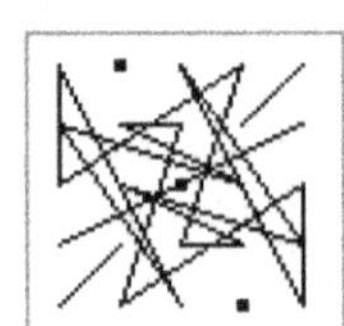

8	16	25	**4**	12
20	3	7	11	24
21	9	**13**	17	5
2	15	19	23	6
14	**22**	1	10	18

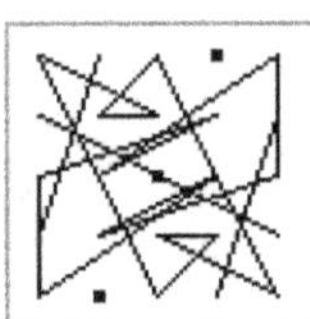

14	22	1	10	18
2	15	19	23	6
21	9	**13**	17	5
20	3	7	11	24
8	16	25	4	12

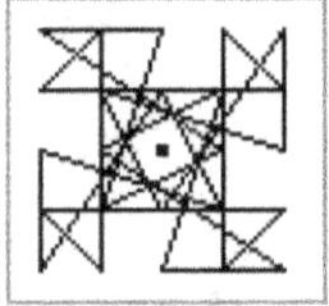

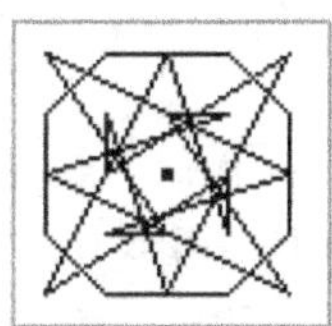

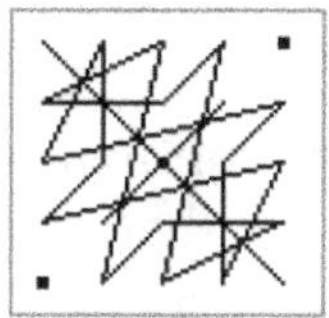

The following regular magic squares shade the odd numbers in the left-side. The right-side draws lines through the numbers in sequence (1 to 2 to 3, etc.).

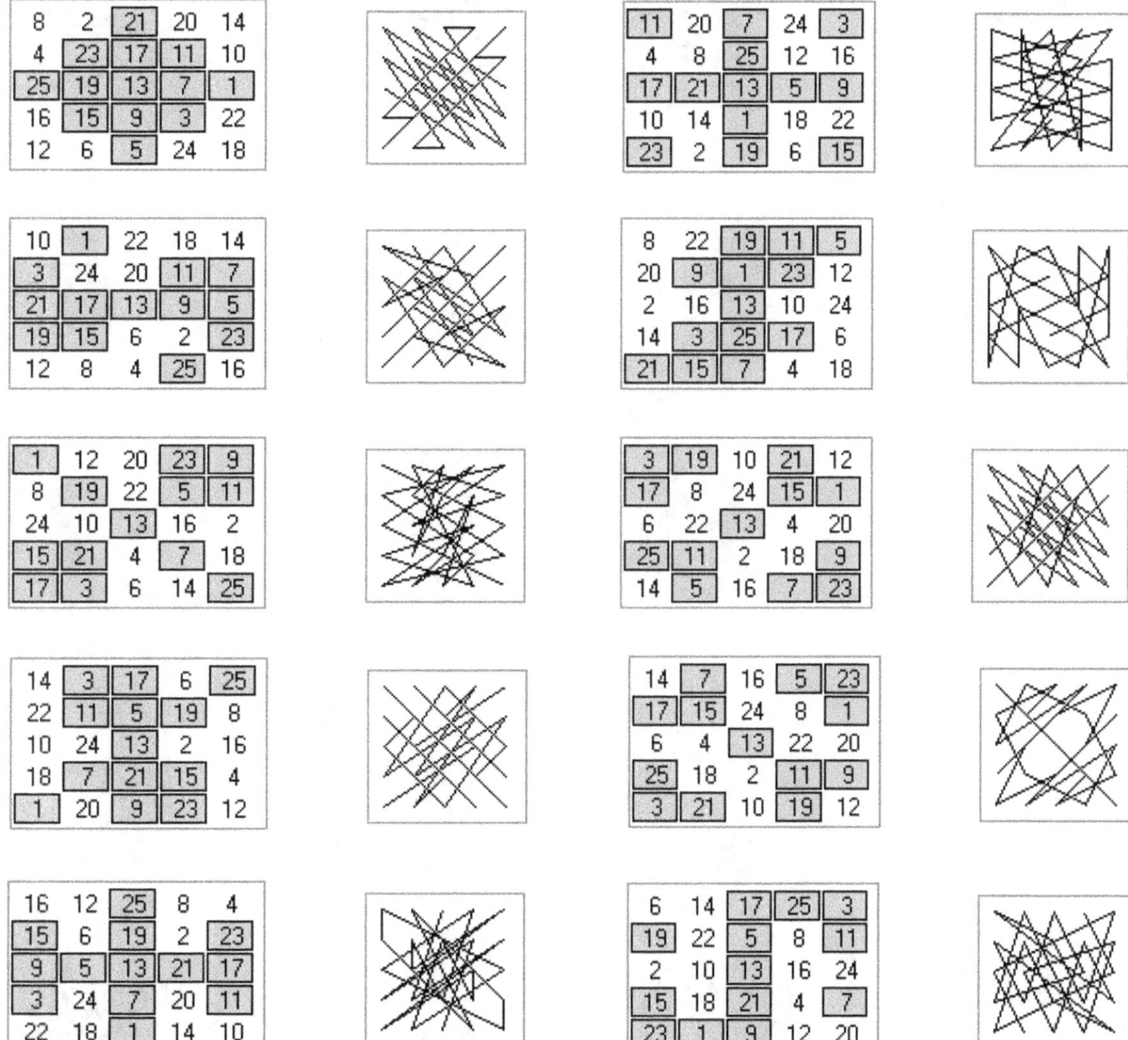

Again the odd numbers are boxed on the left-side. The right-side draws lines through the odd numbers (1 to 3 to 5, etc.), and then the even numbers (2 to 4 to 6, etc.)

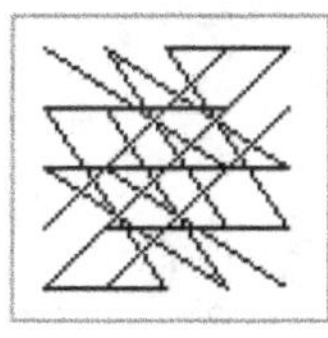

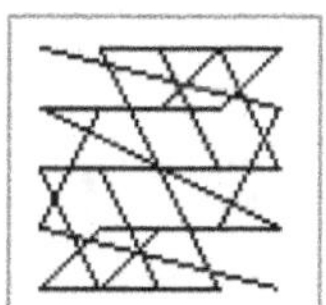

Appendix C – Selected 6th Order Magic Squares

The following are regular 6th order magic squares. The boxed numbers in the left squares are in their original positions. The right square shows the movement of the numbers from their original to their current position. All of these magic squares are unique to each other. Yet, in some cases the differences between them are very subtle.

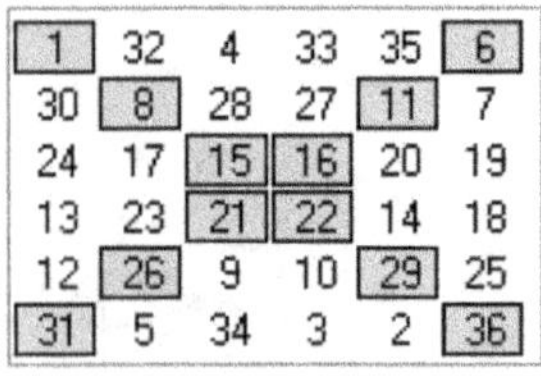

1	32	4	33	35	6
30	8	28	27	11	7
24	17	15	16	20	19
13	23	21	22	14	18
12	26	9	10	29	25
31	5	34	3	2	36

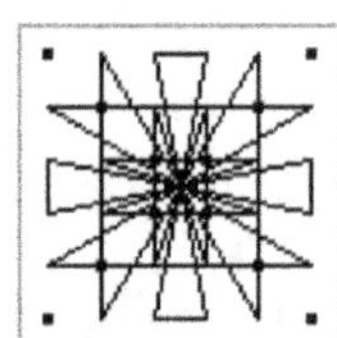

1	35	4	33	32	6
30	8	27	28	11	7
19	17	15	16	20	24
18	23	21	22	14	13
12	26	10	9	29	25
31	2	34	3	5	36

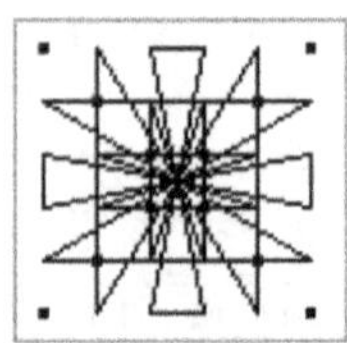

1	35	34	3	32	6
30	8	28	27	11	7
24	23	15	16	14	19
13	17	21	22	20	18
12	26	9	10	29	25
31	2	4	33	5	36

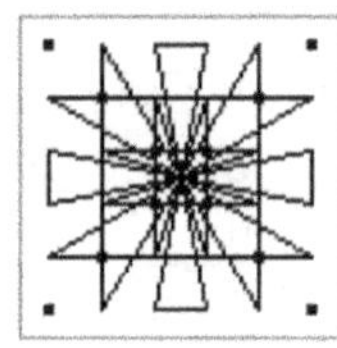

1	35	4	33	32	6
30	8	28	27	11	7
24	23	15	16	14	19
13	17	21	22	20	18
12	26	9	10	29	25
31	2	34	3	5	36

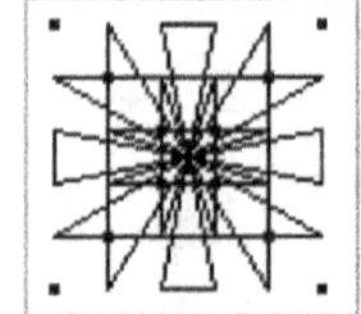

1	12	18	19	30	31
32	8	28	27	11	5
3	17	22	16	20	33
34	23	21	15	14	4
35	26	9	10	29	2
6	25	13	24	7	36

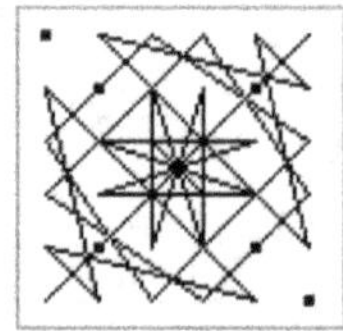

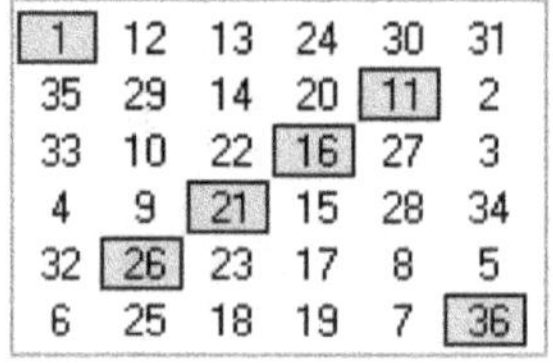

1	12	13	24	30	31
35	29	14	20	11	2
33	10	22	16	27	3
4	9	21	15	28	34
32	26	23	17	8	5
6	25	18	19	7	36

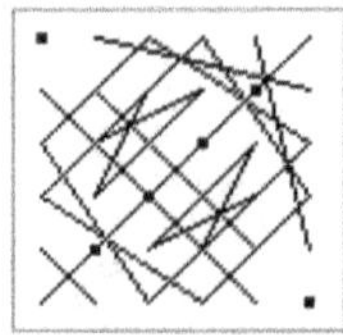

1	30	13	24	12	31
32	8	23	17	26	5
33	27	22	16	10	3
4	28	21	15	9	34
35	11	14	20	29	2
6	7	18	19	25	36

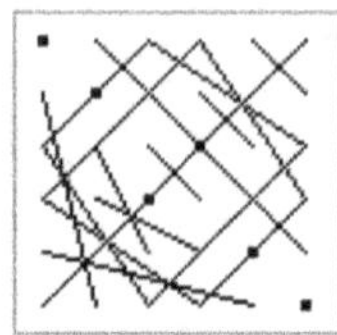

36	2	34	3	5	31
25	8	23	17	26	12
24	28	15	16	9	19
13	27	21	22	10	18
7	11	14	20	29	30
6	35	4	33	32	1

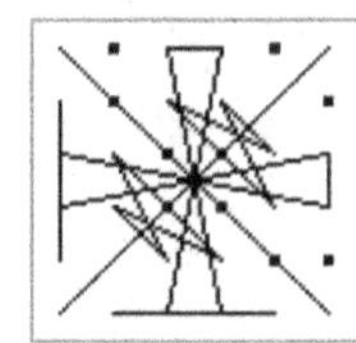

7	8	27	26	22	21
14	13	9	12	31	32
34	36	20	18	1	2
35	33	19	17	4	3
5	6	25	28	24	23
16	15	11	10	29	30

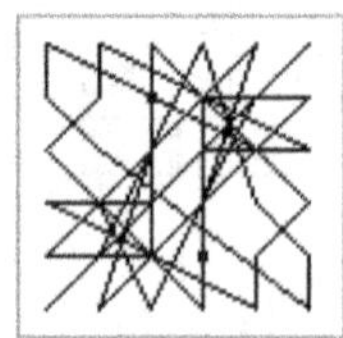

7	25	23	32	21	3
20	11	18	9	31	22
2	29	36	27	13	4
33	24	10	1	8	35
15	6	19	28	26	17
34	16	5	14	12	30

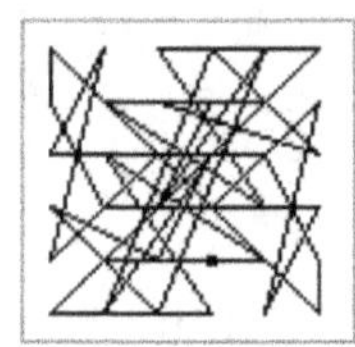

1	34	33	32	9	2
29	11	18	20	25	8
30	22	23	13	16	7
6	17	12	26	19	31
10	24	21	15	14	27
35	3	4	5	28	36

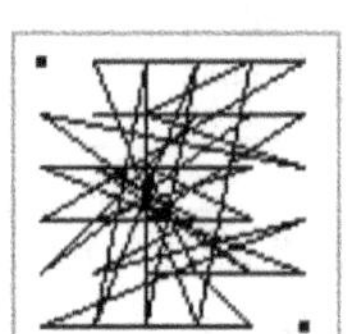

5	35	7	16	34	14
28	17	26	10	18	12
6	33	8	15	36	13
24	4	22	29	1	31
25	19	27	11	20	9
23	3	21	30	2	32

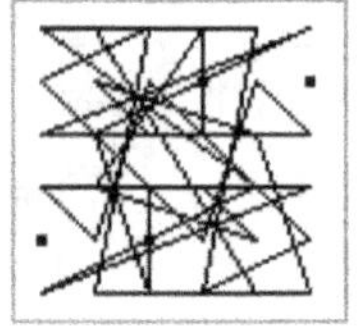

The graphic with the following magic squares show lines drawn in sequence (from 1, to 2, to 3, etc.)

6	36	8	15	33	13
25	20	27	11	19	9
23	2	21	30	3	32
5	34	7	16	35	14
28	18	26	10	17	12
24	1	22	29	4	31

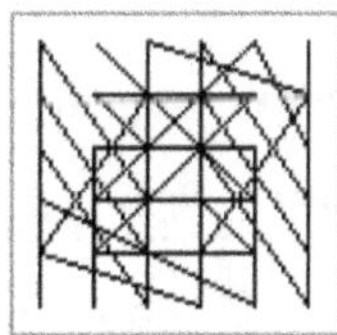

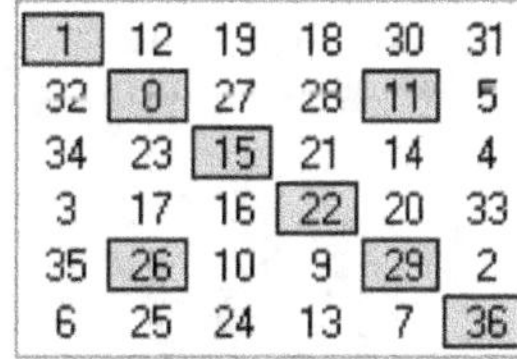

1	12	19	18	30	31
32	0	27	28	11	5
34	23	15	21	14	4
3	17	16	22	20	33
35	26	10	9	29	2
6	25	24	13	7	36

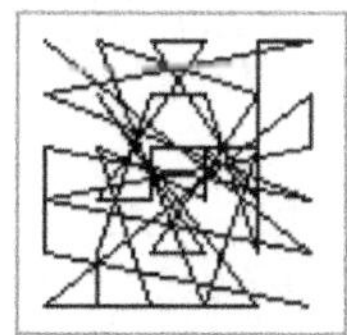

8	6	33	36	13	15
21	23	2	3	32	30
27	25	20	19	9	11
26	28	18	17	12	10
7	5	34	35	14	16
22	24	4	1	31	29

8	7	27	28	30	11
5	36	3	34	31	2
23	18	22	21	13	14
17	19	16	15	24	20
32	6	33	4	1	35
26	25	10	9	12	29

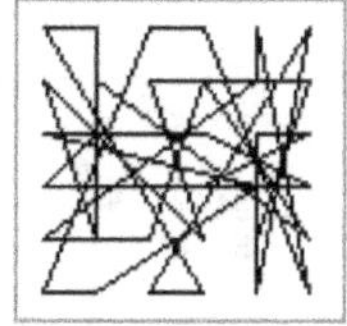

The following magic squares show numbers in their original positions on the left, and lines connecting the sequence pairs on the right.

15	16	11	10	30	29
13	14	9	12	32	31
33	35	19	17	3	4
36	34	20	18	2	1
6	5	25	28	23	24
8	7	27	26	21	22

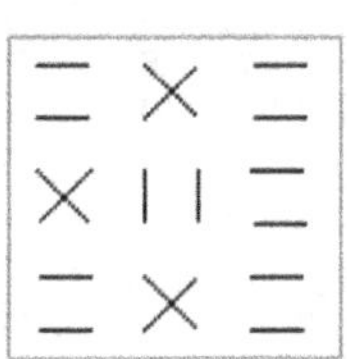

22	21	27	26	7	8
31	32	9	12	14	13
1	2	20	18	34	36
4	3	19	17	35	33
24	23	25	28	5	6
29	30	11	10	16	15

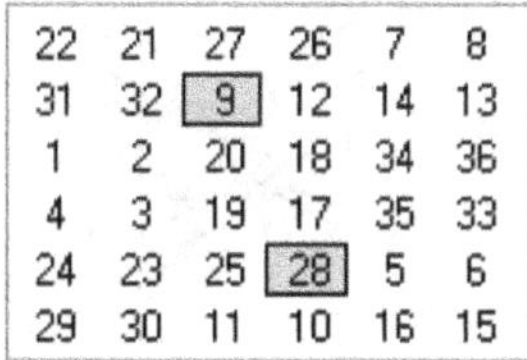

8	26	7	21	27	22
36	17	35	3	19	1
6	28	5	23	25	24
13	12	14	32	9	31
33	18	34	2	20	4
15	10	16	30	11	29

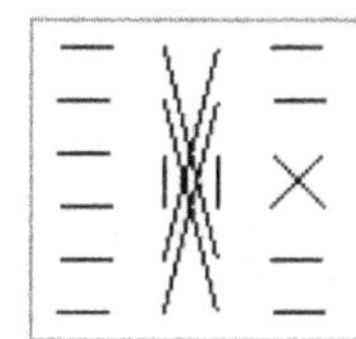

8	28	12	25	27	11
23	15	24	19	16	14
32	34	1	6	3	35
5	4	31	36	33	2
17	21	13	18	22	20
26	9	30	7	10	29

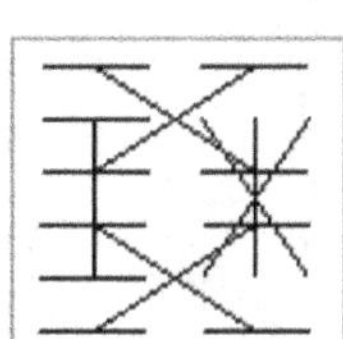

20	25	11	27	9	19
2	23	30	21	32	3
4	24	29	22	31	1
33	6	15	8	13	36
34	5	16	7	14	35
18	28	10	26	12	17

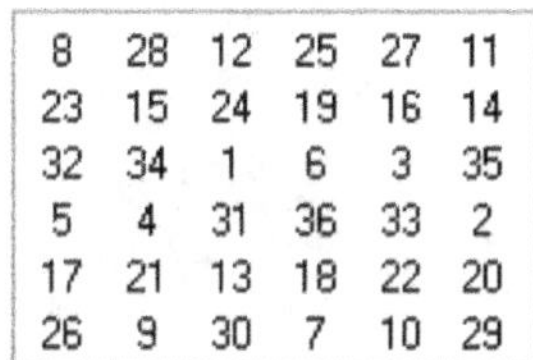

32	12	31	13	9	14
3	17	1	36	19	35
30	10	29	15	11	16
21	26	22	8	27	7
2	18	4	33	20	34
23	28	24	6	25	5

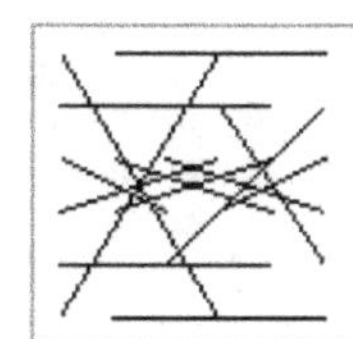

20	25	27	11	9	19
34	5	7	16	14	35
36	6	8	15	13	33
1	24	22	29	31	4
2	23	21	30	32	3
18	28	26	10	12	17

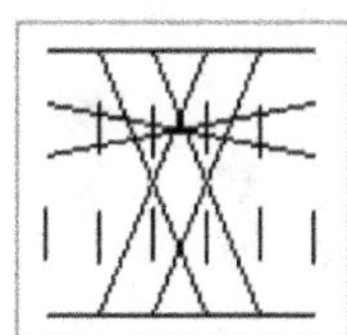

16	35	14	5	34	7
10	17	12	28	18	26
29	4	31	24	1	22
15	33	13	6	36	8
11	19	9	25	20	27
30	3	32	23	2	21

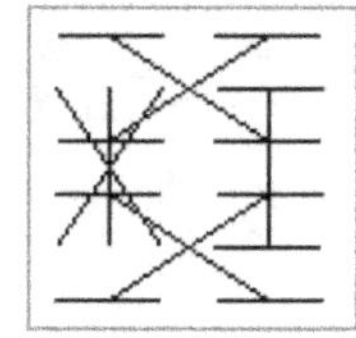

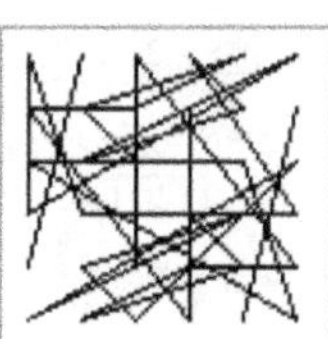

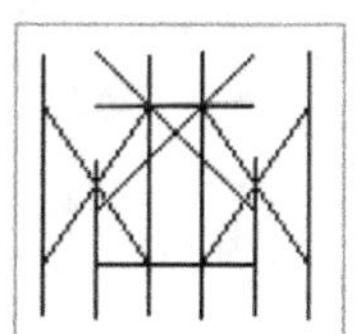

213

Appendix D – Selected 7th Order Magic Squares

The following are regular 7th order magic squares. The boxed numbers in the left squares are in their original positions. The right square shows the movement of the numbers from their original to their current position.

[1]	34	11	37	21	47	24
32	[9]	42	19	45	22	6
14	40	[17]	43	27	4	30
38	15	48	[25]	2	35	12
20	46	23	7	[33]	10	36
44	28	5	31	8	[41]	18
26	3	29	13	39	16	[49]

[1]	40	23	13	45	35	18
26	[9]	48	31	21	4	36
44	34	[17]	7	39	22	12
20	3	42	[25]	8	47	30
38	28	11	43	[33]	16	6
14	46	29	19	2	[41]	24
32	15	5	37	27	10	[49]

18	[2]	10	43	34	42	26
6	39	47	31	15	23	[14]
12	45	4	37	28	29	20
49	33	41	[25]	9	17	1
30	21	22	13	46	5	38
[36]	27	35	19	3	11	44
24	8	16	7	40	[48]	32

18	[2]	34	43	10	42	26
6	39	15	31	47	23	[14]
30	21	46	13	22	5	38
49	33	9	[25]	41	17	1
12	45	28	37	4	29	20
[36]	27	3	19	35	11	44
24	8	40	7	16	[48]	32

11	35	[3]	36	27	44	19
22	46	21	5	38	[13]	30
40	8	32	16	7	24	48
49	17	41	[25]	9	33	1
2	26	43	34	18	42	10
20	[37]	12	45	29	4	28
31	6	23	14	[47]	15	39

23	42	29	17	[5]	48	11
14	26	20	1	38	32	44
[15]	34	28	9	46	40	3
31	43	37	[25]	13	7	19
47	10	4	41	22	16	[35]
6	18	12	49	30	24	36
39	2	[45]	33	21	8	27

24	36	12	49	30	[6]	18
[8]	27	45	33	21	39	2
40	3	28	9	46	15	34
7	19	37	[25]	13	31	43
16	35	4	41	22	47	10
48	11	29	17	5	23	[42]
32	[44]	20	1	38	14	26

24	36	30	49	12	[6]	18
[8]	27	21	33	45	39	2
16	35	22	41	4	47	10
7	19	13	[25]	37	31	43
40	3	46	9	28	15	34
48	11	5	17	29	23	[42]
32	[44]	38	1	20	14	26

39	[2]	21	33	45	8	27
6	18	30	49	[12]	24	36
[15]	34	46	9	28	40	3
31	43	13	[25]	37	7	19
47	10	22	41	4	16	[35]
14	26	[38]	1	20	32	44
23	42	5	17	29	[48]	11

18	14	[3]	48	37	33	22
[8]	4	49	38	34	23	19
5	43	39	35	24	[20]	9
44	40	29	[25]	21	10	6
41	[30]	26	15	11	7	45
31	27	16	12	1	46	[42]
28	17	13	2	[47]	36	32

37	35	26	17	8	[6]	46
28	19	[10]	1	48	39	30
12	3	43	41	32	23	[21]
45	36	34	[25]	16	14	5
[29]	27	18	9	7	47	38
20	11	2	49	[40]	31	22
4	[44]	42	33	24	15	13

39	47	14	31	6	15	23
45	4	20	37	[12]	28	29
8	[16]	32	7	24	40	48
33	41	1	[25]	49	9	17
2	10	26	43	18	[34]	42
21	22	[38]	13	30	46	5
27	35	44	19	36	3	11

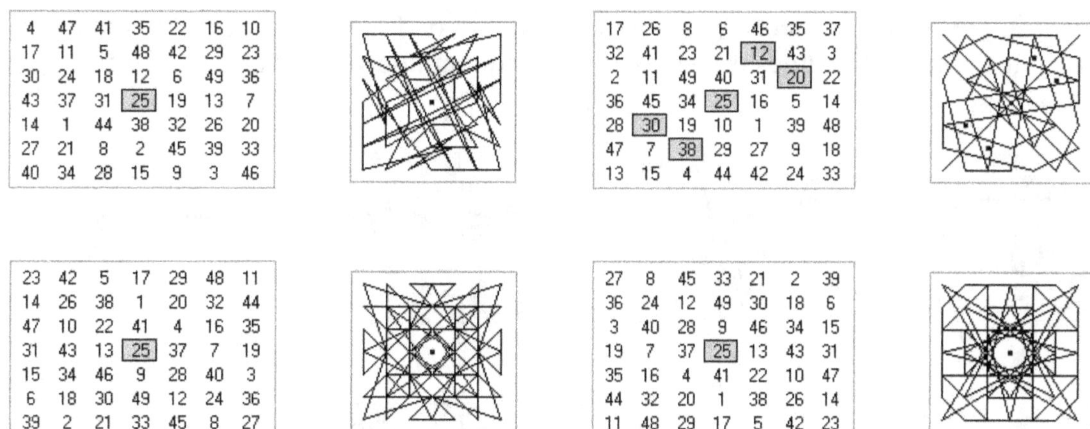

4	47	41	35	22	16	10
17	11	5	48	42	29	23
30	24	18	12	6	49	36
43	37	31	[25]	19	13	7
14	1	44	38	32	26	20
27	21	8	2	45	39	33
40	34	28	15	9	3	46

17	26	8	6	46	35	37
32	41	23	21	[12]	43	3
2	11	49	40	31	[20]	22
36	45	34	[25]	16	5	14
28	[30]	19	10	1	39	48
47	7	[38]	29	27	9	18
13	15	4	44	42	24	23

23	42	5	17	29	48	11
14	26	38	1	20	32	44
47	10	22	41	4	16	35
31	43	13	[25]	37	7	19
15	34	46	9	28	40	3
6	18	30	49	12	24	36
39	2	21	33	45	8	27

27	8	45	33	21	2	39
36	24	12	49	30	18	6
3	40	28	9	46	34	15
19	7	37	[25]	13	43	31
35	16	4	41	22	10	47
44	32	20	1	38	26	14
11	48	29	17	5	42	23

The following regular 7th order magic squares highlight numbers in their original positions in the left square, and the right square shows lines drawn in regular sequence (from 1 to 2 to 3 . . .)

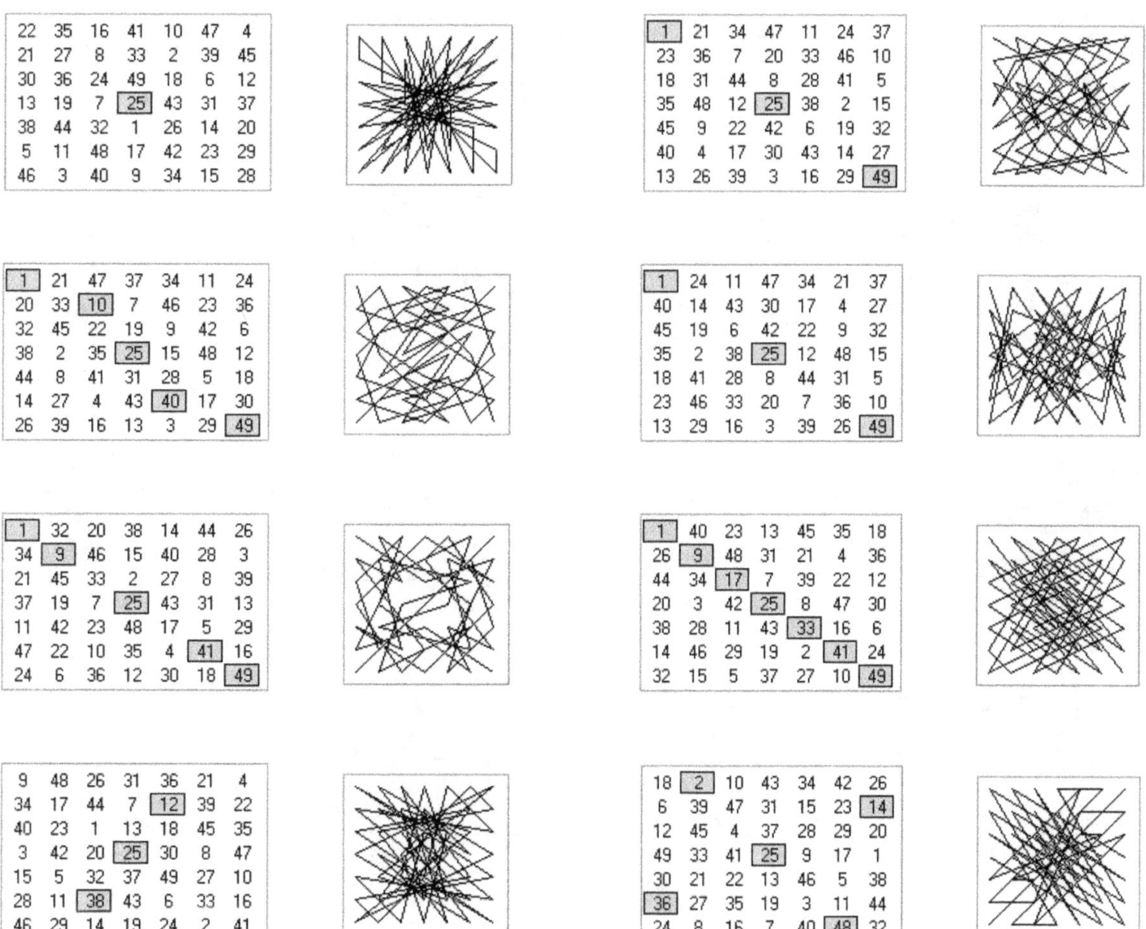

22	35	16	41	10	47	4
21	27	8	33	2	39	45
30	36	24	49	18	6	12
13	19	7	[25]	43	31	37
38	44	32	1	26	14	20
5	11	48	17	42	23	29
46	3	40	9	34	15	28

[1]	21	34	47	11	24	37
23	36	7	20	33	46	10
18	31	44	8	28	41	5
35	48	12	[25]	38	2	15
45	9	22	42	6	19	32
40	4	17	30	43	14	27
13	26	39	3	16	29	[49]

[1]	21	47	37	34	11	24
20	33	[10]	7	46	23	36
32	45	22	19	9	42	6
38	2	35	[25]	15	48	12
44	8	41	31	28	5	18
14	27	4	43	[40]	17	30
26	39	16	13	3	29	[49]

[1]	24	11	47	34	21	37
40	14	43	30	17	4	27
45	19	6	42	22	9	32
35	2	38	[25]	12	48	15
18	41	28	8	44	31	5
23	46	33	20	7	36	10
13	29	16	3	39	26	[49]

[1]	32	20	38	14	44	26
34	[9]	46	15	40	28	3
21	45	33	2	27	8	39
37	19	7	[25]	43	31	13
11	42	23	48	17	5	29
47	22	10	35	4	[41]	16
24	6	36	12	30	18	[49]

[1]	40	23	13	45	35	18
26	[9]	48	31	21	4	36
44	34	[17]	7	39	22	12
20	3	42	[25]	8	47	30
38	28	11	43	[33]	16	6
14	46	29	19	2	[41]	24
32	15	5	37	27	10	[49]

9	48	26	31	36	21	4
34	17	44	7	[12]	39	22
40	23	1	13	18	45	35
3	42	20	[25]	30	8	47
15	5	32	37	49	27	10
28	11	[38]	43	6	33	16
46	29	14	19	24	2	41

18	[2]	10	43	34	42	26
6	39	47	31	15	23	[14]
12	45	4	37	28	29	20
49	33	41	[25]	9	17	1
30	21	22	13	46	5	38
[36]	27	35	19	3	11	44
24	8	16	7	40	[48]	32

215

The Methods, Marvels, and Madness of Magic Squares

The following regular 7th order magic squares highlight odd numbers in the left square, and the right square shows lines drawn in an odd sequential order (from 1 to 3 to 5, etc.)

1	40	23	13	45	35	18
26	9	48	31	21	4	36
44	34	17	7	39	22	12
20	3	42	25	8	47	30
38	28	11	43	33	16	6
14	46	29	19	2	41	24
32	15	5	37	27	10	49

7	38	27	9	47	29	18
24	13	44	33	15	4	42
48	30	19	1	39	28	10
16	5	36	25	14	45	34
40	22	11	49	31	20	2
8	46	35	17	6	37	26
32	21	3	41	23	12	43

4	29	12	37	20	45	28
35	11	36	19	44	27	3
10	42	18	43	26	2	34
41	17	49	25	1	33	9
16	48	24	7	32	8	40
47	23	6	31	14	39	15
22	5	30	13	38	21	46

4	33	13	42	15	44	24
31	11	40	20	49	22	2
9	38	18	47	27	7	29
36	16	45	25	5	34	14
21	43	23	3	32	12	41
48	28	1	30	10	39	19
26	6	35	8	37	17	46

9	32	6	42	22	45	19
26	49	16	3	39	13	29
36	10	33	20	7	23	46
48	15	38	25	12	35	2
4	27	43	30	17	40	14
21	37	11	47	34	1	24
31	5	28	8	44	18	41

9	38	18	29	47	27	7
20	49	22	40	2	31	11
24	4	33	44	13	42	15
5	34	14	25	36	16	45
35	8	37	6	17	46	26
39	19	48	10	28	1	30
43	23	3	21	32	12	41

11	3	44	19	36	35	27
5	46	38	13	30	22	21
48	40	32	7	24	16	8
17	9	1	25	49	41	33
42	34	26	43	18	10	2
29	28	20	37	12	4	45
23	15	14	31	6	47	39

18	10	2	43	42	34	26
12	4	45	37	29	28	20
6	47	39	31	23	15	14
49	41	33	25	17	9	1
36	35	27	19	11	3	44
30	22	21	13	5	46	38
24	16	8	7	48	40	32

18	36	30	49	12	6	24
42	11	5	17	29	23	48
34	3	46	9	28	15	40
43	19	13	25	37	31	7
10	35	22	41	4	47	16
2	27	21	33	45	39	8
26	44	38	1	20	14	32

4	29	12	37	21	45	27
34	11	36	19	44	28	3
10	41	18	43	26	2	35
42	17	48	25	1	33	9
16	49	24	6	32	8	40
47	23	7	31	13	39	15
22	5	30	14	38	20	46

18	40	13	45	35	1	23
38	11	33	16	6	28	43
9	31	4	36	26	48	21
47	20	42	25	8	30	3
29	2	24	14	46	19	41
7	22	44	34	17	39	12
27	49	15	5	37	10	32

18	38	47	9	27	29	7
2	22	31	49	11	20	40
42	13	15	33	44	4	24
34	5	14	25	36	45	16
26	46	6	17	35	37	8
10	30	39	1	19	28	48
43	21	23	41	3	12	32

These are 7th order pan-diagonal magic squares. The boxed numbers are in their original positions. The right square shows the movement of the numbers from their original position to their current position.

1	21	34	47	11	24	37
45	9	22	42	6	19	32
40	4	17	30	43	14	27
35	48	12	25	38	2	15
23	36	7	20	33	46	10
18	31	44	8	28	41	5
13	26	39	3	16	29	49

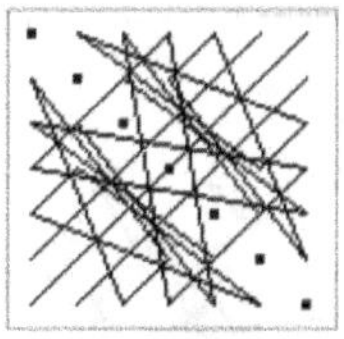

1	38	26	14	44	32	20
28	9	46	34	15	3	40
48	29	17	5	42	23	11
19	7	37	25	13	43	31
39	27	8	45	33	21	2
10	47	35	16	4	41	22
30	18	6	36	24	12	49

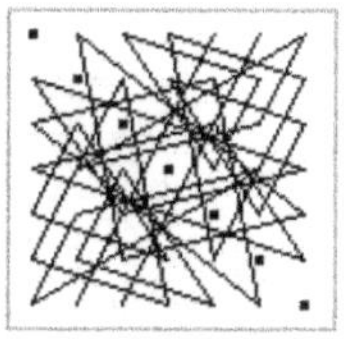

1	39	28	10	48	30	19
27	9	47	29	18	7	38
46	35	17	6	37	26	8
16	5	36	25	14	45	34
42	24	13	44	33	15	4
12	43	32	21	3	41	23
31	20	2	40	22	11	49

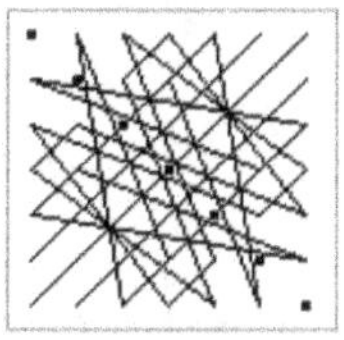

1	48	39	30	28	19	10
18	9	7	47	38	29	27
35	26	17	8	6	46	37
45	36	34	25	16	14	5
13	4	44	42	33	24	15
23	21	12	3	43	41	32
40	31	22	20	11	2	49

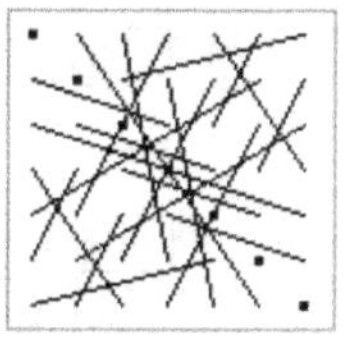

34	2	26	43	18	42	10
28	45	20	37	12	29	4
15	39	14	31	6	23	47
9	33	1	25	49	17	41
3	27	44	19	36	11	35
46	21	38	13	30	5	22
40	8	32	7	24	48	16

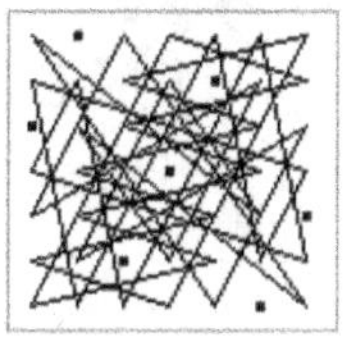

40	46	3	9	15	28	34
8	21	27	33	39	45	2
32	38	44	1	14	20	26
7	13	19	25	31	37	43
24	30	36	49	6	12	18
48	5	11	17	23	29	42
16	22	35	41	47	4	10

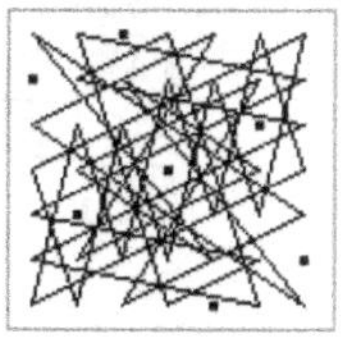

30	22	21	13	5	46	38
6	47	39	31	23	15	14
24	16	8	7	48	40	32
49	41	33	25	17	9	1
18	10	2	43	42	34	26
36	35	27	19	11	3	44
12	4	45	37	29	28	20

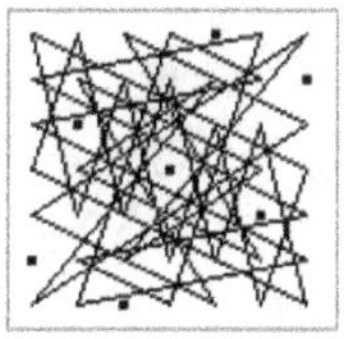

12	36	18	49	24	6	30
4	35	10	41	16	47	22
45	27	2	33	8	39	21
37	19	43	25	7	31	13
29	11	42	17	48	23	5
28	3	34	9	40	15	46
20	44	26	1	32	14	38

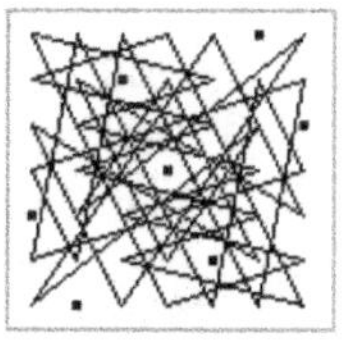

6	47	39	31	23	15	14
18	10	2	43	42	34	26
30	22	21	13	5	46	38
49	41	33	25	17	9	1
12	4	45	37	29	28	20
24	16	8	7	48	40	32
36	35	27	19	11	3	44

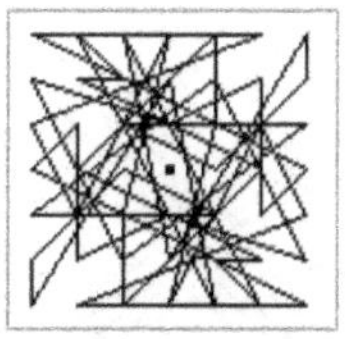

9	6	45	42	32	22	19
36	33	23	20	10	7	46
21	11	1	47	37	34	24
48	38	35	25	15	12	2
26	16	13	3	49	39	29
4	43	40	30	27	17	14
31	28	18	8	5	44	41

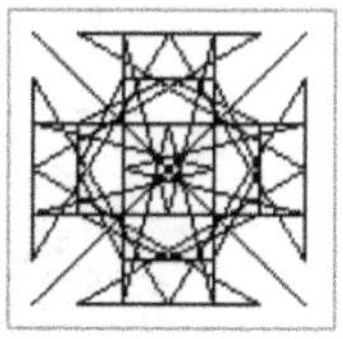

17	34	44	12	22	39	7
40	1	18	35	45	13	23
14	24	41	2	19	29	46
30	47	8	25	42	3	20
4	21	31	48	9	26	36
27	37	5	15	32	49	10
43	11	28	38	6	16	33

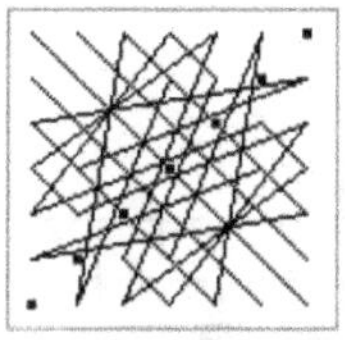

9	18	27	29	38	47	7
24	33	42	44	4	13	15
39	48	1	10	19	28	30
5	14	16	25	34	36	45
20	22	31	40	49	2	11
35	37	46	6	8	17	26
43	3	12	21	23	32	41

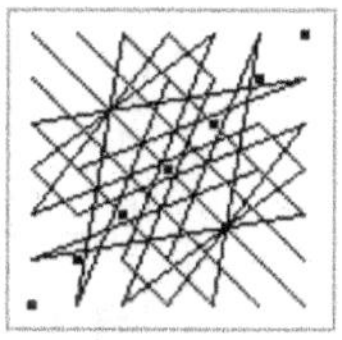

Appendix E – Selected 8th Order Magic Squares

The following are regular 8th order magic squares. The boxed numbers are numbers in their original positions while the right square shows the movement of the numbers from their original to their current positions.

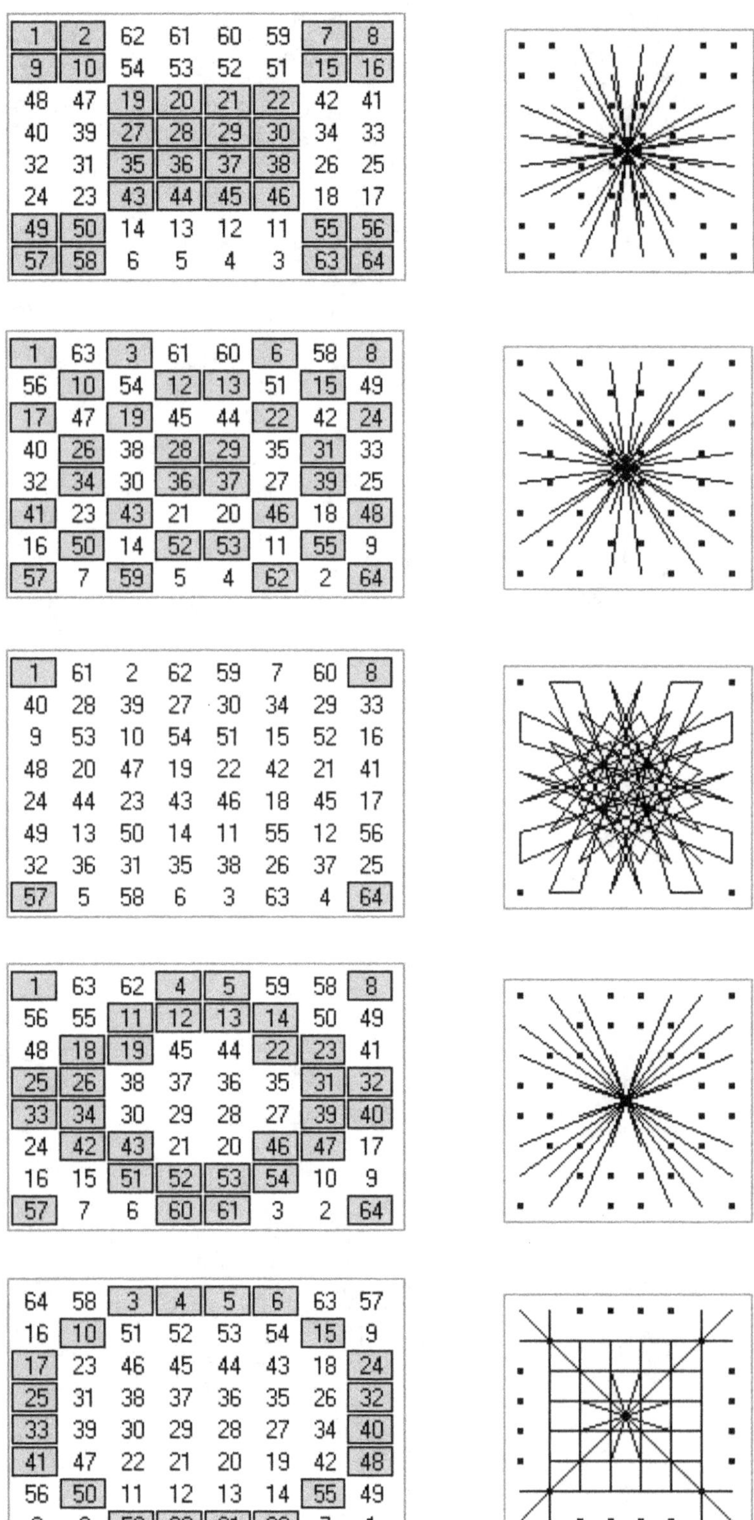

1	2	62	61	60	59	7	8
9	10	54	53	52	51	15	16
48	47	19	20	21	22	42	41
40	39	27	28	29	30	34	33
32	31	35	36	37	38	26	25
24	23	43	44	45	46	18	17
49	50	14	13	12	11	55	56
57	58	6	5	4	3	63	64

1	63	3	61	60	6	58	8
56	10	54	12	13	51	15	49
17	47	19	45	44	22	42	24
40	26	38	28	29	35	31	33
32	34	30	36	37	27	39	25
41	23	43	21	20	46	18	48
16	50	14	52	53	11	55	9
57	7	59	5	4	62	2	64

1	61	2	62	59	7	60	8
40	28	39	27	30	34	29	33
9	53	10	54	51	15	52	16
48	20	47	19	22	42	21	41
24	44	23	43	46	18	45	17
49	13	50	14	11	55	12	56
32	36	31	35	38	26	37	25
57	5	58	6	3	63	4	64

1	63	62	4	5	59	58	8
56	55	11	12	13	14	50	49
48	18	19	45	44	22	23	41
25	26	38	37	36	35	31	32
33	34	30	29	28	27	39	40
24	42	43	21	20	46	47	17
16	15	51	52	53	54	10	9
57	7	6	60	61	3	2	64

64	58	3	4	5	6	63	57
16	10	51	52	53	54	15	9
17	23	46	45	44	43	18	24
25	31	38	37	36	35	26	32
33	39	30	29	28	27	34	40
41	47	22	21	20	19	42	48
56	50	11	12	13	14	55	49
8	2	59	60	61	62	7	1

1	9	53	61	60	52	16	8
2	10	54	59	62	51	15	7
39	47	19	30	27	22	42	34
40	24	44	28	29	45	17	33
32	48	20	36	37	21	41	25
31	23	43	38	35	46	18	26
58	50	14	3	6	11	55	63
57	49	13	5	4	12	56	64

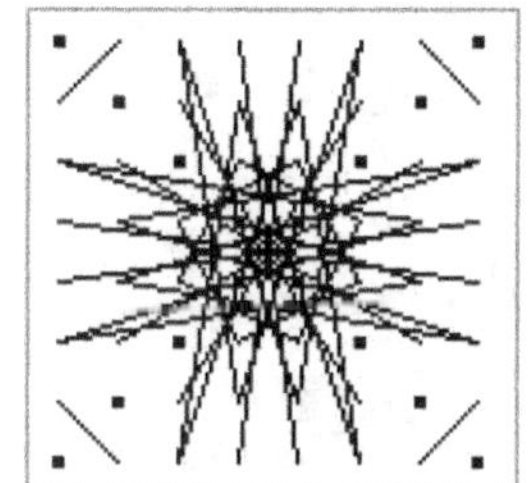

1	63	24	47	42	17	58	8
56	10	25	34	39	32	15	49
59	4	46	20	21	43	5	62
54	13	27	37	36	30	12	51
14	53	35	29	28	38	52	11
3	60	22	44	45	19	61	6
16	50	33	26	31	40	55	9
57	7	48	23	18	41	2	64

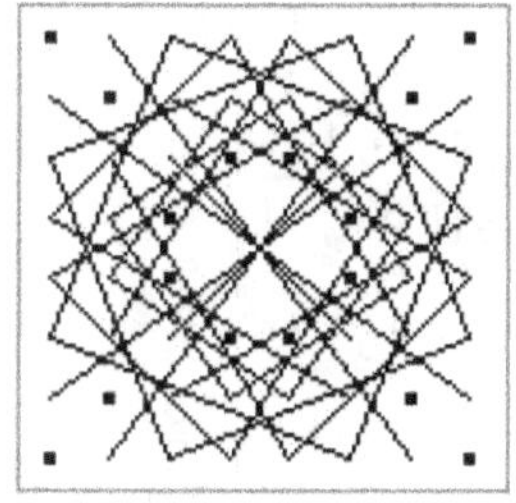

1	63	3	61	14	52	16	50
40	10	38	12	59	21	57	23
17	47	19	45	30	36	32	34
56	26	54	28	43	5	41	7
58	24	60	22	37	11	39	9
31	33	29	35	20	46	18	48
42	8	44	6	53	27	55	25
15	49	13	51	4	62	2	64

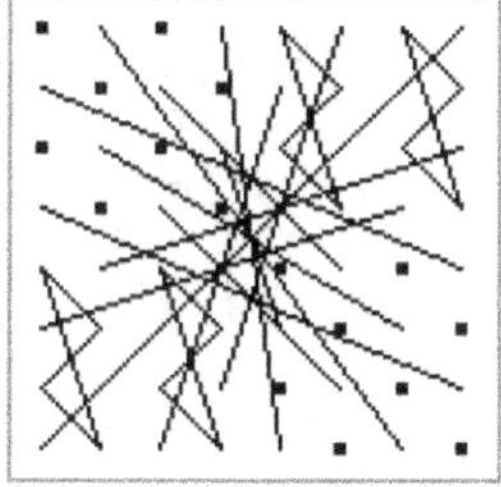

1	9	48	40	32	24	49	57
2	10	47	39	31	23	50	58
62	54	19	27	35	43	14	6
61	53	20	28	36	44	13	5
60	52	21	29	37	45	12	4
59	51	22	30	38	46	11	3
7	15	42	34	26	18	55	63
8	16	41	33	25	17	56	64

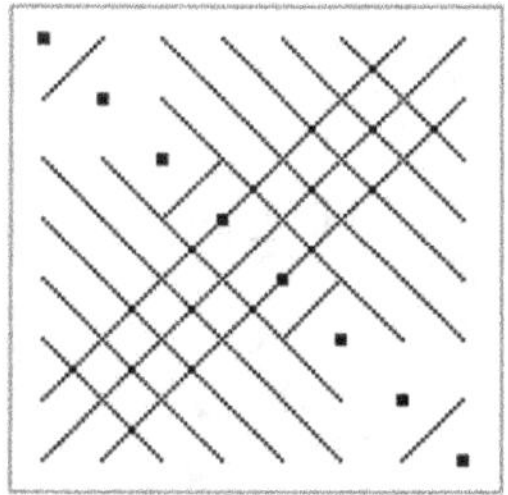

1	9	52	61	32	48	21	36
2	10	54	59	39	23	43	30
31	47	19	38	58	50	14	3
40	24	45	28	57	49	12	5
60	53	16	8	37	20	41	25
62	51	15	7	27	46	18	34
35	22	42	26	6	11	55	63
29	44	17	33	4	13	56	64

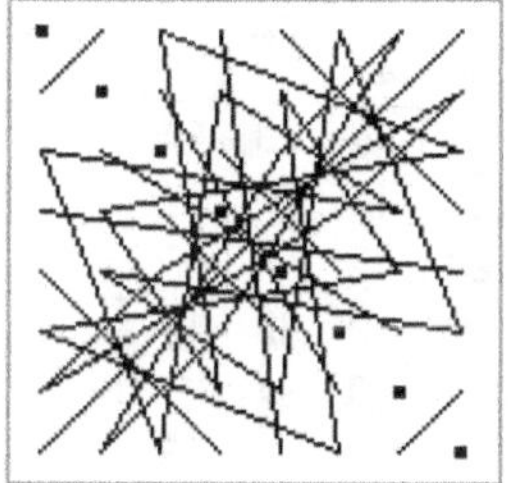

1	20	62	47	5	24	58	43
27	10	40	53	31	14	36	49
48	61	19	2	44	57	23	6
54	39	9	28	50	35	13	32
33	52	30	15	37	56	26	11
59	42	8	21	63	46	4	17
16	29	51	34	12	25	55	38
22	7	41	60	18	3	45	64

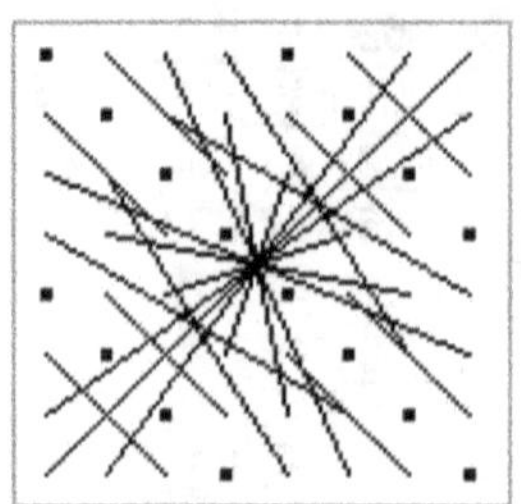

1	41	16	32	40	56	17	57
6	19	54	27	35	14	43	62
58	47	55	34	26	15	23	2
60	20	13	37	29	53	44	4
61	21	12	36	28	52	45	5
63	42	50	39	31	10	18	7
3	22	51	30	38	11	46	59
8	48	9	25	33	49	24	64

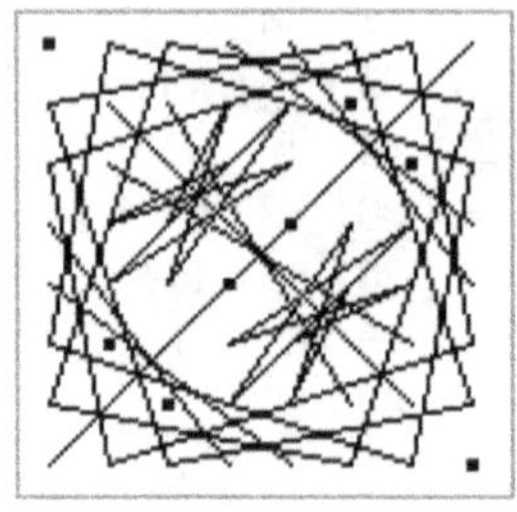

19	60	3	45	44	6	61	22
32	10	56	39	34	49	15	25
17	63	1	42	47	8	58	24
38	53	14	28	29	11	52	35
30	13	54	36	37	51	12	27
41	7	57	18	23	64	2	48
40	50	16	31	26	9	55	33
43	4	59	21	20	62	5	46

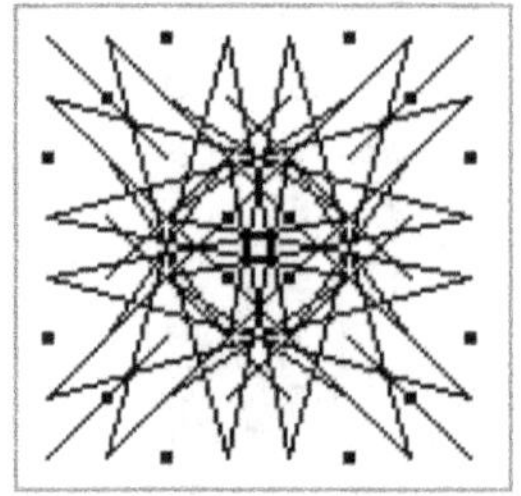

44	22	3	61	14	52	37	27
29	35	54	12	59	5	20	46
39	25	16	50	1	63	42	24
18	48	57	7	56	10	31	33
32	34	55	9	58	8	17	47
41	23	2	64	15	49	40	26
19	45	60	6	53	11	30	36
38	28	13	51	4	62	43	21

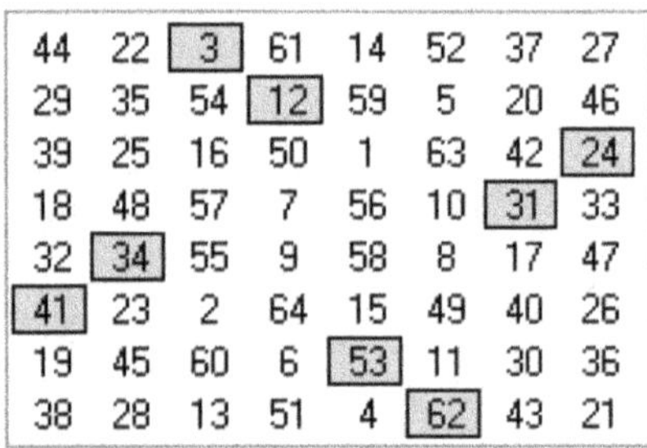

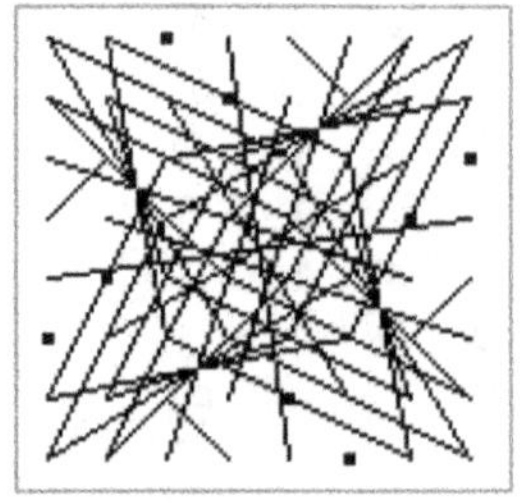

10	63	47	26	34	23	7	50
56	1	17	40	32	41	57	16
54	3	46	27	35	22	59	14
12	61	20	37	29	44	5	52
13	60	21	36	28	45	4	53
51	6	43	30	38	19	62	11
49	8	24	33	25	48	64	9
15	58	42	31	39	18	2	55

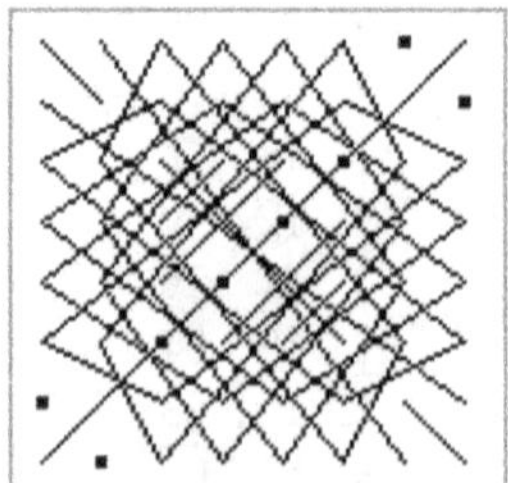

8	7	53	54	60	59	9	10
16	15	61	62	52	51	1	2
17	18	36	35	45	46	32	31
25	26	44	43	37	38	24	23
42	41	27	28	22	21	39	40
34	33	19	20	30	29	47	48
63	64	14	13	3	4	50	49
55	56	6	5	11	12	58	57

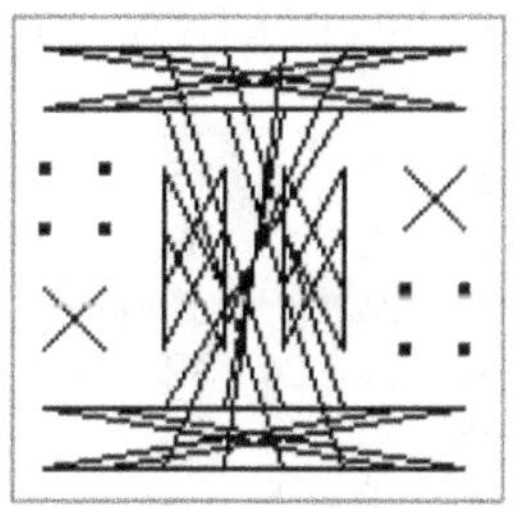

11	12	46	45	49	50	24	23
3	4	38	37	57	58	32	31
55	56	18	17	13	14	44	43
63	64	26	25	5	6	36	35
30	29	59	60	40	39	1	2
22	21	51	52	48	47	9	10
34	33	7	8	28	27	61	62
42	41	15	16	20	19	53	54

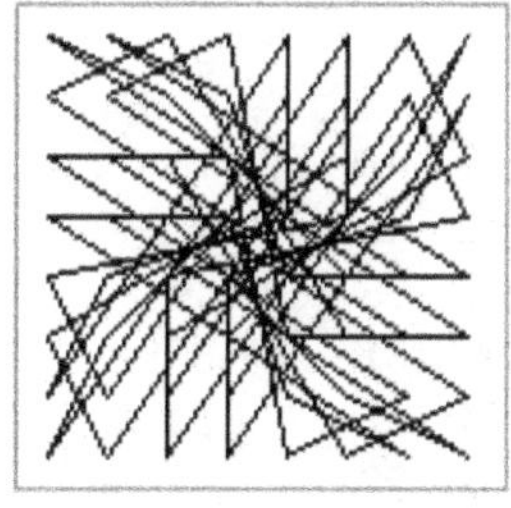

19	45	24	42	61	3	58	8
38	28	33	31	12	54	15	49
59	5	64	2	21	43	18	48
14	52	9	55	36	30	39	25
40	26	35	29	10	56	13	51
17	47	22	44	63	1	60	6
16	50	11	53	34	32	37	27
57	7	62	4	23	41	20	46

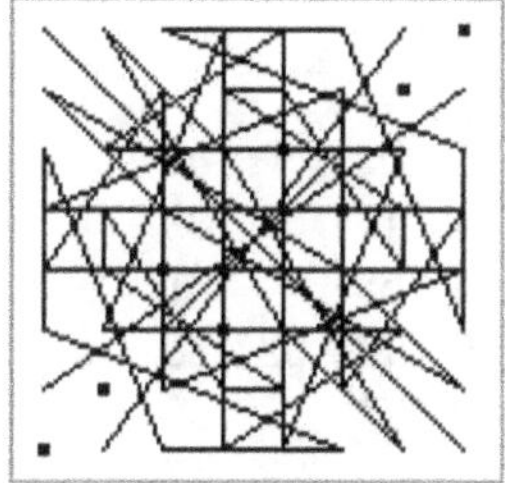

28	27	6	13	53	62	35	36
20	19	14	5	61	54	43	44
41	42	63	64	8	7	18	17
34	33	55	56	16	15	25	26
39	40	50	49	9	10	32	31
48	47	58	57	1	2	23	24
21	22	11	4	60	51	46	45
29	30	3	12	52	59	38	37

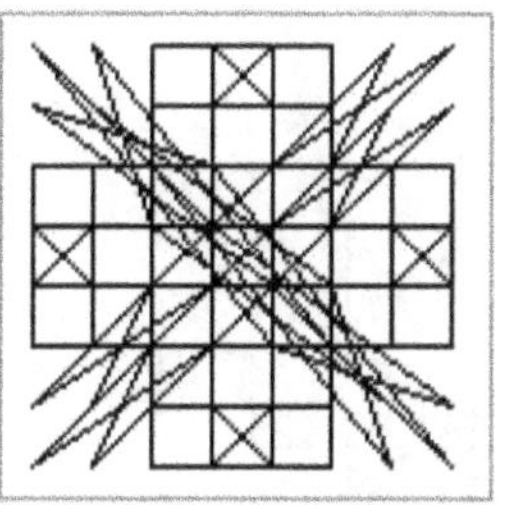

64	56	17	25	33	41	16	8
63	55	18	26	34	42	15	7
3	11	46	38	30	22	51	59
4	12	45	37	29	21	52	60
5	13	44	36	28	20	53	61
6	14	43	35	27	19	54	62
58	50	23	31	39	47	10	2
57	49	24	32	40	48	9	1

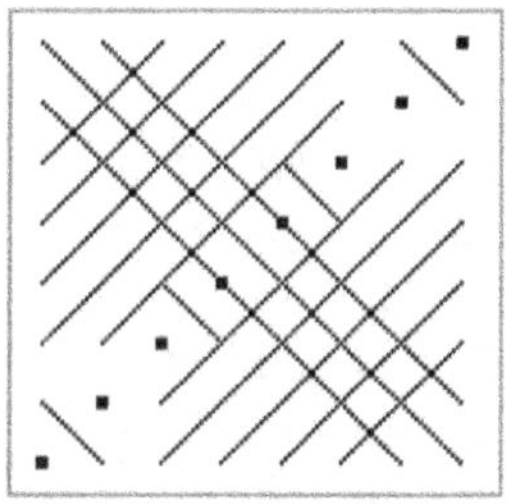

The next five magic squares don't have any numbers in their original position. The right continues to show the original position to their current position in the square.

10	56	12	54	51	13	49	15
63	1	61	3	6	60	8	58
26	40	37	27	30	36	33	31
47	17	20	46	43	21	24	42
23	41	44	22	19	45	48	18
34	32	29	35	38	28	25	39
7	57	5	59	62	4	64	2
50	16	52	14	11	53	9	55

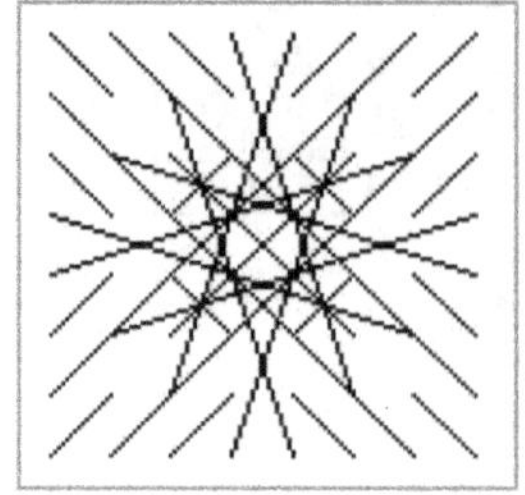

10	54	41	20	21	48	51	15
47	19	16	53	52	9	22	42
6	58	37	32	25	36	63	3
27	39	60	1	8	61	34	30
35	31	4	57	64	5	26	38
62	2	29	40	33	28	7	59
23	43	56	13	12	49	46	18
50	14	17	44	45	24	11	55

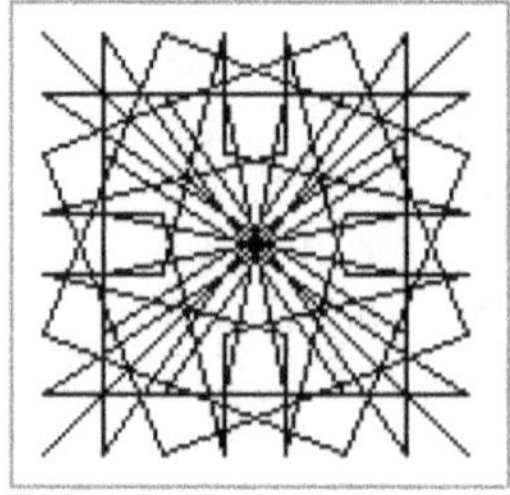

46	4	62	20	21	59	5	43
25	55	9	39	34	16	50	32
48	2	64	18	23	57	7	41
27	53	11	37	36	14	52	30
35	13	51	29	28	54	12	38
24	58	8	42	47	1	63	17
33	15	49	31	26	56	10	40
22	60	6	44	45	3	61	19

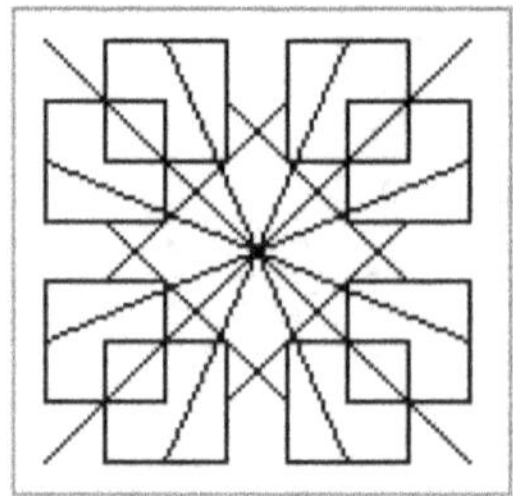

19	27	54	62	6	14	35	43
20	28	53	61	5	13	36	44
47	39	10	2	58	50	31	23
48	40	9	1	57	49	32	24
41	33	16	8	64	56	25	17
42	34	15	7	63	55	26	18
21	29	52	60	4	12	37	45
22	30	51	59	3	11	38	46

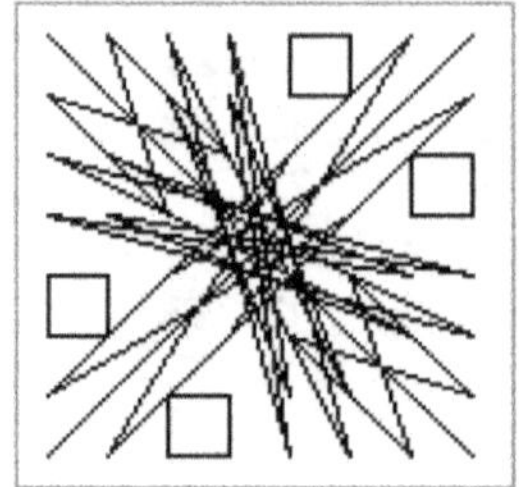

55	63	4	12	13	5	58	50
56	64	3	11	14	6	57	49
25	17	46	38	35	43	24	32
26	18	45	37	36	44	23	31
34	42	21	29	28	20	47	39
33	41	22	30	27	19	48	40
16	8	59	51	54	62	1	9
15	7	60	52	53	61	2	10

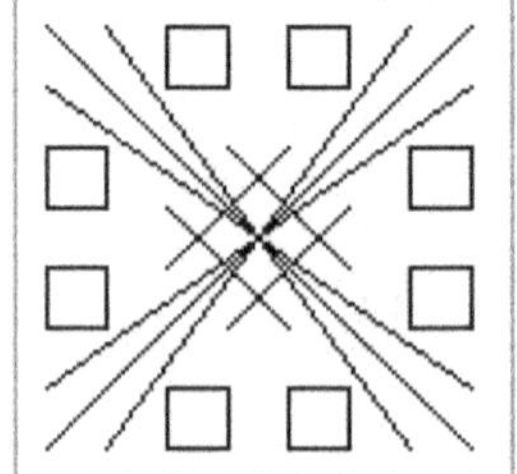

These are regular 8th order magic squares. Odd numbers are highlighted on the left, and the lines are drawn in sequence on the right (from 1 to 2 to 3, etc.)

1	2	61	62	48	47	20	19
32	31	36	35	49	50	13	14
9	10	53	54	40	39	28	27
24	23	44	43	57	58	5	6
59	60	7	8	22	21	42	41
38	37	26	25	11	12	55	56
51	52	15	16	30	29	34	33
46	45	18	17	3	4	63	64

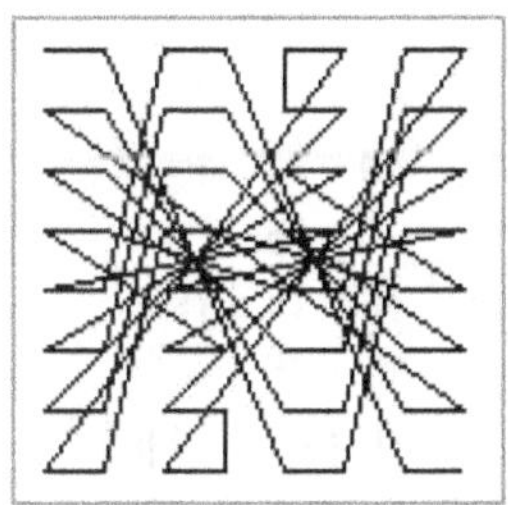

1	3	56	54	57	59	16	14
32	30	41	43	40	38	17	19
45	47	28	26	21	23	36	34
52	50	5	7	12	10	61	63
2	4	55	53	58	60	15	13
31	29	42	44	39	37	18	20
46	48	27	25	22	24	35	33
51	49	6	8	11	9	62	64

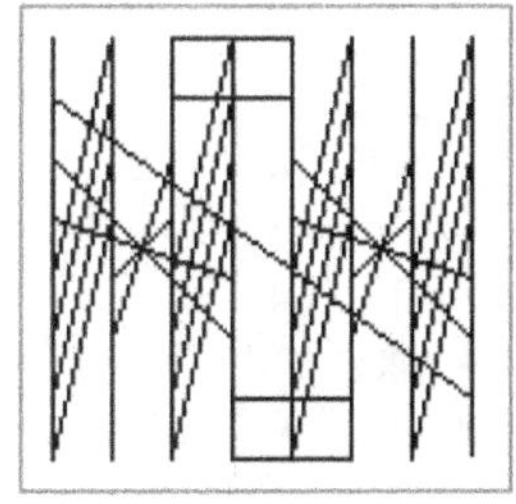

1	3	56	54	63	61	10	12
5	7	52	50	59	57	14	16
48	46	25	27	18	20	39	37
44	42	29	31	22	24	35	33
32	30	41	43	34	36	23	21
28	26	45	47	38	40	19	17
49	51	8	6	15	13	58	60
53	55	4	2	11	9	62	64

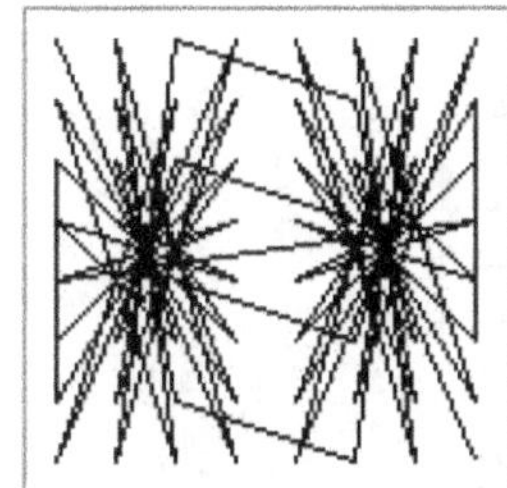

1	3	61	63	58	60	6	8
17	19	45	47	42	44	22	24
40	38	28	26	31	29	35	33
56	54	12	10	15	13	51	49
16	14	52	50	55	53	11	9
32	30	36	34	39	37	27	25
41	43	21	23	18	20	46	48
57	59	5	7	2	4	62	64

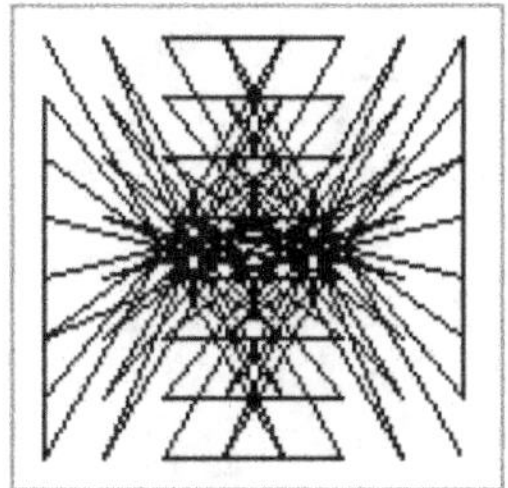

1	17	56	40	32	16	41	57
58	42	15	31	39	55	18	2
3	19	54	38	30	14	43	59
60	44	13	29	37	53	20	4
61	45	12	28	36	52	21	5
6	22	51	35	27	11	46	62
63	47	10	26	34	50	23	7
8	24	49	33	25	9	48	64

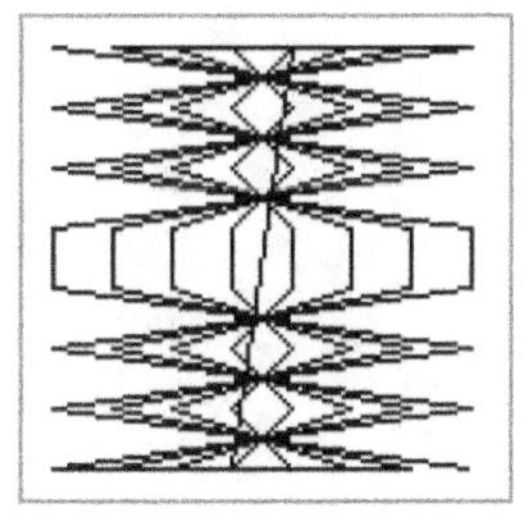

The following are regular magic squares with circular plots.

1	2	61	62	24	23	44	43
9	10	53	54	32	31	36	35
47	48	19	20	58	57	6	5
39	40	27	28	50	49	14	13
52	51	16	15	37	38	25	26
60	59	8	7	45	46	17	18
30	29	34	33	11	12	55	56
22	21	42	41	3	4	63	64

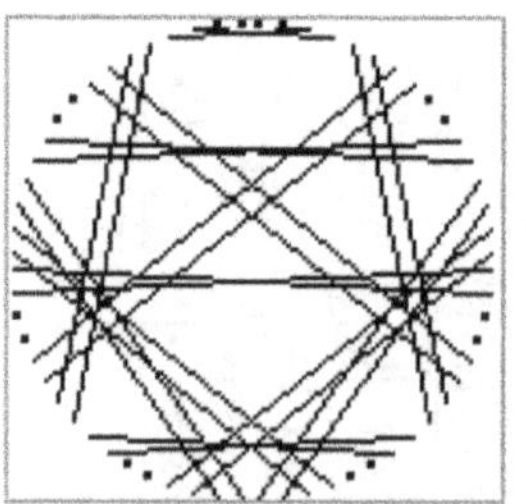

Circular Orig to Curr

1	9	24	32	40	48	49	57
2	10	23	31	39	47	50	58
59	51	46	38	30	22	11	3
60	52	45	37	29	21	12	4
61	53	44	36	28	20	13	5
62	54	43	35	27	19	14	6
7	15	18	26	34	42	55	63
8	16	17	25	33	41	56	64

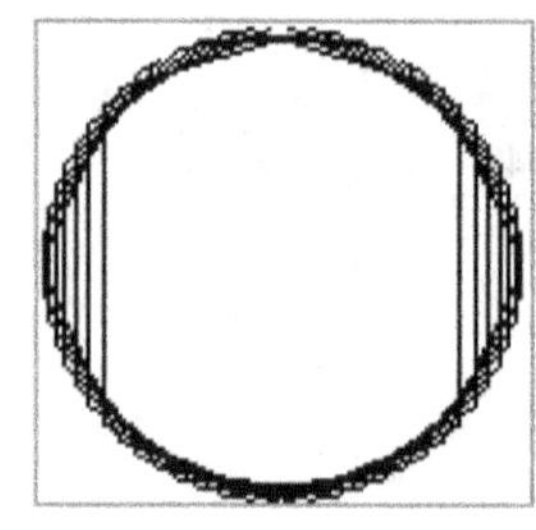

Circular Std Sequence

1	9	48	40	32	24	49	57
2	10	47	39	31	23	50	58
62	54	19	27	35	43	14	6
61	53	20	28	36	44	13	5
60	52	21	29	37	45	12	4
59	51	22	30	38	46	11	3
7	15	42	34	26	18	55	63
8	16	41	33	25	17	56	64

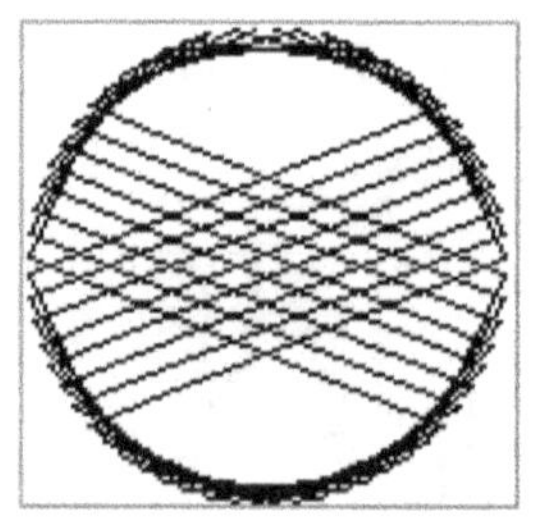

Circular Std Sequence

1	17	32	16	63	47	34	50
3	19	30	14	61	45	36	52
60	44	37	53	6	22	27	11
58	42	39	55	8	24	25	9
56	40	41	57	10	26	23	7
54	38	43	59	12	28	21	5
13	29	20	4	51	35	46	62
15	31	18	2	49	33	48	64

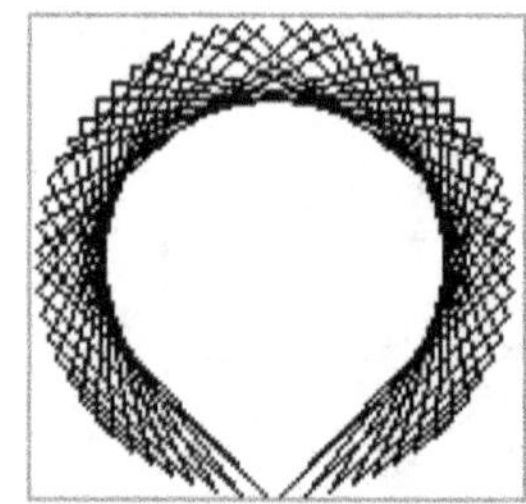

Circular Std Sequence

Appendix F – Selected 9th Order Magic Squares

The following are regular 9th order magic squares. The boxed numbers in the left squares are numbers in their original positions. The right square shows the movement of the numbers from their original to their current position.

1	80	12	23	34	45	47	69	58
61	50	72	74	4	15	26	39	28
31	20	42	53	55	66	77	18	7
68	57	79	9	11	22	33	46	44
17	6	19	30	41	52	63	76	65
38	36	49	60	71	73	3	25	14
75	64	5	16	27	29	40	62	51
54	43	56	67	78	8	10	32	21
24	13	35	37	48	59	70	2	81

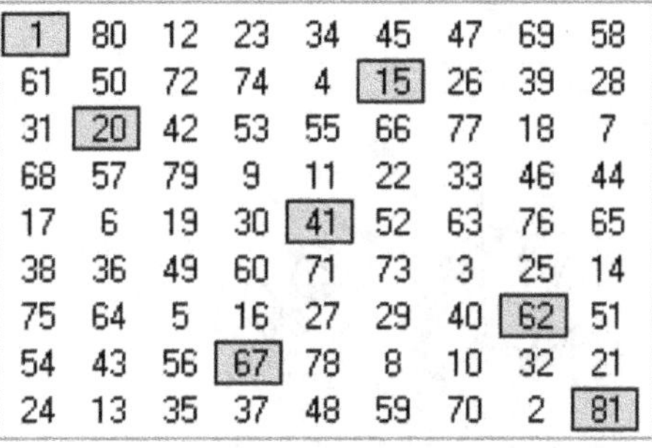

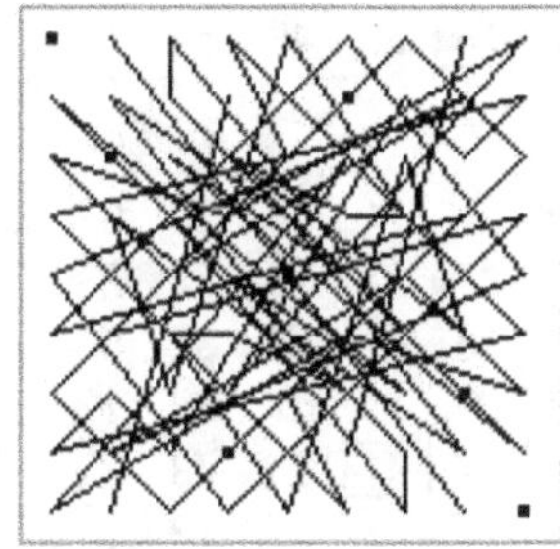

54	2	49	67	9	47	65	4	72
14	79	12	30	77	16	34	75	32
63	38	58	22	45	56	20	40	27
18	74	13	31	81	11	29	76	36
59	43	57	21	41	61	25	39	23
46	6	53	71	1	51	69	8	64
55	42	62	26	37	60	24	44	19
50	7	48	66	5	52	70	3	68
10	78	17	35	73	15	33	80	28

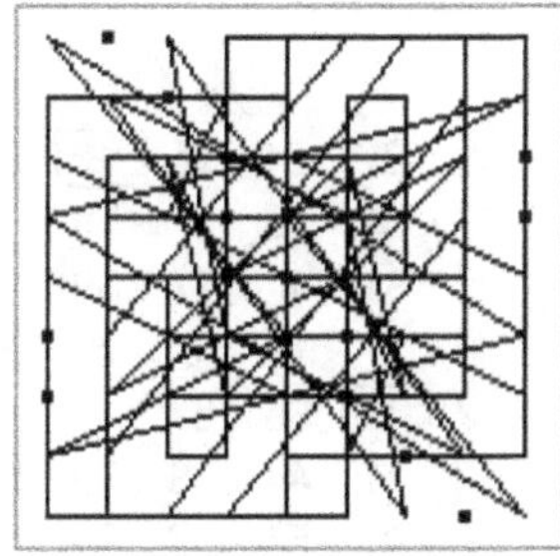

26	10	3	77	33	70	63	47	40
18	2	76	69	25	62	46	39	32
1	75	68	61	17	54	38	31	24
74	67	60	53	9	37	30	23	16
34	27	11	4	41	78	71	55	48
66	59	52	45	73	29	22	15	8
58	51	44	28	65	21	14	7	81
50	43	36	20	57	13	6	80	64
42	35	19	12	49	5	79	72	56

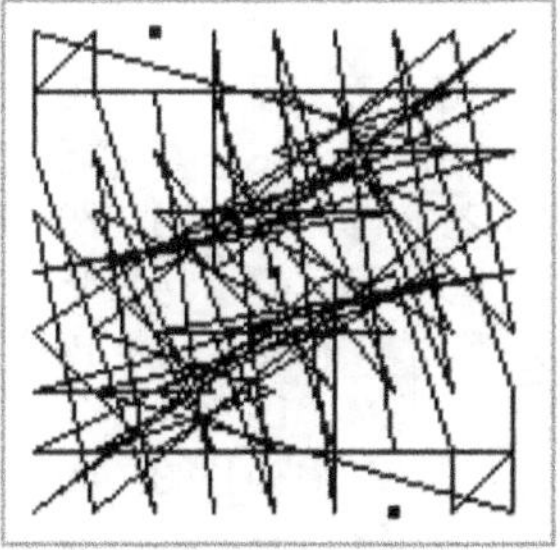

52	70	3	68	5	50	7	48	66
15	33	80	28	73	10	78	17	35
56	20	40	27	45	63	38	58	22
11	29	76	36	81	18	74	13	31
61	25	39	23	41	59	43	57	21
51	69	8	64	1	46	6	53	71
60	24	44	19	37	55	42	62	26
47	65	4	72	9	54	2	49	67
16	34	75	32	77	14	79	12	30

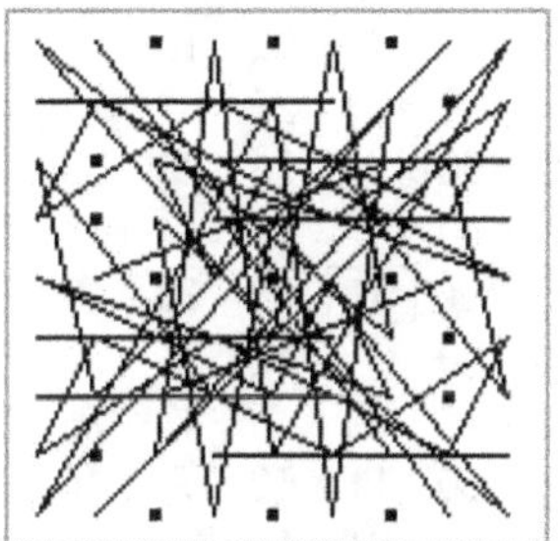

225

77	67	26	36	57	6	16	47	37
69	59	18	19	49	79	8	39	29
20	10	50	60	9	30	40	80	70
28	27	58	68	17	38	48	7	78
61	51	1	11	41	71	81	31	21
4	75	34	44	65	14	24	55	54
12	2	42	52	73	22	32	72	62
53	43	74	3	33	63	64	23	13
45	35	66	76	25	46	56	15	5

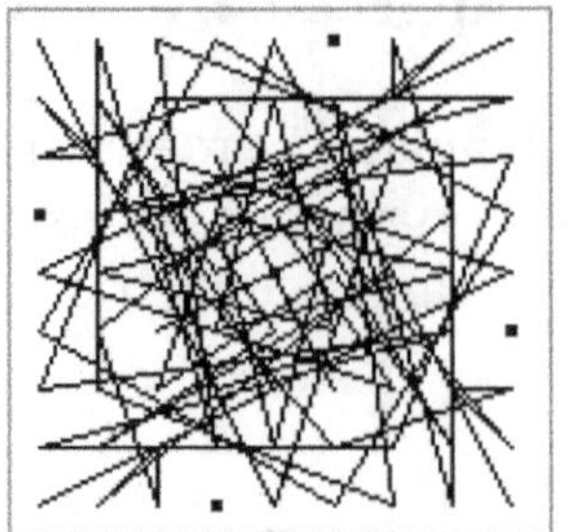

38	79	30	6	71	46	22	63	14
7	39	80	47	31	15	72	23	55
48	8	40	16	81	56	32	64	24
78	29	70	37	21	5	62	13	54
17	49	9	57	41	25	73	33	65
28	69	20	77	61	45	12	53	4
58	18	50	26	1	66	42	74	34
27	59	10	67	51	35	2	43	75
68	19	60	36	11	76	52	3	44

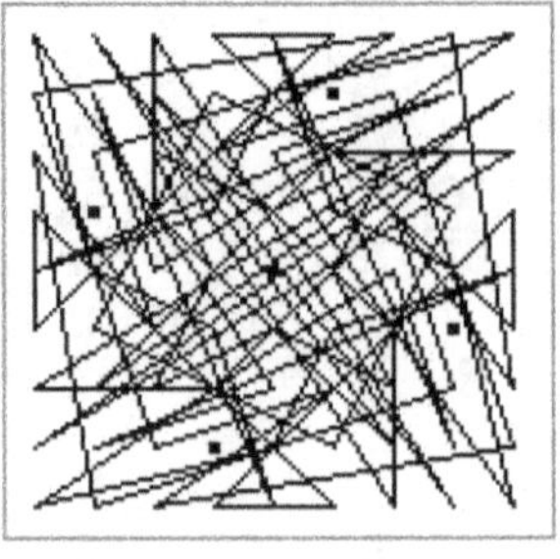

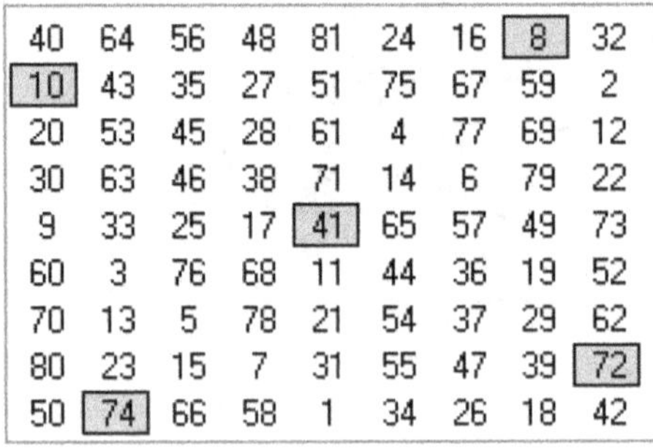

40	64	56	48	81	24	16	8	32
10	43	35	27	51	75	67	59	2
20	53	45	28	61	4	77	69	12
30	63	46	38	71	14	6	79	22
9	33	25	17	41	65	57	49	73
60	3	76	68	11	44	36	19	52
70	13	5	78	21	54	37	29	62
80	23	15	7	31	55	47	39	72
50	74	66	58	1	34	26	18	42

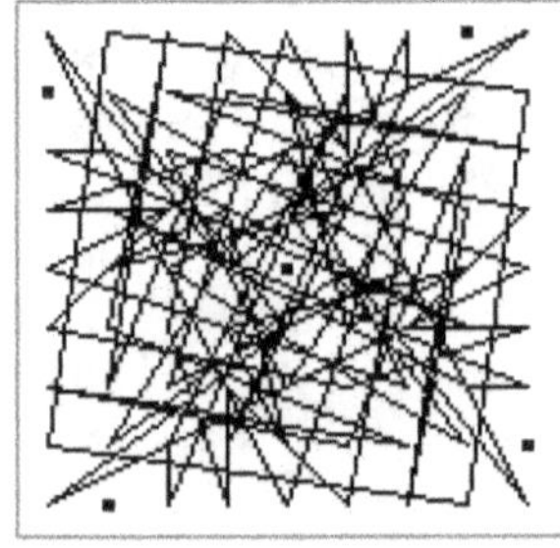

42	20	31	18	7	77	55	66	53
72	50	61	39	28	26	4	15	74
46	33	44	22	11	9	68	79	57
12	80	1	69	58	47	34	45	23
76	63	65	52	41	30	17	19	6
59	37	48	35	24	13	81	2	70
25	3	14	73	71	60	38	49	36
8	67	78	56	54	43	21	32	10
29	16	27	5	75	64	51	62	40

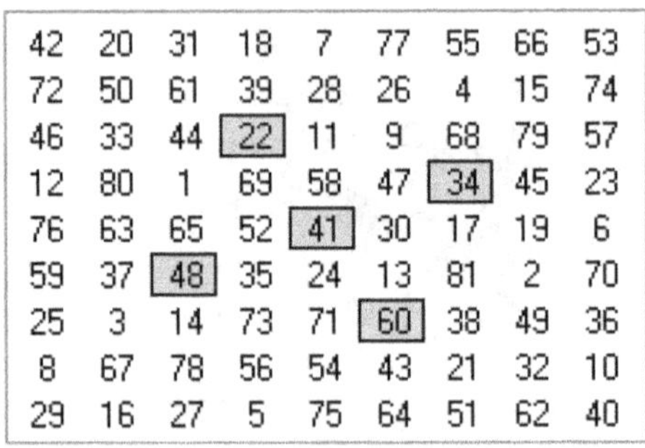

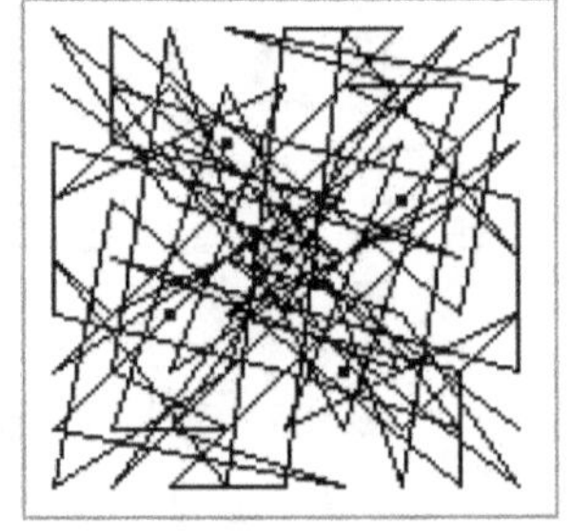

37	78	29	70	21	62	13	54	5
6	38	79	30	71	22	63	14	46
47	7	39	80	31	72	23	55	15
16	48	8	40	81	32	64	24	56
57	17	49	9	41	73	33	65	25
26	58	18	50	1	42	74	34	66
67	27	59	10	51	2	43	75	35
36	68	19	60	11	52	3	44	76
77	28	69	20	61	12	53	4	45

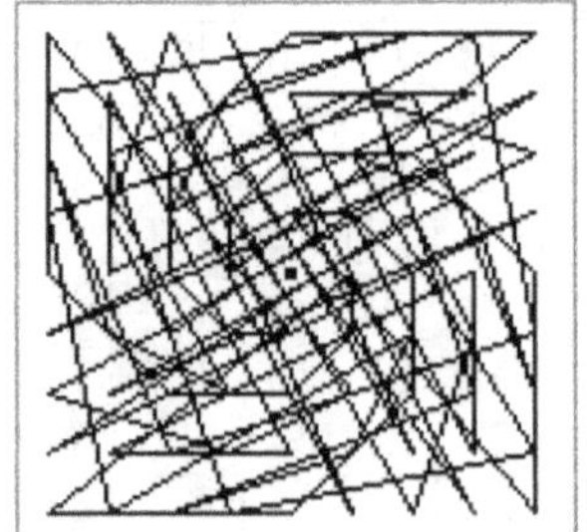

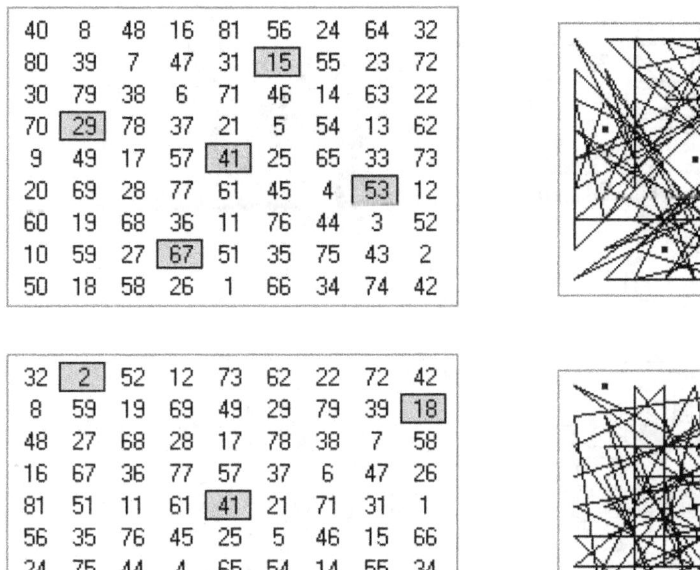

```
40   8  48  16  81  56  24  64  32
80  39   7  47  31 [15] 55  23  72
30  79  38   6  71  46  14  63  22
70 [29] 78  37  21   5  54  13  62
 9  49  17  57 [41] 25  65  33  73
20  69  28  77  61  45   4 [53] 12
60  19  68  36  11  76  44   3  52
10  59  27 [67] 51  35  75  43   2
50  18  58  26   1  66  34  74  42
```

```
32  [2] 52  12  73  62  22  72  42
 8  59  19  69  49  29  79  39 [18]
48  27  68  28  17  78  38   7  58
16  67  36  77  57  37   6  47  26
81  51  11  61 [41] 21  71  31   1
56  35  76  45  25   5  46  15  66
24  75  44   4  65  54  14  55  34
[64] 43   3  53  33  13  63  23  74
40  10  60  20   9  70  30 [80] 50
```

These are regular 9th order magic squares. Odd numbers are highlighted on the left, and the lines are drawn in sequence on the right (from 1 to 2 to 3, etc.).

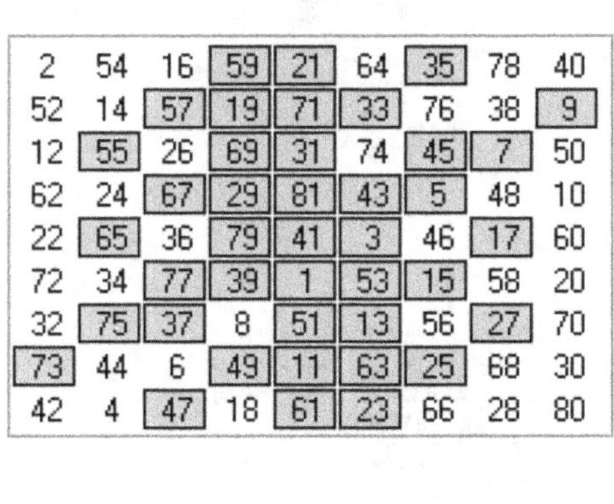

```
 2  54  16 [59][21] 64 [35] 78  40
52  14 [57][19][71][33] 76  38  [9]
12 [55] 26 [69][31] 74 [45][7]  50
62  24 [67][29][81][43][5]  48  10
22 [65] 36 [79][41][3]  46 [17] 60
72  34 [77][39][1][53][15] 58  20
32 [75][37]  8 [51][13] 56 [27] 70
[73] 44   6 [49][11][63][25] 68  30
42   4 [47] 18 [61][23] 66  28  80
```

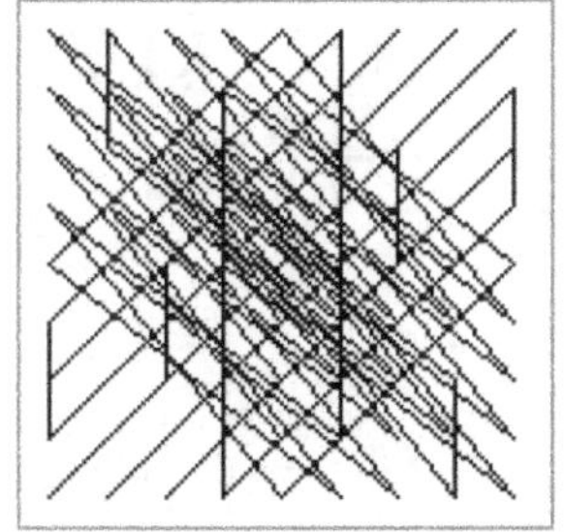

```
[37][51] 56  70 [75]  8 [13][27] 32
[33] 38  52 [57][71] 76  [9] 14 [19]
20  34 [39][53] 58  72 [77][1] [15]
16 [21][35] 40  54 [59] 64  78   2
[3][17] 22  36 [41] 46  60 [65][79]
80   4  18 [23] 28  42 [47][61] 66
[67][81][5] 10  24 [29][43] 48  62
[63] 68 [73]  6 [11][25] 30  44 [49]
50 [55][69] 74 [7]  12  26 [31][45]
```

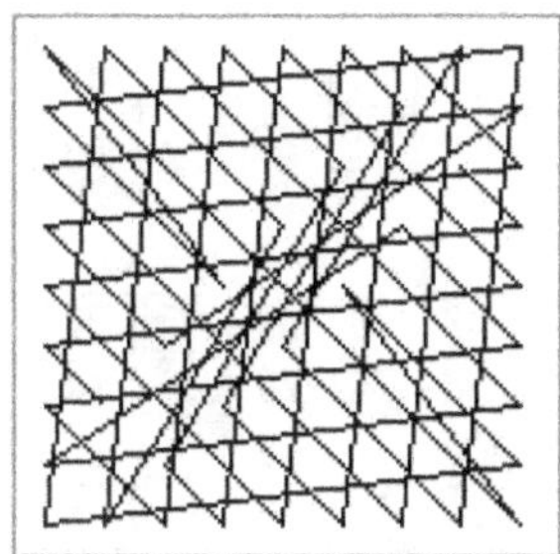

227

37	70	29	54	21	78	13	62	5
16	40	8	24	81	48	64	32	56
67	10	59	75	51	27	43	2	35
36	60	19	44	11	68	3	52	76
57	9	49	65	41	17	33	73	25
6	30	79	14	71	38	63	22	46
47	80	39	55	31	7	23	72	15
26	50	18	34	1	58	74	42	66
77	20	69	4	61	28	53	12	45

11	18	13	74	81	76	29	36	31
16	14	12	79	77	75	34	32	30
15	10	17	78	73	80	33	28	35
56	63	58	38	45	40	20	27	22
61	59	57	43	41	39	25	23	21
60	55	62	42	37	44	24	19	26
47	54	49	2	9	4	65	72	67
52	50	48	7	5	3	70	68	66
51	46	53	6	1	8	69	64	71

5	78	13	70	21	62	29	54	37
76	68	3	60	11	52	19	44	36
15	7	23	80	31	72	39	55	47
66	58	74	50	1	42	18	34	26
25	17	33	9	41	73	49	65	57
56	48	64	40	81	32	8	24	16
35	27	43	10	51	2	59	75	67
46	38	63	30	71	22	79	14	6
45	28	53	20	61	12	69	4	77

45	28	69	20	61	12	53	4	77
46	38	79	30	71	22	63	14	6
15	7	39	80	31	72	23	55	47
56	48	8	40	81	32	64	24	16
25	17	49	9	41	73	33	65	57
66	58	18	50	1	42	74	34	26
35	27	59	10	51	2	43	75	67
76	68	19	60	11	52	3	44	36
5	78	29	70	21	62	13	54	37

10	24	29	43	48	62	67	81	5
33	38	52	57	71	76	9	14	19
47	61	66	80	4	18	23	28	42
70	75	8	13	27	32	37	51	56
3	17	22	36	41	46	60	65	79
26	31	45	50	55	69	74	7	12
40	54	59	64	78	2	16	21	35
63	68	73	6	11	25	30	44	49
77	1	15	20	34	39	53	58	72

14	6	63	30	71	22	79	46	38
4	77	53	20	61	12	69	45	28
55	47	23	80	31	72	39	15	7
34	26	74	50	1	42	18	66	58
65	57	33	9	41	73	49	25	17
24	16	64	40	81	32	8	56	48
75	67	43	10	51	2	59	35	27
54	37	13	70	21	62	29	5	78
44	36	3	60	11	52	19	76	68

32	69	25	62	76	18	46	2	39
67	23	60	16	30	53	9	37	74
21	58	14	51	65	7	44	81	28
56	12	49	5	19	42	79	35	72
78	34	71	27	41	55	11	48	4
10	47	3	40	63	77	33	70	26
54	1	38	75	17	31	68	24	61
8	45	73	29	52	66	22	59	15
43	80	36	64	6	20	57	13	50

39	7	80	47	31	15	72	55	23
29	78	70	37	21	5	62	54	13
19	68	60	36	11	76	52	44	3
18	58	50	26	1	66	42	34	74
49	17	9	57	41	25	73	65	33
8	48	40	16	81	56	32	24	64
79	38	30	6	71	46	22	14	63
69	28	20	77	61	45	12	4	53
59	27	10	67	51	35	2	75	43

40	8	16	48	81	24	56	64	32
80	39	47	7	31	55	15	23	72
70	29	37	78	21	54	5	13	62
30	79	6	38	71	14	46	63	22
9	49	57	17	41	65	25	33	73
60	19	36	68	11	44	76	3	52
20	69	77	28	61	4	45	53	12
10	59	67	27	51	75	35	43	2
50	18	26	58	1	34	66	74	42

39	47	55	72	31	80	7	15	23
29	37	54	62	21	70	78	5	13
19	36	44	52	11	60	68	76	3
18	26	34	42	1	50	58	66	74
49	57	65	73	41	9	17	25	33
8	16	24	32	81	40	48	56	64
79	6	14	22	71	30	38	46	63
69	77	4	12	61	20	28	45	53
59	67	75	2	51	10	27	35	43

The following are regular 9th order magic squares with circular plots.

3	44	52	76	11	36	60	68	19
64	24	32	56	81	16	40	48	8
13	54	62	5	21	37	70	78	29
23	55	72	15	31	47	80	7	39
33	65	73	25	41	57	9	17	49
43	75	2	35	51	67	10	27	59
53	4	12	45	61	77	20	28	69
74	34	42	66	1	26	50	58	18
63	14	22	46	71	6	30	38	79

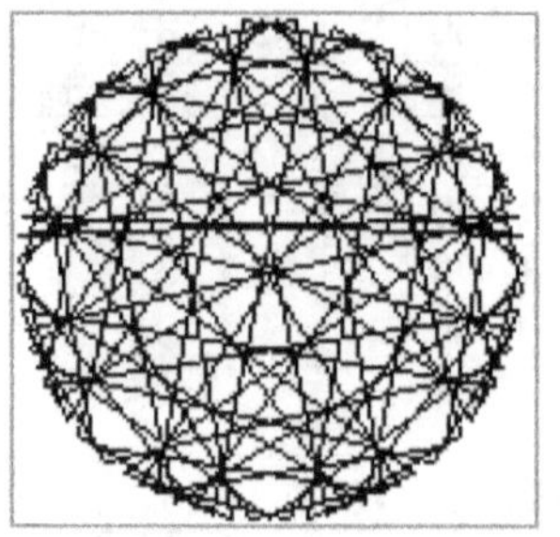

Circular Std Sequence

5	78	13	70	21	62	29	54	37
76	68	3	60	11	52	19	44	36
15	7	23	80	31	72	39	55	47
66	58	74	50	1	42	18	34	26
25	17	33	9	41	73	49	65	57
56	48	64	40	81	32	8	24	16
35	27	43	10	51	2	59	75	67
46	38	63	30	71	22	79	14	6
45	28	53	20	61	12	69	4	77

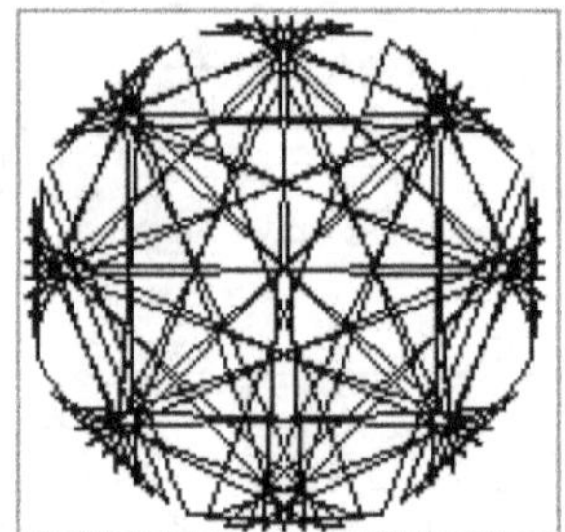

Circular Std Sequence

37	27	8	70	51	32	13	75	56
57	38	19	9	71	52	33	14	76
77	58	39	20	1	72	53	34	15
16	78	59	40	21	2	64	54	35
36	17	79	60	41	22	3	65	46
47	28	18	80	61	42	23	4	66
67	48	29	10	81	62	43	24	5
6	68	49	30	11	73	63	44	25
26	7	69	50	31	12	74	55	45

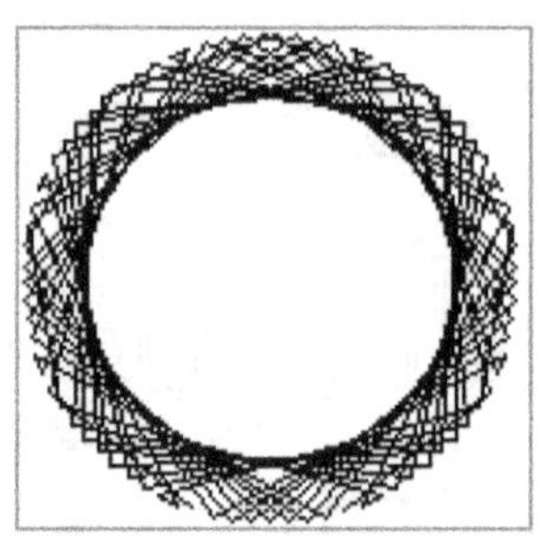

Circular Std Sequence

37	51	56	70	75	8	13	27	32
33	38	52	57	71	76	9	14	19
20	34	39	53	58	72	77	1	15
16	21	35	40	54	59	64	78	2
3	17	22	36	41	46	60	65	79
80	4	18	23	28	42	47	61	66
67	81	5	10	24	29	43	48	62
63	68	73	6	11	25	30	44	49
50	55	69	74	7	12	26	31	45

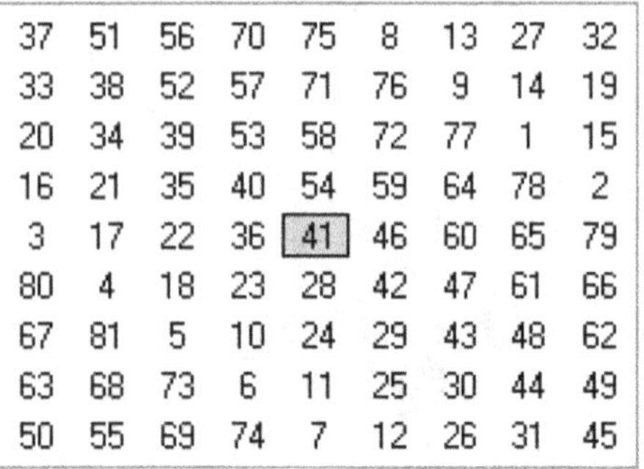

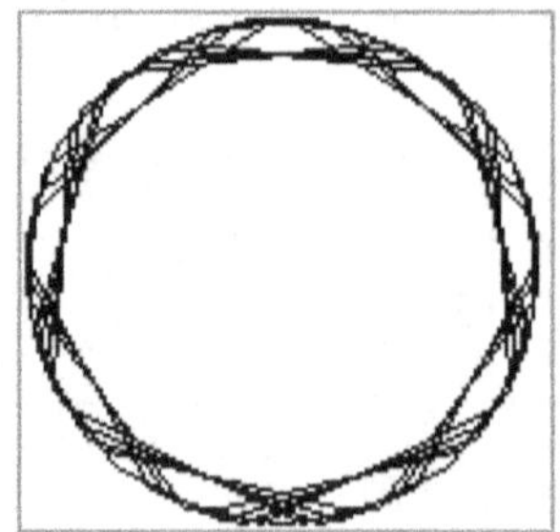

Circular Std Sequence

Appendix G – Selected 10th Order Magic Squares

The following are regular 10th order magic squares. The boxed numbers in the left squares are in their original positions. The right square shows the movement of the numbers from their original to their current position.

1	99	3	97	96	5	94	8	92	10
20	12	88	17	86	85	14	83	19	81
21	29	23	77	75	76	74	28	72	30
40	62	38	34	66	65	37	63	39	61
51	59	48	47	45	46	54	43	52	60
50	49	58	57	55	56	44	53	42	41
70	32	68	64	35	36	67	33	69	31
71	79	73	24	26	25	27	78	22	80
90	82	13	84	15	16	87	18	89	11
91	2	93	4	6	95	7	98	9	100

1	20	21	70	50	60	40	71	81	91
2	12	29	39	59	49	69	72	82	92
8	83	23	68	58	48	38	73	13	93
97	87	77	34	44	54	64	27	14	7
96	86	76	35	45	55	65	26	16	5
95	85	75	36	46	56	66	25	15	6
94	84	74	37	47	57	67	24	17	4
3	18	28	63	53	43	33	78	88	98
99	19	22	62	52	42	32	79	89	9
10	11	80	61	51	41	31	30	90	100

1	2	3	94	95	96	97	8	99	10
20	12	88	84	85	86	87	13	19	11
71	79	78	24	25	26	27	73	72	30
40	69	33	67	66	65	64	38	32	31
60	49	43	57	56	55	54	48	42	41
50	59	53	47	46	45	44	58	52	51
70	39	63	37	36	35	34	68	62	61
21	22	28	74	75	76	77	23	29	80
81	82	18	17	15	16	14	83	89	90
91	92	98	4	6	5	7	93	9	100

45	35	26	16	96	5	86	76	65	55
44	34	27	14	97	7	87	77	64	54
53	63	78	88	3	98	18	28	33	43
52	62	79	89	99	9	19	22	32	42
50	70	71	81	1	91	20	21	40	60
51	61	30	90	10	100	11	80	31	41
59	39	72	82	2	92	12	29	69	49
58	68	73	13	8	93	83	23	38	48
47	37	24	17	94	4	84	74	67	57
46	36	25	15	95	6	85	75	66	56

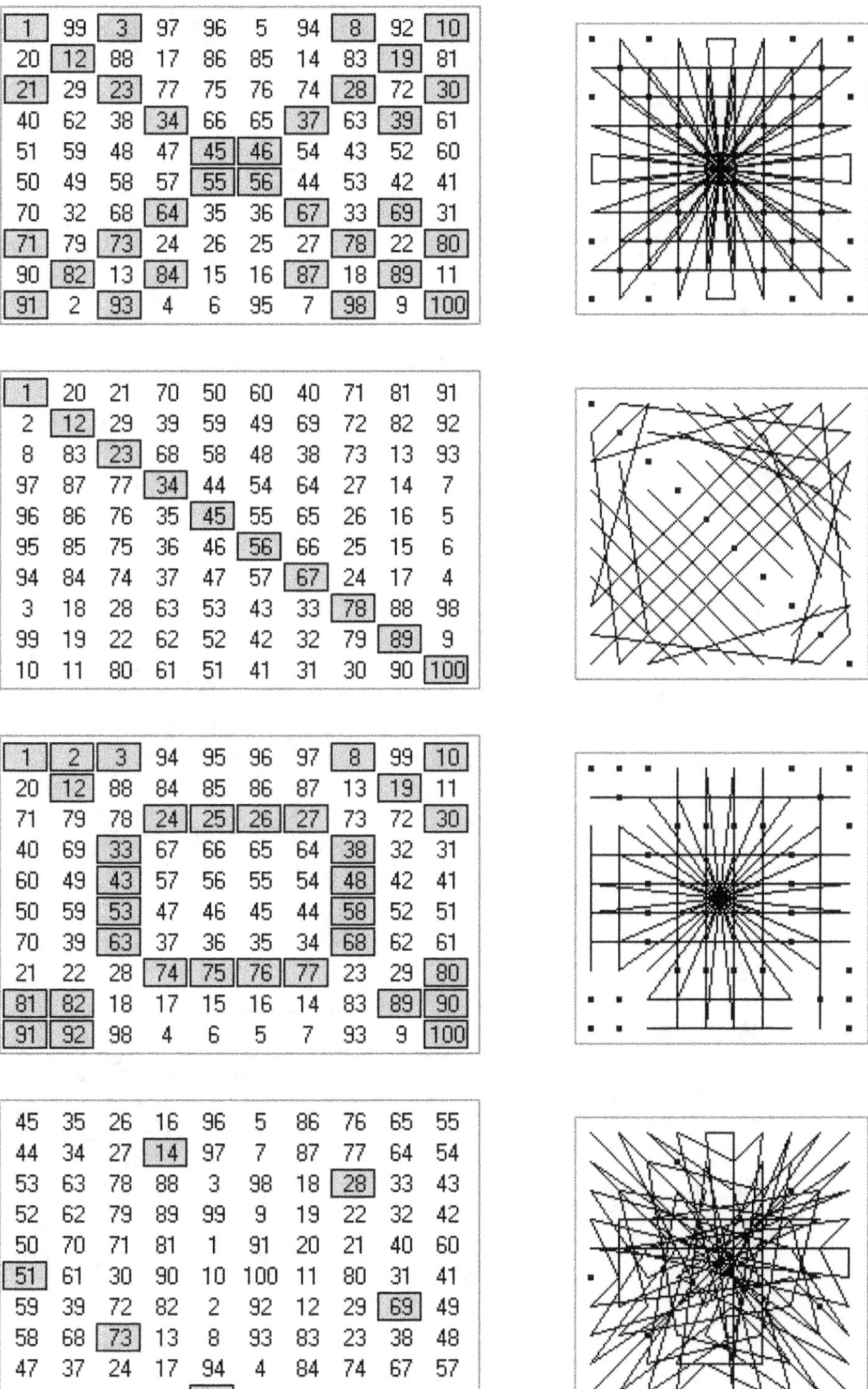

45	44	58	59	41	60	52	53	47	46
35	34	68	69	61	70	32	63	37	36
76	77	23	29	30	71	22	28	74	75
86	84	83	12	20	11	19	18	87	85
95	97	3	92	10	1	9	8	94	96
6	7	98	2	100	91	99	93	4	5
16	17	13	82	81	90	89	88	14	15
26	27	73	72	80	21	79	78	24	25
65	64	38	39	31	40	62	33	67	66
55	54	48	49	51	50	42	43	57	56

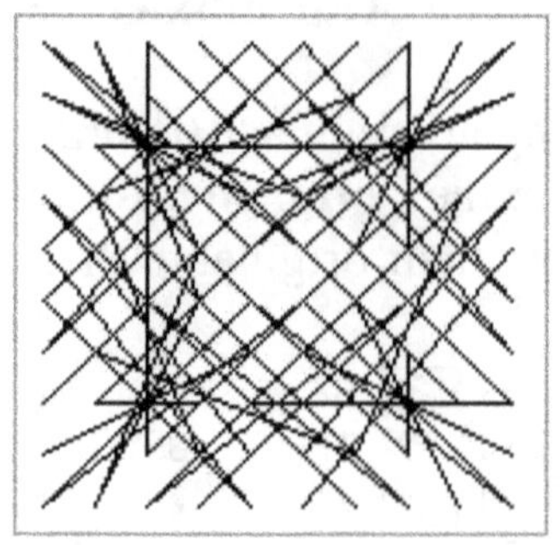

1	9	8	97	95	96	94	3	92	10
90	89	88	17	16	15	14	13	82	81
21	79	78	27	26	25	24	73	72	80
70	32	63	34	35	36	37	68	69	61
60	52	53	44	45	46	47	58	59	41
50	42	43	54	55	56	57	48	49	51
40	62	33	64	65	66	67	38	39	31
71	22	28	77	76	75	74	23	29	30
11	19	18	84	86	85	87	83	12	20
91	99	93	7	6	5	4	98	2	100

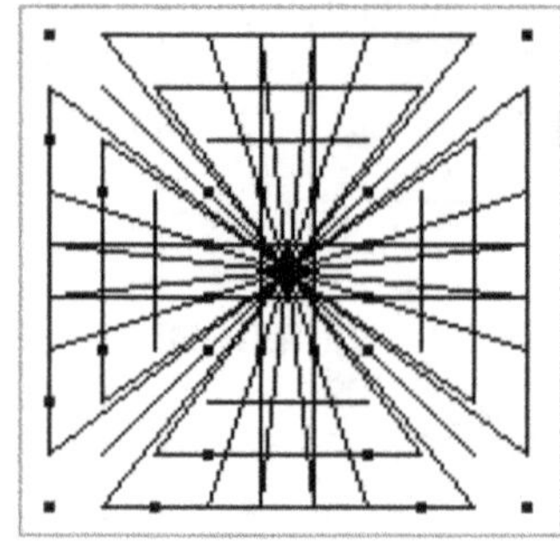

34	63	39	70	35	36	61	62	68	37
27	78	72	71	26	25	30	79	73	24
84	88	12	11	86	85	20	19	13	87
97	8	2	1	95	96	10	99	3	94
44	53	59	50	45	46	51	52	58	47
54	43	49	60	55	56	41	42	48	57
7	93	92	91	6	5	100	9	98	4
17	18	82	90	16	15	81	89	83	14
77	28	29	21	76	75	80	22	23	74
64	33	69	40	65	66	31	32	38	67

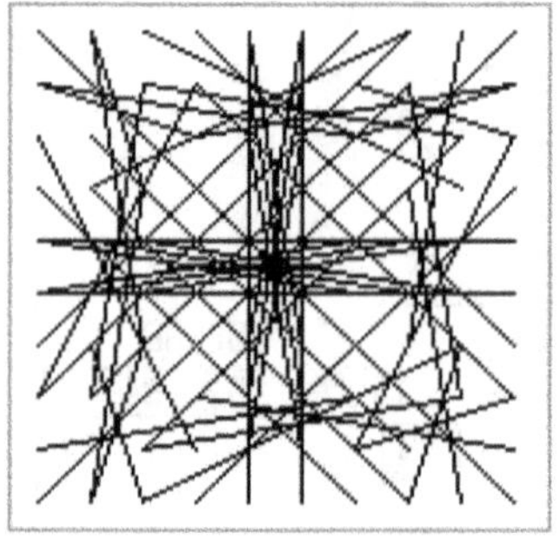

1	96	99	3	94	97	8	2	95	10
60	56	42	48	57	54	43	49	55	41
90	15	89	83	14	17	18	82	16	81
21	75	22	23	74	77	28	29	76	80
40	66	32	38	67	64	33	69	65	31
70	36	62	68	37	34	63	39	35	61
71	25	79	73	24	27	78	72	26	30
11	85	19	13	87	84	88	12	86	20
50	46	52	58	47	44	53	59	45	51
91	5	9	98	4	7	93	92	6	100

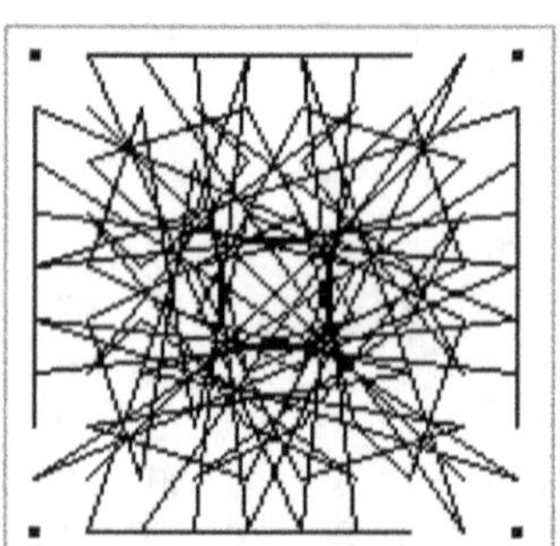

The next magic squares highlight the odd numbers on the left side, and on the right side, the lines connect the numbers in a standard sequence.

1	3	57	59	94	95	32	30	68	66
2	4	58	60	96	93	31	29	67	65
89	92	25	26	64	61	20	18	56	54
90	91	27	28	62	63	19	17	55	53
79	77	15	14	49	50	87	88	22	24
78	80	13	16	51	52	85	86	23	21
48	46	84	82	37	39	73	74	10	12
47	45	83	81	40	38	75	76	9	11
36	34	72	70	7	6	41	43	97	99
35	33	71	69	5	8	42	44	98	100

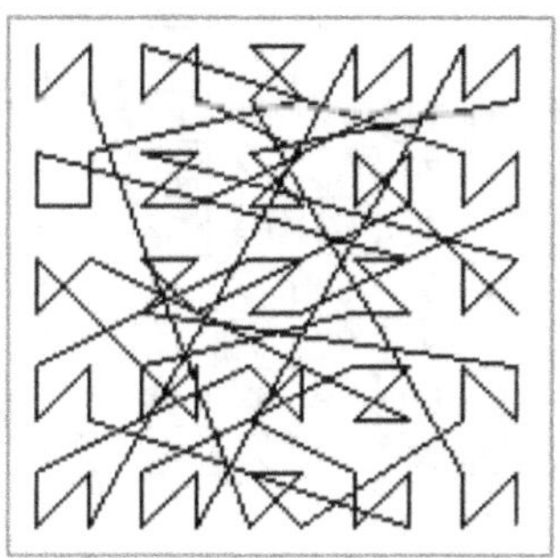

1	20	71	70	60	50	40	21	81	91
92	12	22	69	59	49	39	79	82	2
8	83	23	68	58	48	38	73	13	93
97	87	77	34	44	54	64	27	14	7
96	86	76	35	45	55	65	26	16	5
95	85	75	36	46	56	66	25	15	6
94	84	74	37	47	57	67	24	17	4
3	18	28	63	53	43	33	78	88	98
9	19	29	32	52	42	62	72	89	99
10	11	30	61	41	51	31	80	90	100

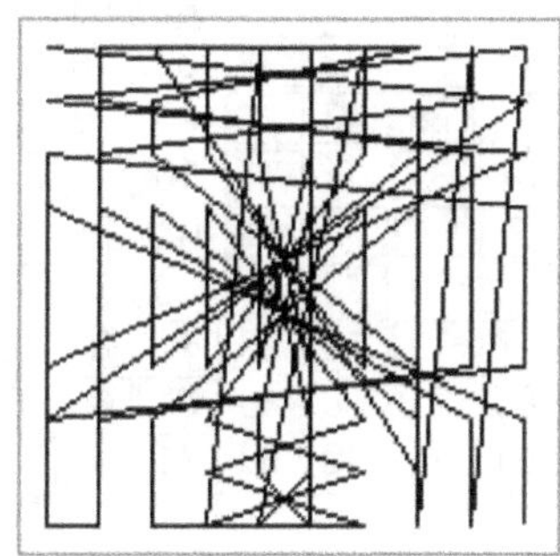

1	92	93	4	95	6	7	98	99	10
20	12	88	84	16	15	87	83	19	81
30	29	23	77	25	76	74	28	72	71
31	39	68	34	66	65	37	63	62	40
50	42	43	57	56	55	54	48	59	41
51	52	58	47	46	45	44	53	49	60
61	69	38	64	36	35	67	33	32	70
80	79	73	27	75	26	24	78	22	21
90	82	13	17	85	86	14	18	89	11
91	9	8	94	5	96	97	3	2	100

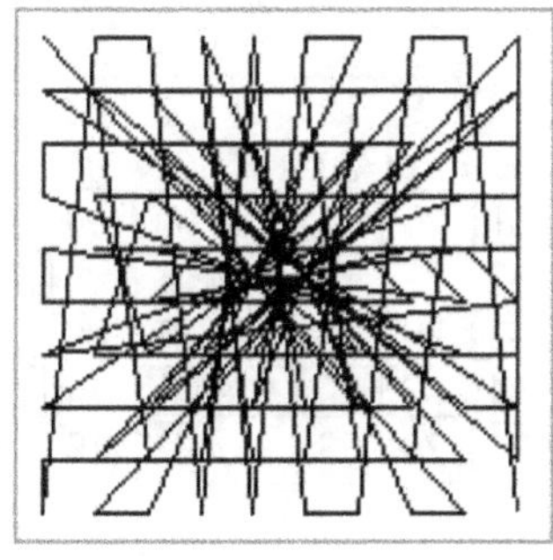

23	73	6	56	94	19	77	27	90	40
48	98	31	81	69	44	52	2	65	15
10	60	18	68	51	26	89	39	97	47
35	85	43	93	1	76	64	14	72	22
92	42	80	30	13	63	46	21	34	84
17	67	5	55	38	88	71	96	59	9
79	29	87	37	25	100	8	58	16	66
54	4	62	12	75	50	33	83	41	91
86	36	99	49	57	32	20	70	3	53
61	11	74	24	82	7	45	95	28	78

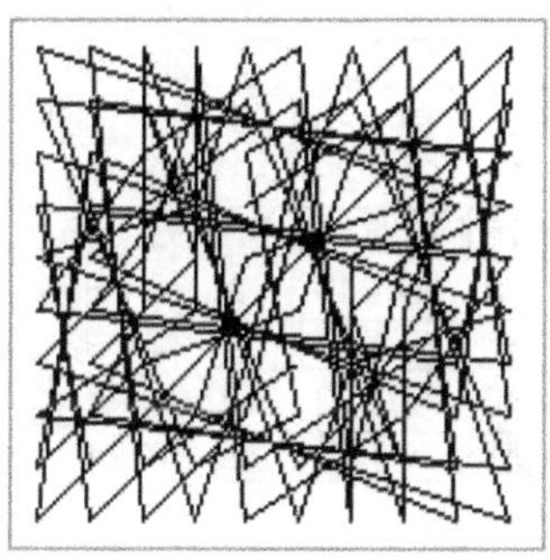

23	75	21	29	77	74	22	80	76	28
48	56	60	49	54	57	42	41	55	43
3	96	1	2	97	94	99	10	95	8
13	85	11	12	84	87	19	20	86	88
68	36	70	39	34	37	62	61	35	63
38	66	40	69	64	67	32	31	65	33
83	15	90	82	17	14	89	81	16	18
98	5	91	92	7	4	9	100	6	93
58	46	50	59	44	47	52	51	45	53
73	25	71	72	27	24	79	30	26	78

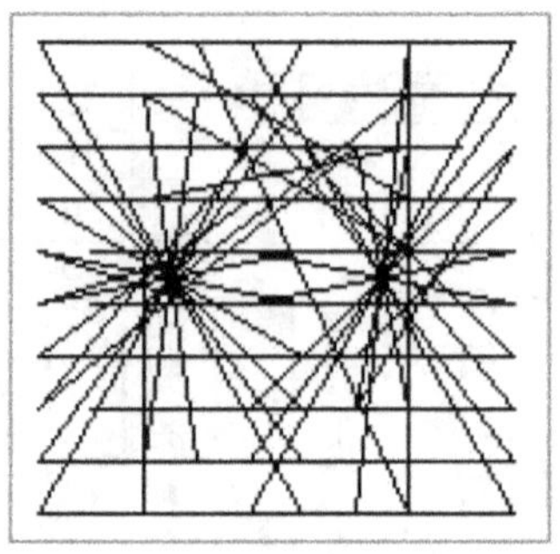

45	44	52	58	50	51	53	59	47	46
35	34	62	68	70	61	63	39	37	36
16	14	89	83	81	90	18	82	17	15
76	77	29	23	21	80	28	22	74	75
96	97	99	8	1	10	3	2	94	95
5	7	9	93	91	100	98	92	4	6
26	27	72	73	71	30	78	79	24	25
86	87	19	13	20	11	88	12	84	85
65	64	32	38	40	31	33	69	67	66
55	54	42	48	60	41	43	49	57	56

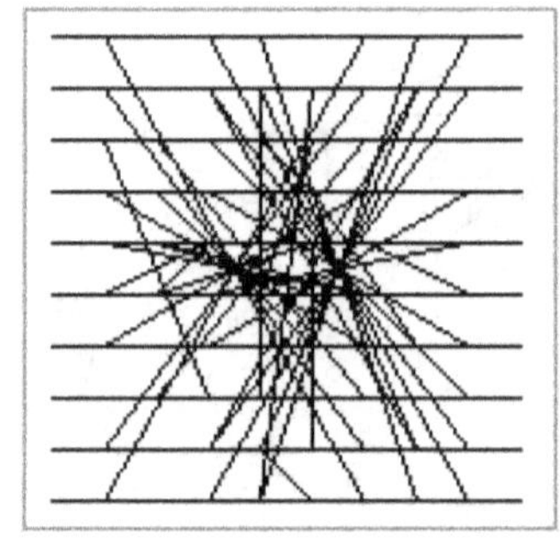

56	66	75	85	95	6	15	25	36	46
57	67	74	84	94	4	17	24	37	47
48	38	23	83	8	93	13	73	68	58
49	69	29	12	2	92	82	72	39	59
60	40	21	20	1	91	81	71	70	50
41	31	80	11	10	100	90	30	61	51
42	32	22	19	99	9	89	79	62	52
43	33	28	18	3	98	88	78	63	53
54	64	77	87	97	7	14	27	34	44
55	65	76	86	96	5	16	26	35	45

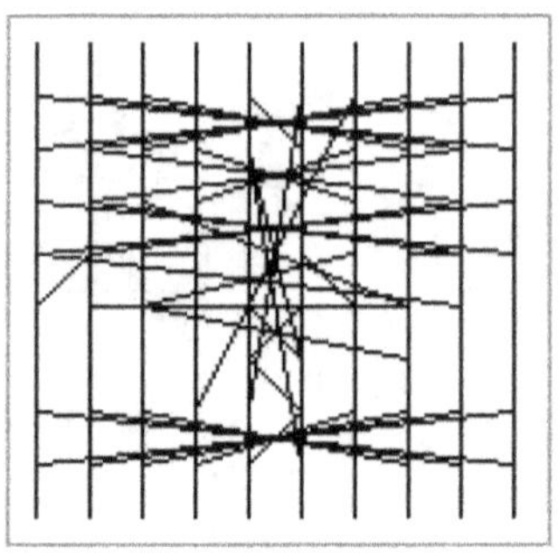

65	67	93	95	2	3	32	30	60	58
66	68	94	96	4	1	31	29	59	57
89	91	17	19	27	26	56	54	64	62
90	92	18	20	25	28	55	53	63	61
16	14	24	22	49	50	78	80	87	85
13	15	21	23	51	52	79	77	86	88
40	38	48	46	73	76	81	83	9	11
39	37	47	45	75	74	82	84	10	12
44	42	72	70	100	97	5	7	33	35
43	41	71	69	99	98	6	8	34	36

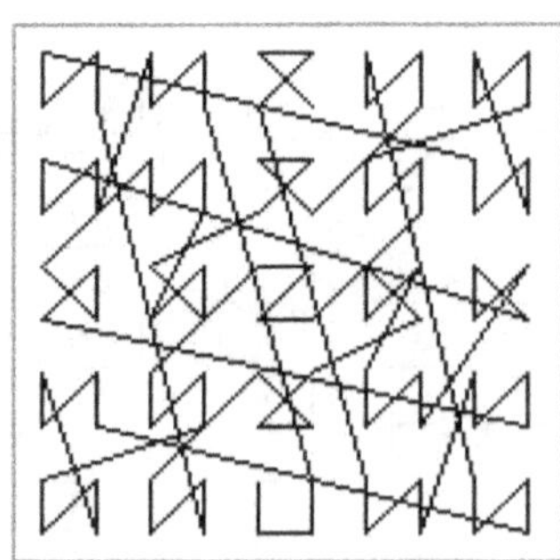

The following magic squares continue highlighting the odd numbers on the left, and on the right, the lines connect numbers in sequence pairs.

1	3	57	59	65	68	96	94	32	30
2	4	58	60	67	66	95	93	31	29
53	55	89	91	20	17	28	26	64	62
54	56	90	92	18	19	27	25	63	61
86	88	22	24	49	50	77	78	16	15
87	85	23	21	51	52	80	79	13	14
40	37	76	74	84	82	9	11	45	47
39	38	75	73	81	83	10	12	46	48
72	70	8	6	36	33	41	43	97	99
71	69	7	5	34	35	42	44	98	100

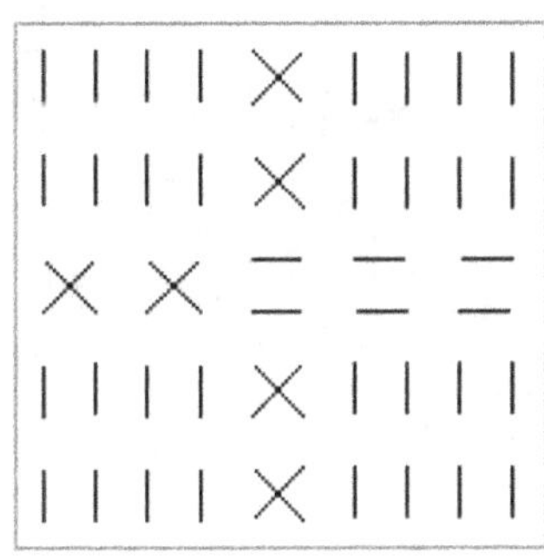

23	75	21	29	77	74	22	80	76	28
48	56	60	49	54	57	42	41	55	43
3	96	1	2	97	94	99	10	95	8
13	85	11	12	84	87	19	20	86	88
68	36	70	39	34	37	62	61	35	63
38	66	40	69	64	67	32	31	65	33
83	15	90	82	17	14	89	81	16	18
98	5	91	92	7	4	9	100	6	93
58	46	50	59	44	47	52	51	45	53
73	25	71	72	27	24	79	30	26	78

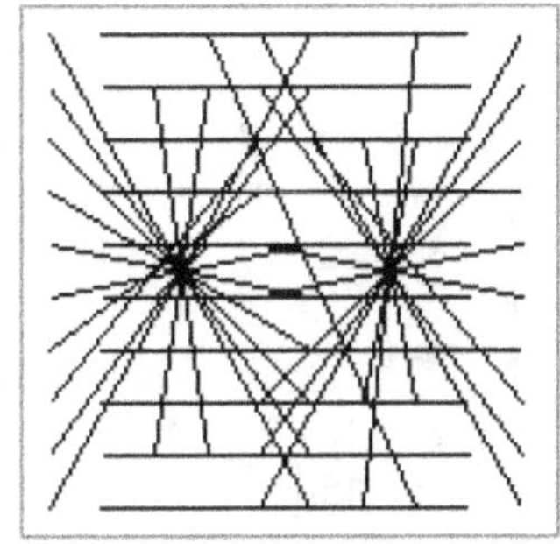

12	11	85	13	87	84	88	86	20	19
2	1	96	3	94	97	8	95	10	99
49	60	56	48	57	54	43	55	41	42
29	21	75	23	74	77	28	76	80	22
69	40	66	38	67	64	33	65	31	32
39	70	36	68	37	34	63	35	61	62
72	71	25	73	24	27	78	26	30	79
59	50	46	58	47	44	53	45	51	52
92	91	5	98	4	7	93	6	100	9
82	90	15	83	14	17	18	16	81	89

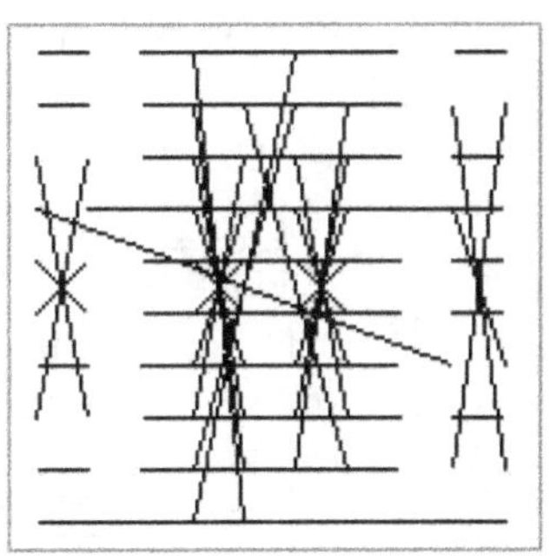

23	73	6	56	94	19	77	27	90	40
48	98	31	81	69	44	52	2	65	15
10	60	18	68	51	26	89	39	97	47
35	85	43	93	1	76	64	14	72	22
92	42	80	30	13	63	46	21	34	84
17	67	5	55	38	88	71	96	59	9
79	29	87	37	25	100	8	58	16	66
54	4	62	12	75	50	33	83	41	91
86	36	99	49	57	32	20	70	3	53
61	11	74	24	82	7	45	95	28	78

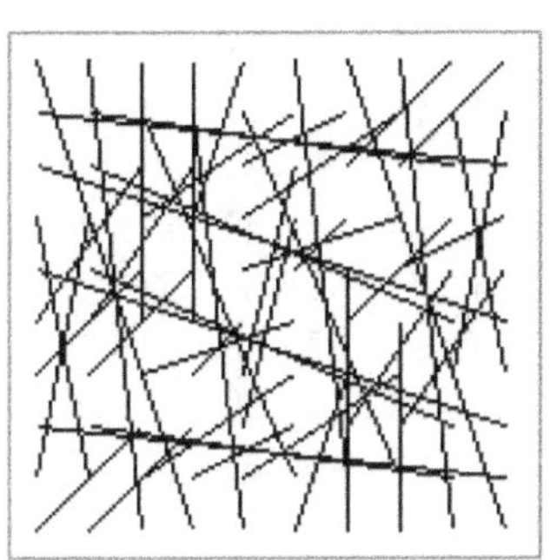

The following are regular 10th order magic squares with circular plots.

1	92	3	97	95	96	94	8	9	10
11	12	83	84	86	85	87	18	19	20
71	29	23	77	76	75	74	28	22	30
70	69	68	34	35	36	37	63	32	61
60	59	58	44	45	46	47	53	52	41
50	49	48	54	55	56	57	43	42	51
40	39	38	64	65	66	67	33	62	31
21	72	73	27	26	25	24	78	79	80
90	82	13	17	16	15	14	88	89	81
91	2	98	7	6	5	4	93	99	100

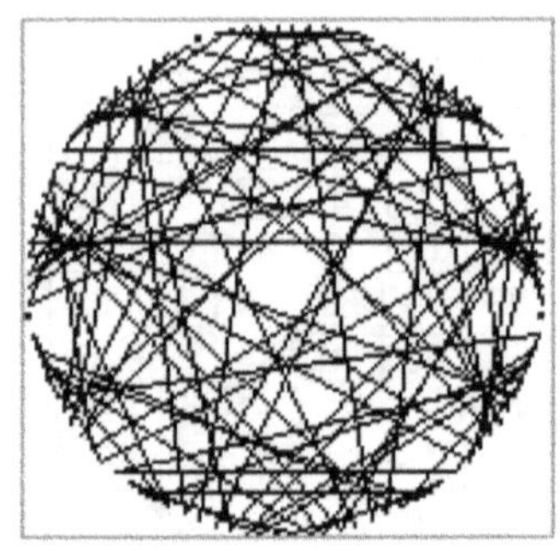

Circular Orig to Curr

45	35	26	16	96	5	86	76	65	55
44	34	27	14	97	7	87	77	64	54
53	63	78	88	3	98	18	28	33	43
52	62	79	89	99	9	19	22	32	42
50	70	71	81	1	91	20	21	40	60
51	61	30	90	10	100	11	80	31	41
59	39	72	82	2	92	12	29	69	49
58	68	73	13	8	93	83	23	38	48
47	37	24	17	94	4	84	74	67	57
46	36	25	15	95	6	85	75	66	56

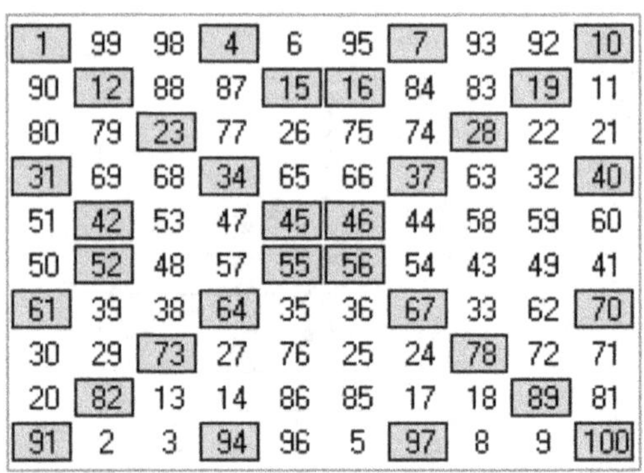

Circular Orig to Curr

1	99	98	4	6	95	7	93	92	10
90	12	88	87	15	16	84	83	19	11
80	79	23	77	26	75	74	28	22	21
31	69	68	34	65	66	37	63	32	40
51	42	53	47	45	46	44	58	59	60
50	52	48	57	55	56	54	43	49	41
61	39	38	64	35	36	67	33	62	70
30	29	73	27	76	25	24	78	72	71
20	82	13	14	86	85	17	18	89	81
91	2	3	94	96	5	97	8	9	100

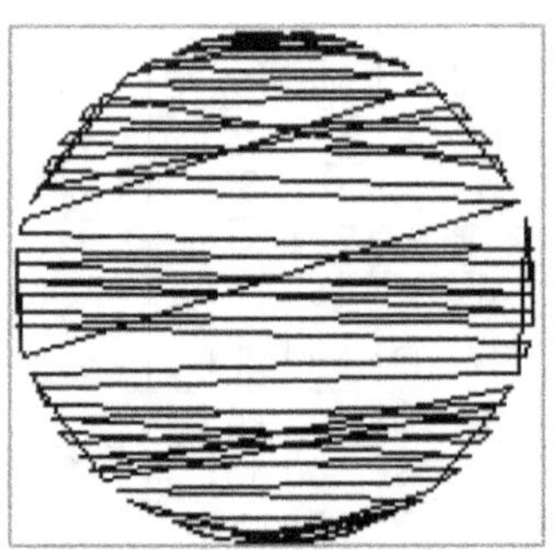

Circular Std Sequence

1	3	94	96	65	68	59	57	32	30
2	4	93	95	67	66	60	58	31	29
39	38	12	10	81	83	73	75	46	48
40	37	11	9	84	82	74	76	45	47
86	88	78	77	49	50	24	22	16	15
87	85	79	80	51	52	21	23	13	14
54	56	25	27	18	19	92	90	63	61
53	55	26	28	20	17	91	89	64	62
72	70	43	41	36	33	6	8	97	99
71	69	44	42	34	35	5	7	98	100

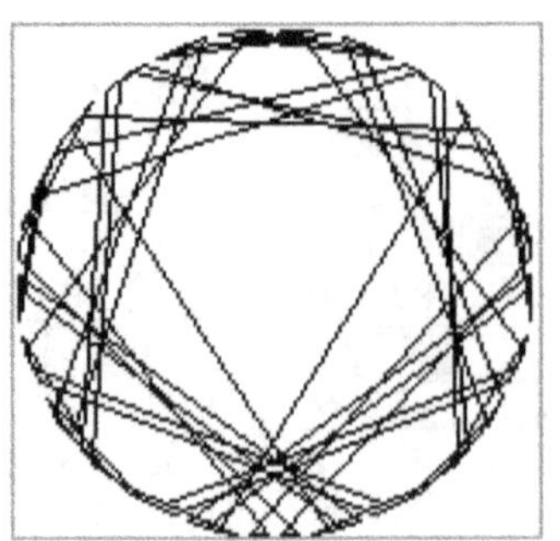

Circular Std Sequence

Appendix H – Selected 11th Order Magic Squares

The following are regular 11th order magic squares. Boxed numbers in the left squares are numbers in their original positions. The right square shows the movement of the numbers from their original to their current position.

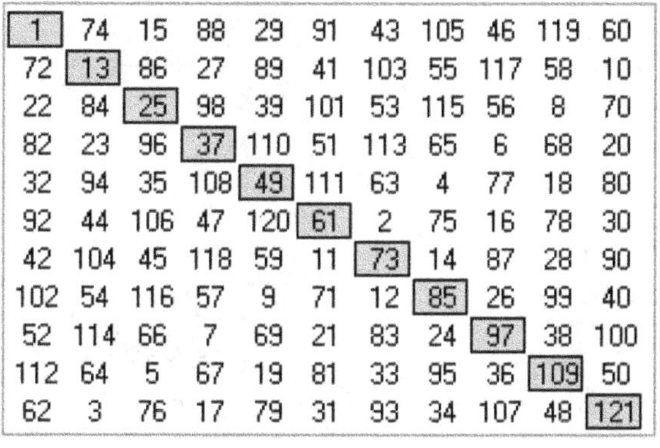

1	74	15	88	29	91	43	105	46	119	60
72	13	86	27	89	41	103	55	117	58	10
22	84	25	98	39	101	53	115	56	8	70
82	23	96	37	110	51	113	65	6	68	20
32	94	35	108	49	111	63	4	77	18	80
92	44	106	47	120	61	2	75	16	78	30
42	104	45	118	59	11	73	14	87	28	90
102	54	116	57	9	71	12	85	26	99	40
52	114	66	7	69	21	83	24	97	38	100
112	64	5	67	19	81	33	95	36	109	50
62	3	76	17	79	31	93	34	107	48	121

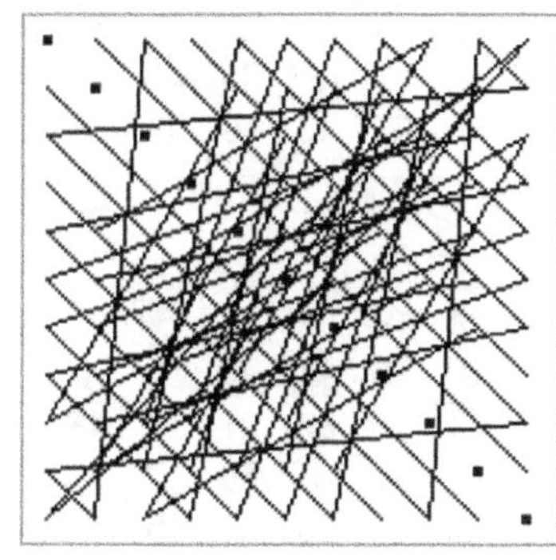

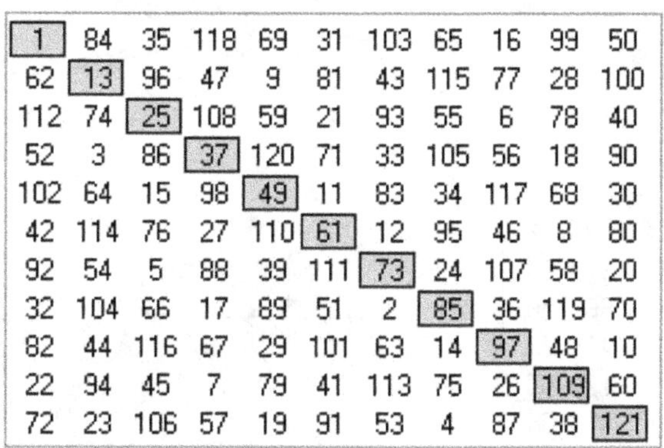

1	84	35	118	69	31	103	65	16	99	50
62	13	96	47	9	81	43	115	77	28	100
112	74	25	108	59	21	93	55	6	78	40
52	3	86	37	120	71	33	105	56	18	90
102	64	15	98	49	11	83	34	117	68	30
42	114	76	27	110	61	12	95	46	8	80
92	54	5	88	39	111	73	24	107	58	20
32	104	66	17	89	51	2	85	36	119	70
82	44	116	67	29	101	63	14	97	48	10
22	94	45	7	79	41	113	75	26	109	60
72	23	106	57	19	91	53	4	87	38	121

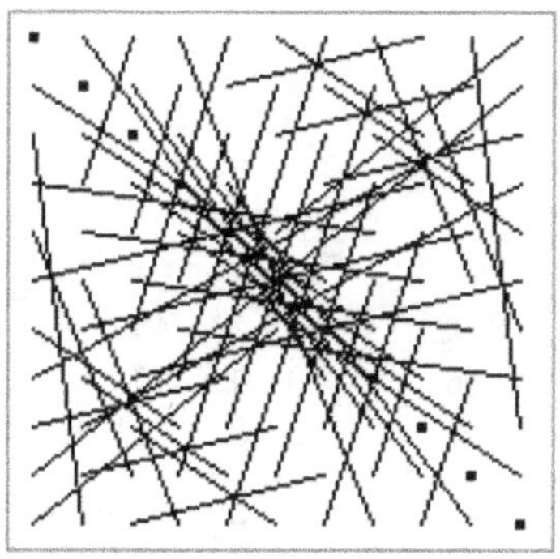

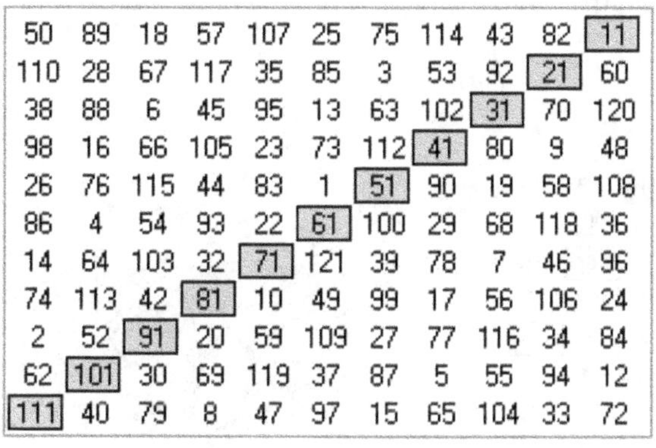

50	89	18	57	107	25	75	114	43	82	11
110	28	67	117	35	85	3	53	92	21	60
38	88	6	45	95	13	63	102	31	70	120
98	16	66	105	23	73	112	41	80	9	48
26	76	115	44	83	1	51	90	19	58	108
86	4	54	93	22	61	100	29	68	118	36
14	64	103	32	71	121	39	78	7	46	96
74	113	42	81	10	49	99	17	56	106	24
2	52	91	20	59	109	27	77	116	34	84
62	101	30	69	119	37	87	5	55	94	12
111	40	79	8	47	97	15	65	104	33	72

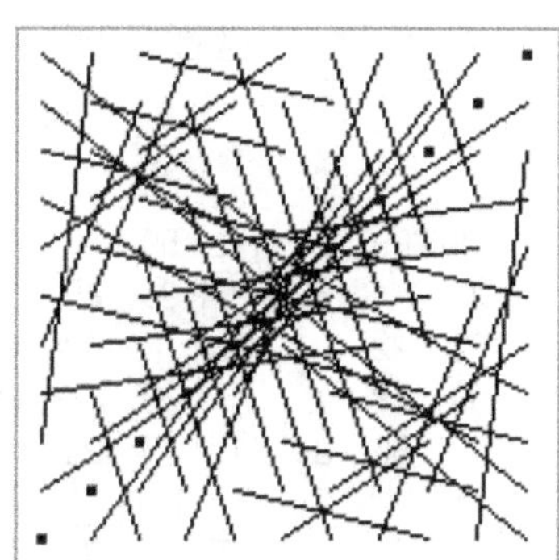

1	28	114	98	63	36	20	106	79	55	71
58	85	50	23	120	93	77	42	15	101	7
44	60	25	9	95	68	52	17	111	87	103
108	3	89	73	38	22	116	81	65	30	46
83	110	75	48	13	118	91	56	40	5	32
26	53	18	112	88	61	34	10	104	69	96
90	117	82	66	31	4	109	74	47	12	39
76	92	57	41	6	100	84	49	33	119	14
19	35	11	105	70	54	27	113	97	62	78
115	21	107	80	45	29	2	99	72	37	64
51	67	43	16	102	86	59	24	8	94	121

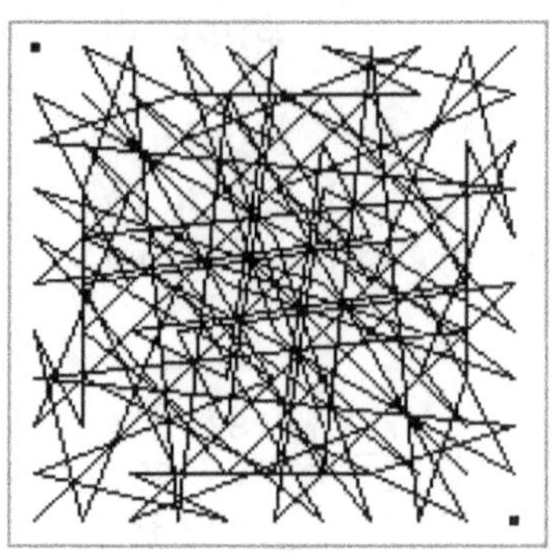

59	18	98	46	5	85	44	113	72	31	100
107	66	14	94	53	1	81	40	120	68	27
23	103	62	21	90	49	8	88	36	116	75
71	30	110	58	17	97	45	4	84	43	112
119	67	26	106	65	13	93	52	11	80	39
35	115	74	33	102	61	20	89	48	7	87
83	42	111	70	29	109	57	16	96	55	3
10	79	38	118	77	25	105	64	12	92	51
47	6	86	34	114	73	32	101	60	19	99
95	54	2	82	41	121	69	28	108	56	15
22	91	50	9	78	37	117	76	24	104	63

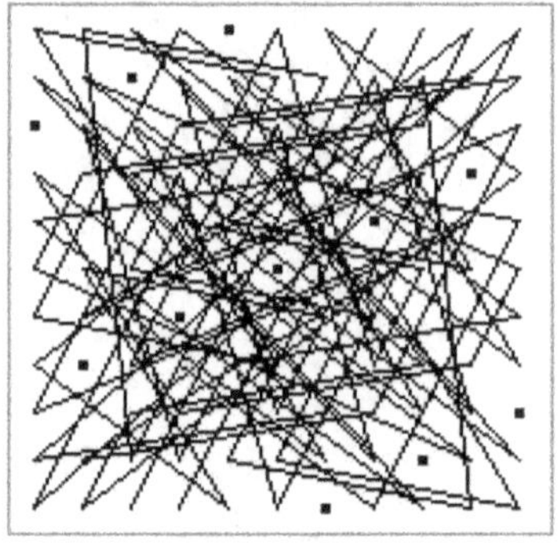

121	32	46	78	64	96	7	103	14	39	71
102	13	38	70	45	88	120	95	6	31	63
16	48	73	105	80	112	23	9	41	66	98
8	40	65	97	72	104	15	111	33	47	79
94	5	30	62	37	69	101	87	119	12	55
86	118	22	54	29	61	93	68	100	4	36
67	110	3	35	21	53	85	60	92	117	28
43	75	89	11	107	18	50	25	57	82	114
24	56	81	113	99	10	42	17	49	74	106
59	91	116	27	2	34	77	52	84	109	20
51	83	108	19	115	26	58	44	76	90	1

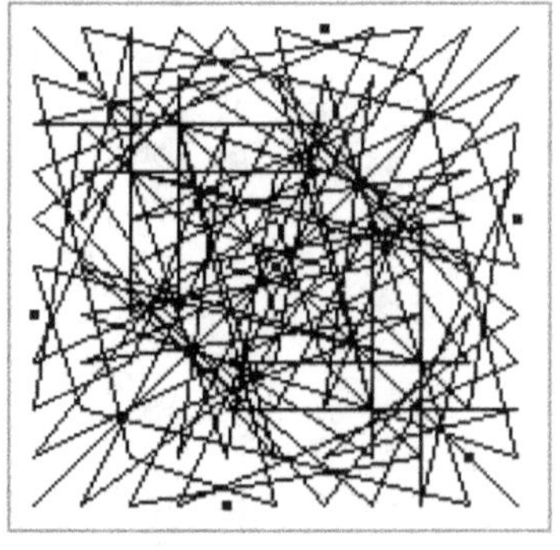

120	69	29	110	59	19	89	49	9	79	39
25	106	66	15	96	45	5	86	35	116	76
62	22	92	52	1	82	42	112	72	32	102
99	48	8	78	38	119	68	28	109	58	18
4	85	34	115	75	24	105	65	14	95	55
41	111	71	31	101	61	21	91	51	11	81
67	27	108	57	17	98	47	7	88	37	118
104	64	13	94	54	3	84	44	114	74	23
20	90	50	10	80	40	121	70	30	100	60
46	6	87	36	117	77	26	107	56	16	97
83	43	113	73	33	103	63	12	93	53	2

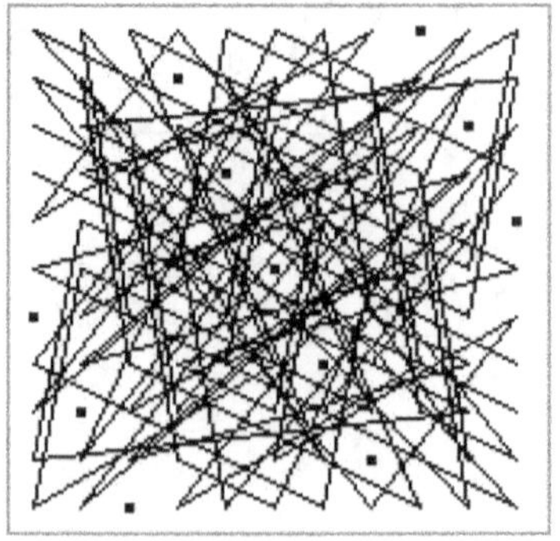

49	119	57	6	76	100	14	84	33	92	41
99	37	107	45	115	29	64	2	72	21	80
17	87	25	95	44	68	103	52	111	60	9
56	5	75	13	83	118	32	91	40	110	48
106	55	114	63	1	36	71	20	79	28	98
10	69	18	88	26	61	96	34	104	53	112
24	94	43	102	51	86	121	59	8	67	16
74	12	82	31	90	4	39	109	47	117	66
113	62	11	70	19	54	78	27	97	35	105
42	101	50	120	58	93	7	77	15	85	23
81	30	89	38	108	22	46	116	65	3	73

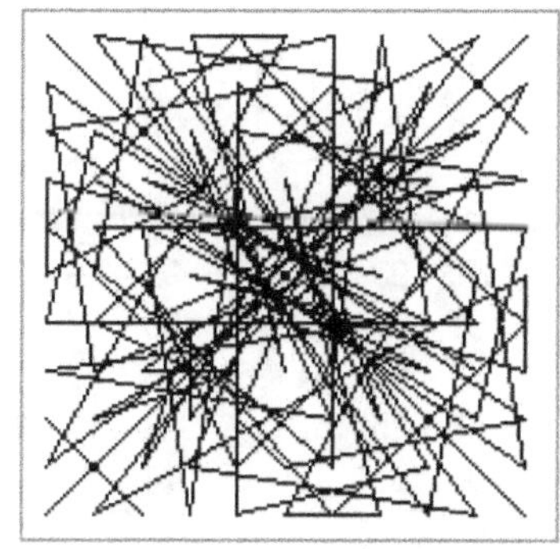

121	79	107	93	3	76	17	48	34	62	31
19	109	5	112	33	95	36	67	64	81	50
87	45	73	59	90	42	104	14	11	28	118
53	22	39	25	56	8	70	101	98	115	84
23	113	20	6	37	110	51	82	68	96	65
106	75	92	78	120	61	2	44	30	47	16
57	26	54	40	71	12	85	116	102	9	99
38	7	24	21	52	114	66	97	83	100	69
4	94	111	108	18	80	32	63	49	77	35
72	41	58	55	86	27	89	10	117	13	103
91	60	88	74	105	46	119	29	15	43	1

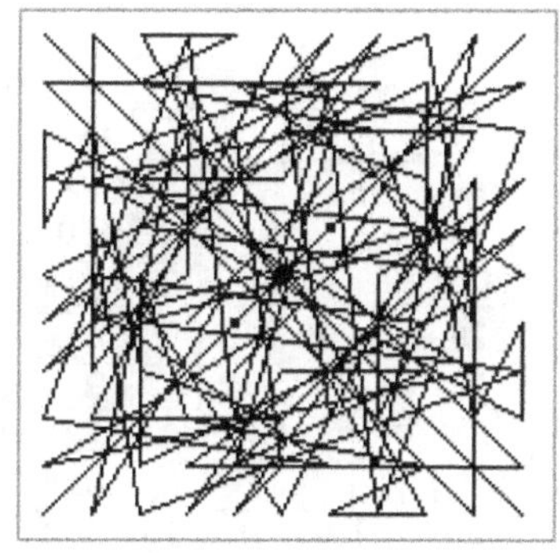

54	42	30	18	6	115	103	91	79	67	66
41	29	17	5	114	102	90	78	77	65	53
28	16	4	113	101	89	88	76	64	52	40
15	3	112	100	99	87	75	63	51	39	27
2	111	110	98	86	74	62	50	38	26	14
121	109	97	85	73	61	49	37	25	13	1
108	96	84	72	60	48	36	24	12	11	120
95	83	71	59	47	35	23	22	10	119	107
82	70	58	46	34	33	21	9	118	106	94
69	57	45	44	32	20	8	117	105	93	81
56	55	43	31	19	7	116	104	92	80	68

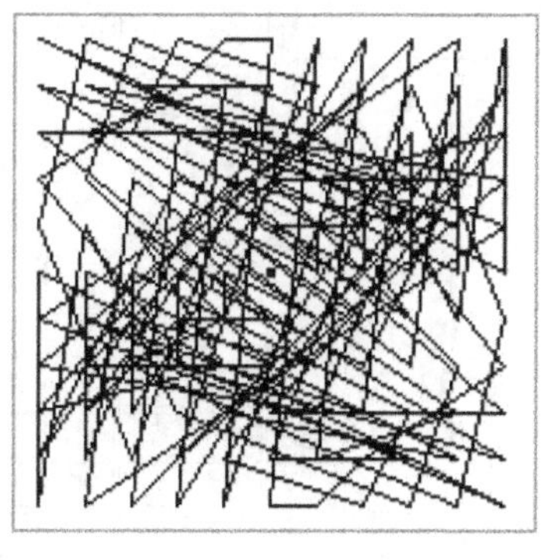

60	9	68	17	87	25	95	44	103	52	111
120	58	7	77	15	85	23	93	42	101	50
48	118	56	5	75	13	83	32	91	40	110
108	46	116	65	3	73	22	81	30	89	38
36	106	55	114	63	1	71	20	79	28	98
96	34	104	53	112	61	10	69	18	88	26
24	94	43	102	51	121	59	8	67	16	86
84	33	92	41	100	49	119	57	6	76	14
12	82	31	90	39	109	47	117	66	4	74
72	21	80	29	99	37	107	45	115	64	2
11	70	19	78	27	97	35	105	54	113	62

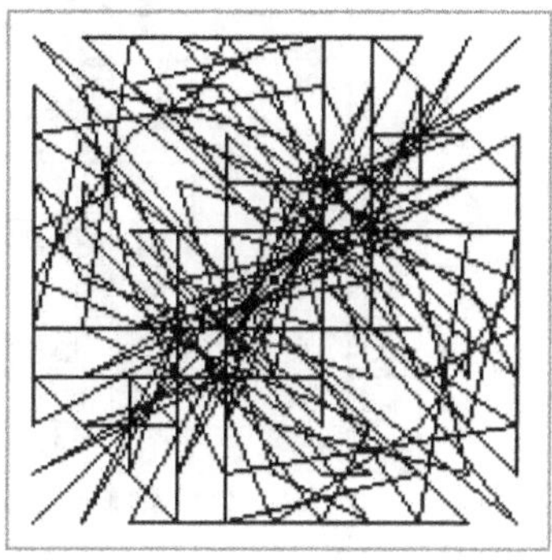

239

54	41	28	15	2	121	108	95	82	69	56
42	29	16	3	111	109	96	83	70	57	55
30	17	4	112	110	97	84	71	58	45	43
18	5	113	100	98	85	72	59	46	44	31
6	114	101	99	86	73	60	47	34	32	19
115	102	89	87	74	61	48	35	33	20	7
103	90	88	75	62	49	36	23	21	8	116
91	78	76	63	50	37	24	22	9	117	104
79	77	64	51	38	25	12	10	118	105	92
67	65	52	39	26	13	11	119	106	93	80
66	53	40	27	14	1	120	107	94	81	68

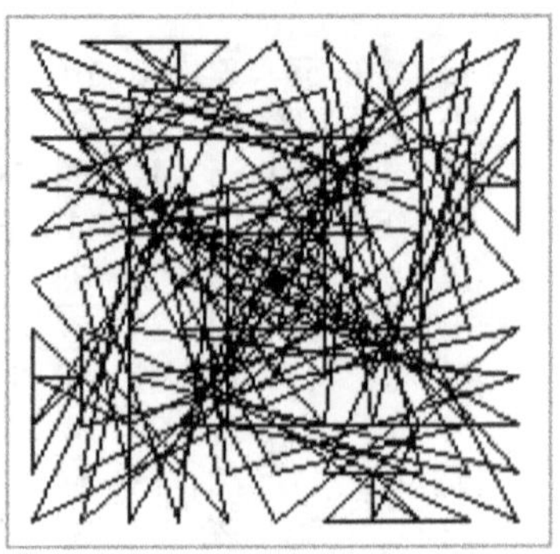

These are regular 11th order magic squares. Odd numbers are highlighted on the left, and the lines are drawn in sequence on the right (from 1 to 2 to 3, etc.)

1	63	114	106	55	36	28	98	79	20	71
83	13	75	56	5	118	110	48	40	91	32
44	95	25	17	87	68	60	9	111	52	103
76	6	57	49	119	100	92	41	33	84	14
115	45	107	99	37	29	21	80	72	2	64
26	88	18	10	69	61	53	112	104	34	96
58	120	50	42	101	93	85	23	15	77	7
108	38	89	81	30	22	3	73	65	116	46
19	70	11	113	62	54	35	105	97	27	78
90	31	82	74	12	4	117	66	47	109	39
51	102	43	24	94	86	67	16	8	59	121

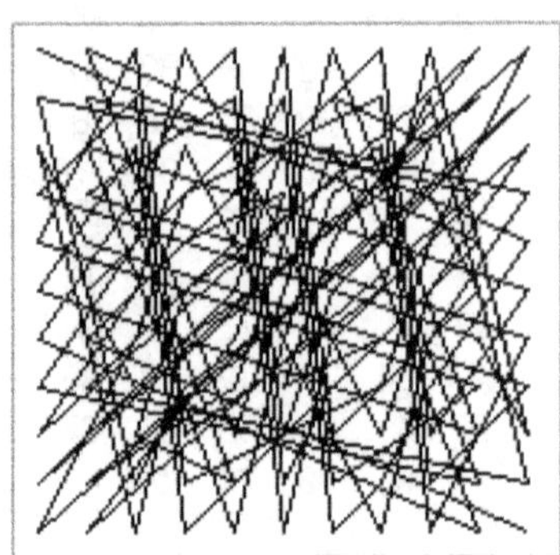

1	74	15	88	29	91	43	105	46	119	60
72	13	86	27	89	41	103	55	117	58	10
22	84	25	98	39	101	53	115	56	8	70
82	23	96	37	110	51	113	65	6	68	20
32	94	35	108	49	111	63	4	77	18	80
92	44	106	47	120	61	2	75	16	78	30
42	104	45	118	59	11	73	14	87	28	90
102	54	116	57	9	71	12	85	26	99	40
52	114	66	7	69	21	83	24	97	38	100
112	64	5	67	19	81	33	95	36	109	50
62	3	76	17	79	31	93	34	107	48	121

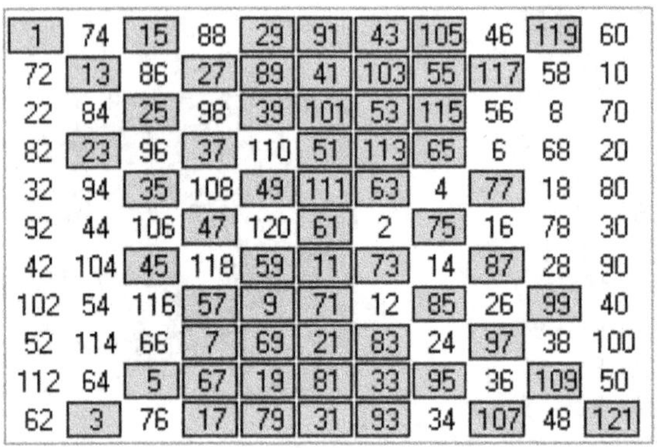

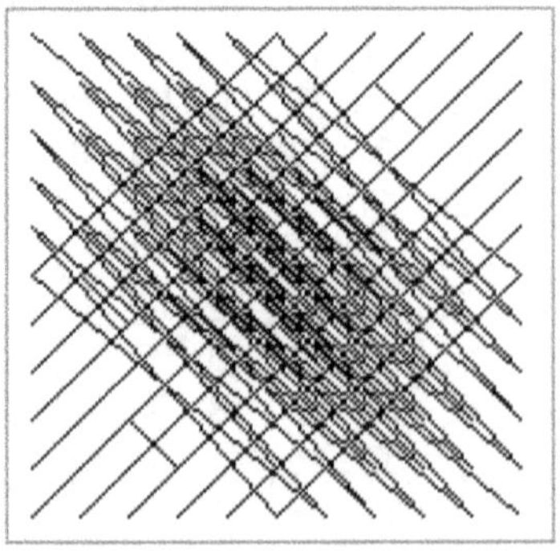

50	100	40	90	30	121	80	20	70	10	60
110	39	89	29	79	49	19	69	9	59	120
38	99	28	78	18	109	68	8	58	119	48
98	27	88	17	67	37	7	57	118	47	108
26	87	16	77	6	97	56	117	46	107	36
111	51	101	41	91	61	31	81	21	71	11
86	15	76	5	66	25	116	45	106	35	96
14	75	4	65	115	85	55	105	34	95	24
74	3	64	114	54	13	104	44	94	23	84
2	63	113	53	103	73	43	93	33	83	12
62	112	52	102	42	1	92	32	82	22	72

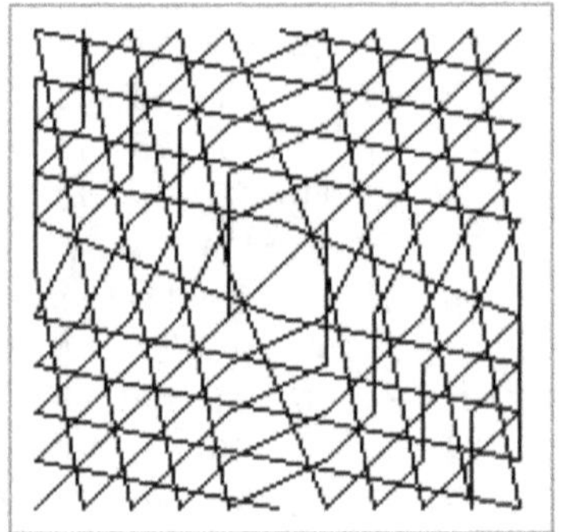

60	9	103	44	95	25	87	17	68	52	111
120	58	42	93	23	85	15	77	7	101	50
48	118	91	32	83	13	75	5	56	40	110
108	46	30	81	22	73	3	65	116	89	38
36	106	79	20	71	1	63	114	55	28	98
96	34	18	69	10	61	112	53	104	88	26
24	94	67	8	59	121	51	102	43	16	86
84	33	6	57	119	49	100	41	92	76	14
12	82	66	117	47	109	39	90	31	4	74
72	21	115	45	107	37	99	29	80	64	2
11	70	54	105	35	97	27	78	19	113	62

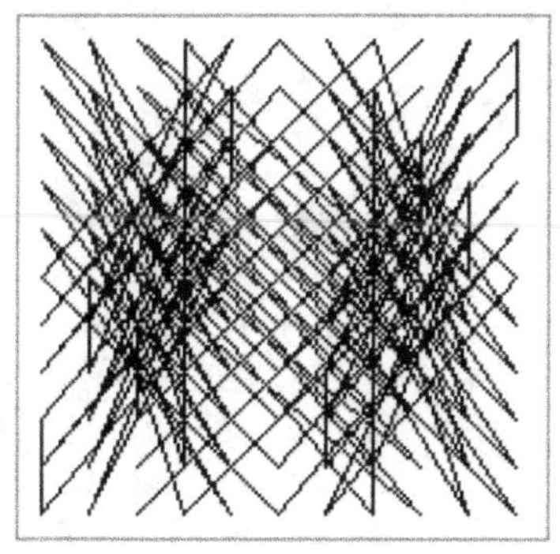

60	114	101	47	6	73	19	99	34	32	86
11	65	52	119	67	13	80	39	106	93	26
12	77	64	10	79	25	92	51	118	105	38
72	5	113	59	18	85	31	100	46	44	98
120	53	40	107	66	1	68	27	94	81	14
48	102	89	35	115	61	7	87	33	20	74
108	41	28	95	54	121	56	15	82	69	2
24	78	76	22	91	37	104	63	9	117	50
84	17	4	71	30	97	43	112	58	45	110
96	29	16	83	42	109	55	3	70	57	111
36	90	88	23	103	49	116	75	21	8	62

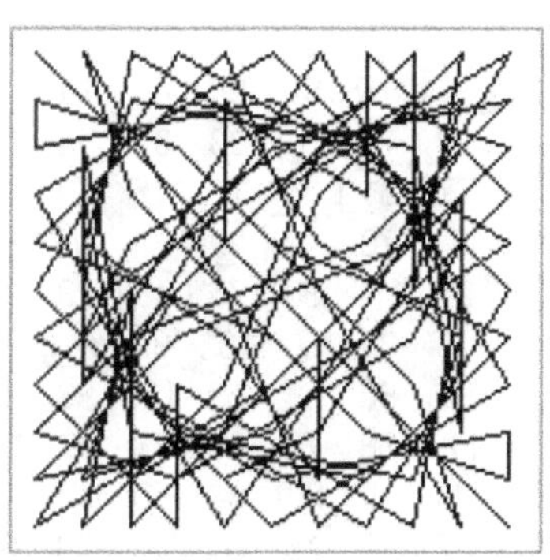

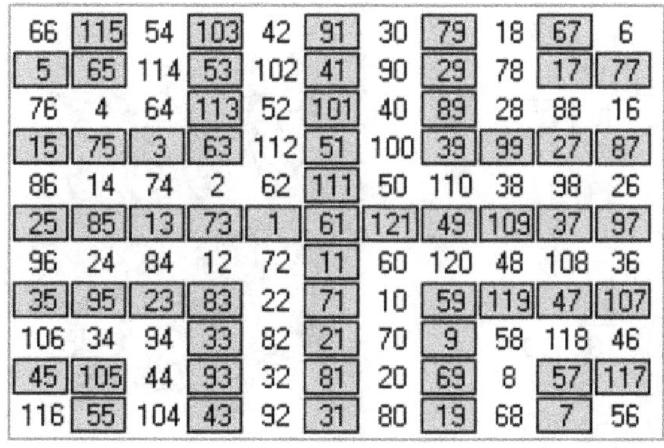

66	115	54	103	42	91	30	79	18	67	6
5	65	114	53	102	41	90	29	78	17	77
76	4	64	113	52	101	40	89	28	88	16
15	75	3	63	112	51	100	39	99	27	87
86	14	74	2	62	111	50	110	38	98	26
25	85	13	73	1	61	121	49	109	37	97
96	24	84	12	72	11	60	120	48	108	36
35	95	23	83	22	71	10	59	119	47	107
106	34	94	33	82	21	70	9	58	118	46
45	105	44	93	32	81	20	69	8	57	117
116	55	104	43	92	31	80	19	68	7	56

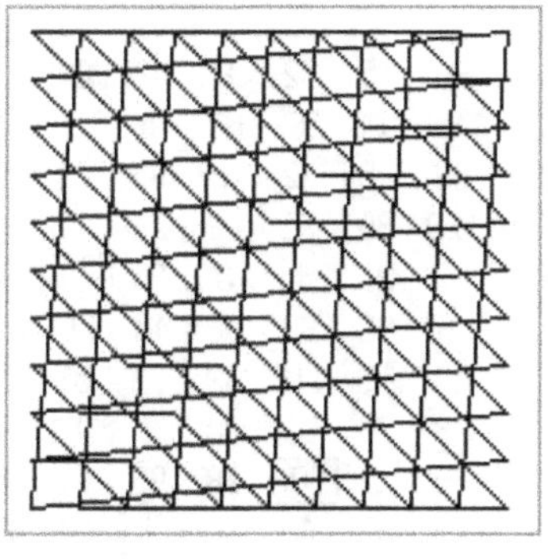

83	33	12	43	93	73	53	103	2	113	63
23	94	84	104	44	13	114	54	74	64	3
22	82	72	92	32	1	102	42	62	52	112
35	106	96	116	45	25	5	66	86	76	15
95	34	24	55	105	85	65	115	14	4	75
71	21	11	31	81	61	41	91	111	101	51
47	118	108	7	57	37	17	67	98	88	27
107	46	36	56	117	97	77	6	26	16	87
10	70	60	80	20	121	90	30	50	40	100
119	58	48	68	8	109	78	18	38	28	99
59	9	120	19	69	49	29	79	110	89	39

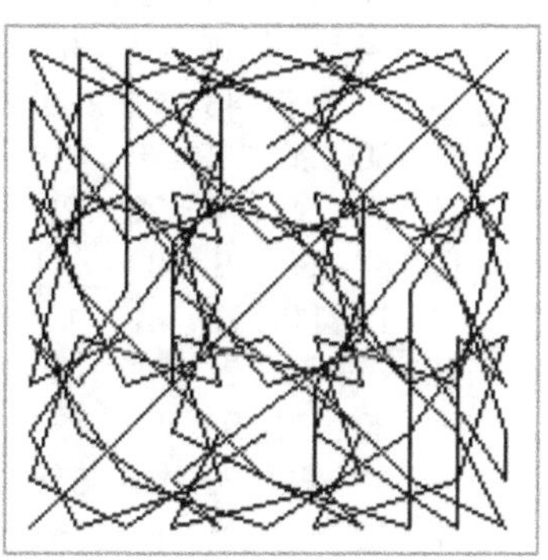

The following 11th order magic squares are pan-diagonal

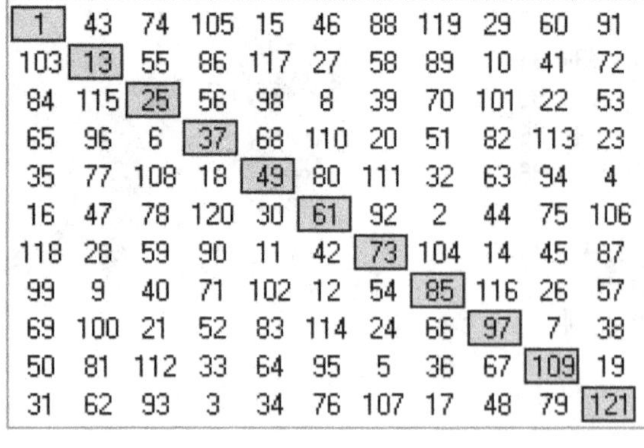

1	43	74	105	15	46	88	119	29	60	91
103	13	55	86	117	27	58	89	10	41	72
84	115	25	56	98	8	39	70	101	22	53
65	96	6	37	68	110	20	51	82	113	23
35	77	108	18	49	80	111	32	63	94	4
16	47	78	120	30	61	92	2	44	75	106
118	28	59	90	11	42	73	104	14	45	87
99	9	40	71	102	12	54	85	116	26	57
69	100	21	52	83	114	24	66	97	7	38
50	81	112	33	64	95	5	36	67	109	19
31	62	93	3	34	76	107	17	48	79	121

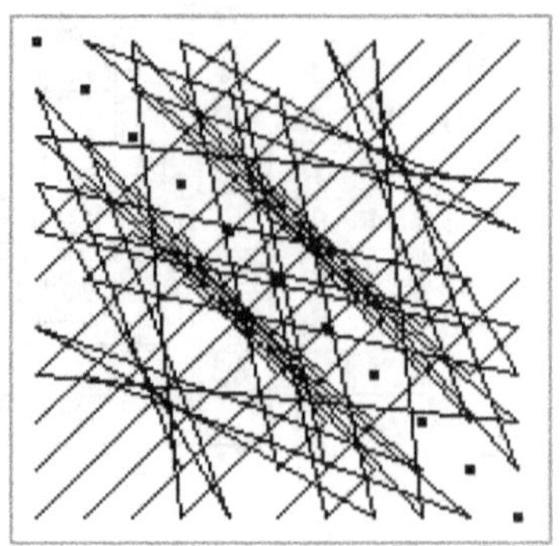

1	107	81	66	40	14	120	94	68	53	27
39	13	119	93	67	52	26	11	106	80	65
77	51	25	10	105	79	64	38	12	118	92
104	78	63	37	22	117	91	76	50	24	9
21	116	90	75	49	23	8	103	88	62	36
48	33	7	102	87	61	35	20	115	89	74
86	60	34	19	114	99	73	47	32	6	101
113	98	72	46	31	5	100	85	59	44	18
30	4	110	84	58	43	17	112	97	71	45
57	42	16	111	96	70	55	29	3	109	83
95	69	54	28	2	108	82	56	41	15	121

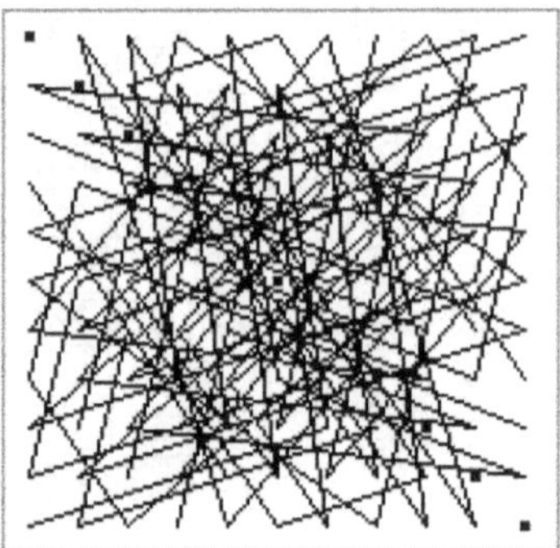

19	116	92	68	55	31	7	104	80	56	43
69	45	32	8	105	81	57	44	20	117	93
9	106	82	58	34	21	118	94	70	46	33
59	35	22	119	95	71	47	23	10	107	83
120	96	72	48	24	11	108	84	60	36	12
49	25	1	109	85	61	37	13	121	97	73
110	86	62	38	14	111	98	74	50	26	2
39	15	112	99	75	51	27	3	100	87	63
89	76	52	28	4	101	88	64	40	16	113
29	5	102	78	65	41	17	114	90	77	53
79	66	42	18	115	91	67	54	30	6	103

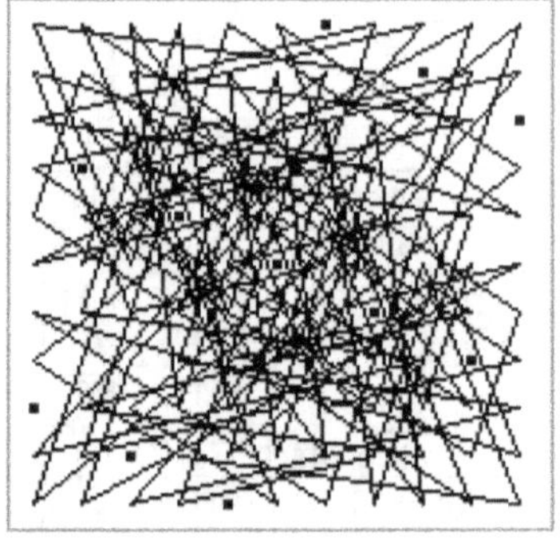

22	103	74	34	5	97	57	28	120	80	51
83	54	14	106	77	37	8	89	60	31	112
23	115	86	46	17	109	69	40	11	92	63
95	66	26	118	78	49	20	101	72	43	3
35	6	98	58	29	121	81	52	12	104	75
107	67	38	9	90	61	32	113	84	55	15
47	18	110	70	41	1	93	64	24	116	87
119	79	50	21	102	73	44	4	96	56	27
59	30	111	82	53	13	105	76	36	7	99
10	91	62	33	114	85	45	16	108	68	39
71	42	2	94	65	25	117	88	48	19	100

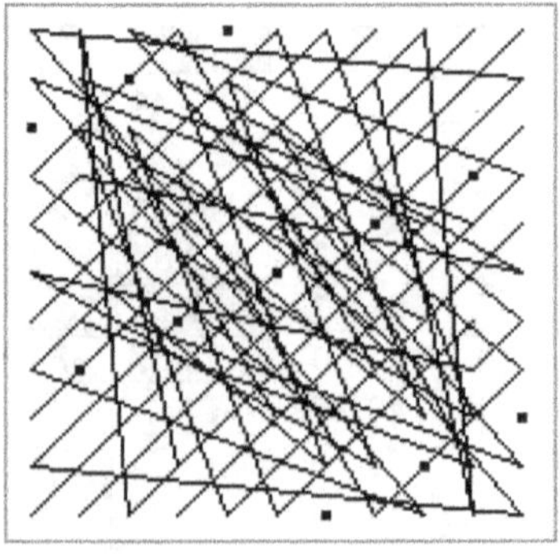

The following are regular magic squares with circular plots.

1	84	35	118	69	31	103	65	16	99	50
62	13	96	47	9	81	43	115	77	28	100
112	74	25	108	59	21	93	55	6	78	40
52	3	86	37	120	71	33	105	56	18	90
102	64	15	98	49	11	83	34	117	68	30
42	114	76	27	110	61	12	95	46	8	80
92	54	5	88	39	111	73	24	107	58	20
32	104	66	17	89	51	2	85	36	119	70
82	44	116	67	29	101	63	14	97	48	10
22	94	45	7	79	41	113	75	26	109	60
72	23	106	57	19	91	53	4	87	38	121

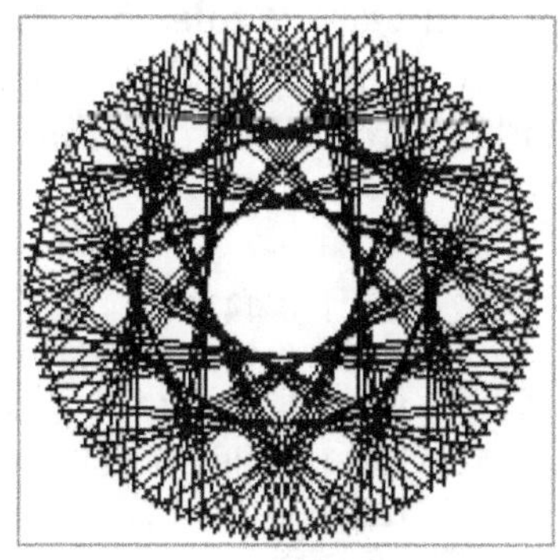

Circular Std Sequence

1	72	22	82	32	92	42	102	52	112	62
74	13	84	23	94	44	104	54	114	64	3
15	86	25	96	35	106	45	116	66	5	76
88	27	98	37	108	47	118	57	7	67	17
29	89	39	110	49	120	59	9	69	19	79
91	41	101	51	111	61	11	71	21	81	31
43	103	53	113	63	2	73	12	83	33	93
105	55	115	65	4	75	14	85	24	95	34
46	117	56	6	77	16	87	26	97	36	107
119	58	8	68	18	78	28	99	38	109	48
60	10	70	20	80	30	90	40	100	50	121

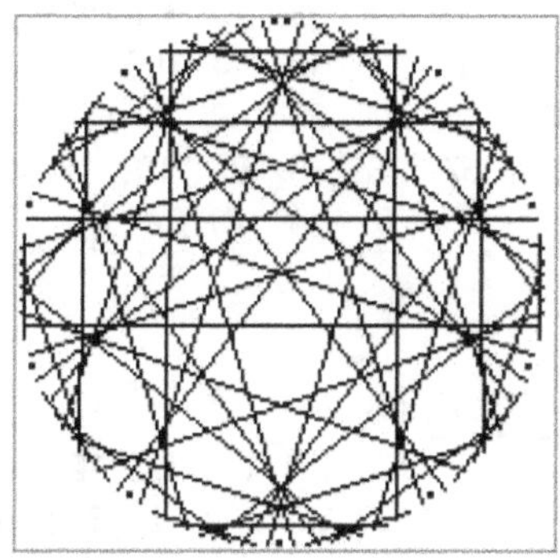

Circular Orig to Curr

50	41	32	12	3	115	106	97	88	68	59
37	28	19	10	111	102	93	84	75	66	46
24	15	6	118	109	89	80	71	62	53	44
22	2	114	105	96	87	67	58	49	40	31
9	121	101	92	83	74	65	45	36	27	18
117	108	99	79	70	61	52	43	23	14	5
104	95	86	77	57	48	39	30	21	1	113
91	82	73	64	55	35	26	17	8	120	100
78	69	60	51	42	33	13	4	116	107	98
76	56	47	38	29	20	11	112	103	94	85
63	54	34	25	16	7	119	110	90	81	72

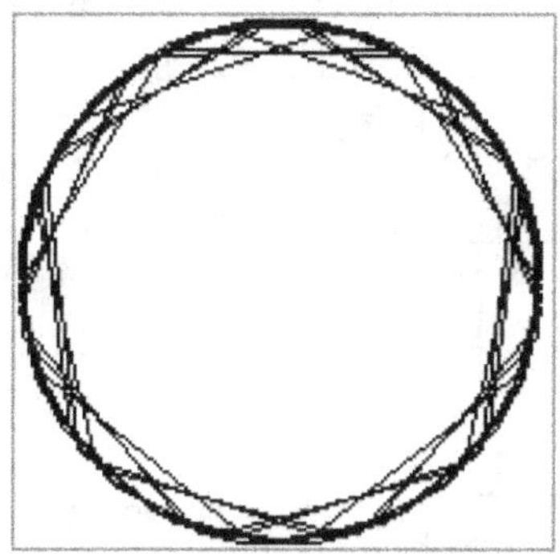

Circular Std Sequence

39	87	3	51	99	15	63	100	27	75	112
68	116	43	80	7	55	92	19	56	104	31
108	24	72	120	36	84	11	48	96	12	60
16	64	101	28	76	113	40	88	4	52	89
45	93	20	57	105	32	69	117	44	81	8
85	1	49	97	13	61	109	25	73	121	37
114	41	78	5	53	90	17	65	102	29	77
33	70	118	34	82	9	46	94	21	58	106
62	110	26	74	111	38	86	2	50	98	14
91	18	66	103	30	67	115	42	79	6	54
10	47	95	22	59	107	23	71	119	35	83

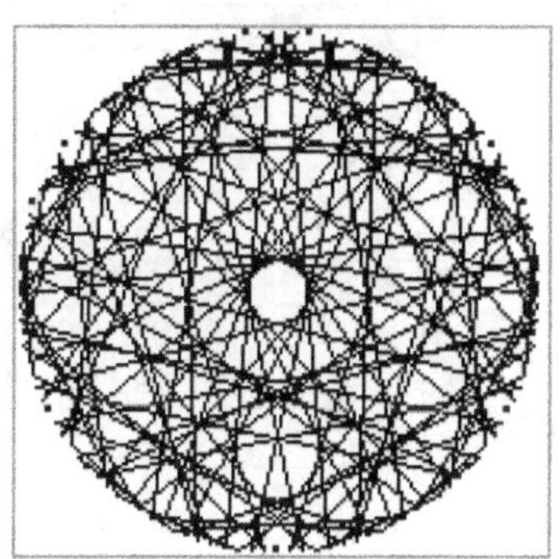

Circular Orig to Curr

Appendix I – Selected 12th Order Magic Squares

The following are regular 12th order magic squares. The boxed numbers in the left squares are numbers in their original positions. The right square shows the movement of the numbers from their original to their current position.

1	2	3	72	140	141	70	71	139	76	77	78
13	14	15	60	128	129	58	59	127	88	89	90
25	26	27	115	47	117	46	116	48	100	101	102
108	107	106	109	110	40	111	41	42	33	32	31
96	95	94	121	53	52	123	122	54	21	20	19
84	83	82	66	65	64	135	134	133	9	8	7
138	137	136	12	11	10	81	80	79	63	62	61
126	125	124	91	23	22	93	92	24	51	50	49
114	113	112	103	104	34	105	35	36	39	38	37
43	44	45	97	29	99	28	98	30	118	119	120
55	56	57	18	86	87	16	17	85	130	131	132
67	68	69	6	74	75	4	5	73	142	143	144

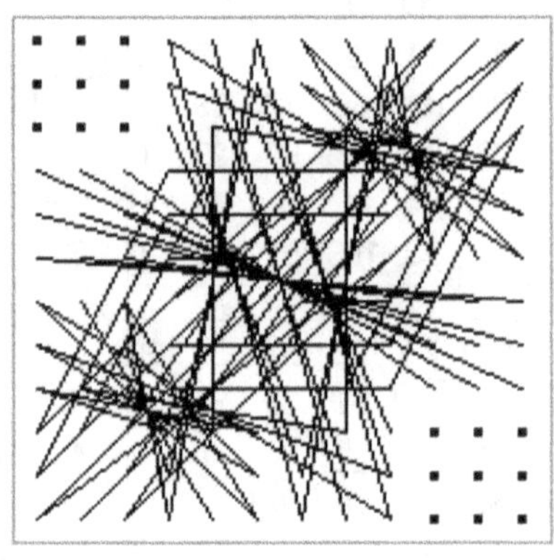

1	2	3	135	134	133	126	125	124	28	29	30
4	5	6	132	131	130	123	122	121	31	32	33
7	8	9	129	128	127	120	119	118	34	35	36
102	101	100	52	53	54	61	62	63	75	74	73
108	107	106	46	47	48	55	56	57	81	80	79
105	104	103	49	50	51	58	59	60	78	77	76
69	68	67	85	86	87	94	95	96	42	41	40
66	65	64	88	89	90	97	98	99	39	38	37
72	71	70	82	83	84	91	92	93	45	44	43
109	110	111	27	26	25	18	17	16	136	137	138
112	113	114	24	23	22	15	14	13	139	140	141
115	116	117	21	20	19	12	11	10	142	143	144

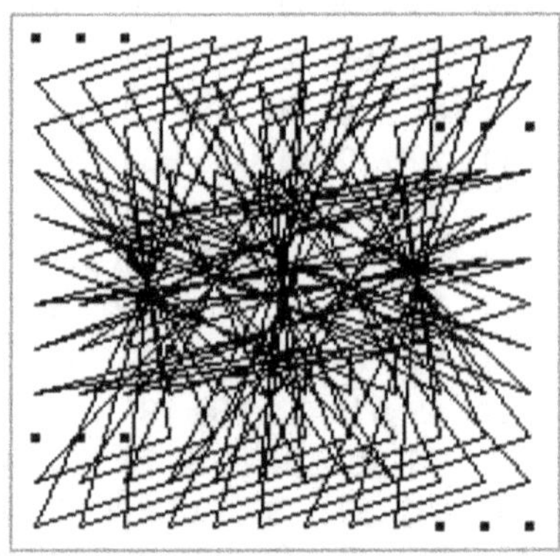

1	2	3	141	137	139	138	140	136	10	11	12
13	14	15	129	125	127	126	128	124	22	23	24
25	26	27	117	116	115	114	113	112	34	35	36
108	107	106	40	41	42	43	44	45	99	98	97
60	59	94	52	92	90	91	89	57	87	50	49
84	83	82	64	68	66	67	65	69	75	74	73
72	71	70	76	80	78	79	77	81	63	62	61
96	95	58	88	56	54	55	53	93	51	86	85
48	47	46	100	101	102	103	104	105	39	38	37
109	110	111	33	32	31	30	29	28	118	119	120
121	122	123	21	17	19	18	20	16	130	131	132
133	134	135	9	5	7	6	8	4	142	143	144

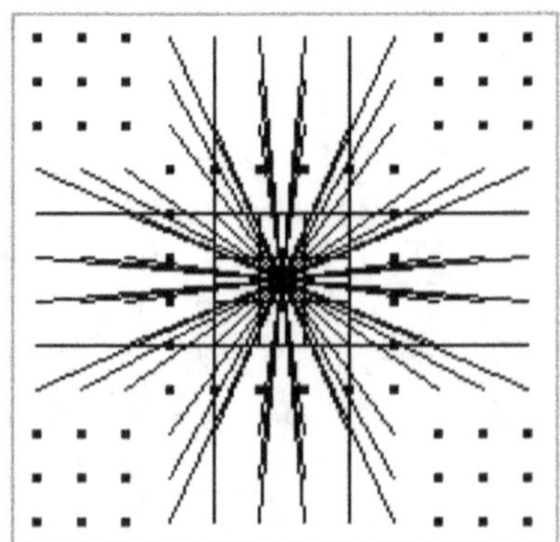

1	2	3	141	140	139	72	71	70	76	77	78
13	14	15	129	128	127	60	59	58	88	89	90
25	26	27	117	116	115	48	47	46	100	101	102
108	107	106	40	41	42	109	110	111	33	32	31
96	95	94	52	53	54	121	122	123	21	20	19
84	83	82	64	65	66	133	134	135	9	8	7
138	137	136	10	11	12	79	80	81	63	62	61
126	125	124	22	23	24	91	92	93	51	50	49
114	113	112	34	35	36	103	104	105	39	38	37
43	44	45	99	98	97	30	29	28	118	119	120
55	56	57	87	86	85	18	17	16	130	131	132
67	68	69	75	74	73	6	5	4	142	143	144

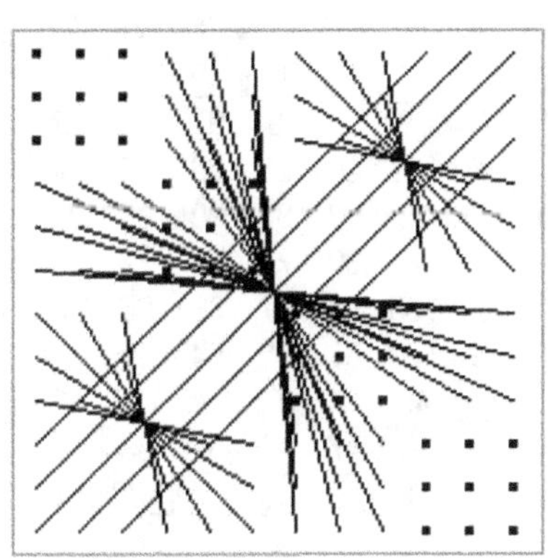

1	2	76	70	83	72	84	71	9	135	134	133
13	14	88	58	95	60	96	59	21	123	122	121
43	44	118	28	29	30	108	107	33	111	110	109
114	113	39	105	104	103	25	26	100	46	47	48
126	125	51	93	56	91	55	92	130	16	17	18
138	137	63	81	68	79	67	80	142	4	5	6
139	140	141	3	65	78	66	77	64	82	8	7
127	128	129	15	53	90	54	89	52	94	20	19
97	98	99	45	119	120	42	41	40	106	32	31
36	35	34	112	38	37	115	116	117	27	101	102
24	23	22	124	86	49	85	50	87	57	131	132
12	11	10	136	74	61	73	62	75	69	143	144

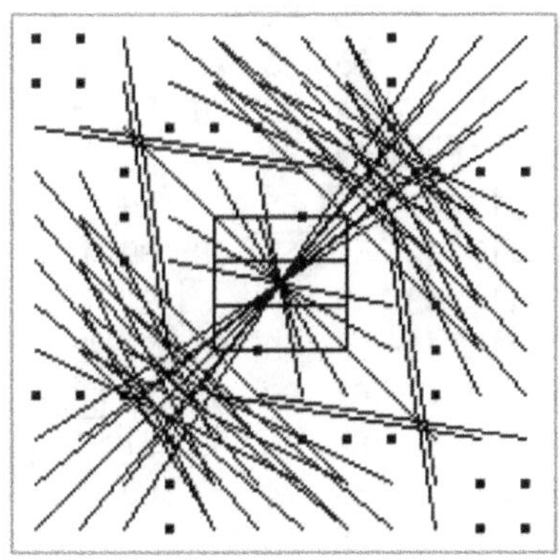

1	2	76	70	140	139	72	71	141	3	77	78
13	14	88	58	128	127	60	59	129	15	89	90
43	44	118	28	98	97	30	29	99	45	119	120
114	113	39	105	35	36	103	104	34	112	38	37
96	95	21	123	53	54	121	122	52	94	20	19
84	83	9	135	65	66	133	134	64	82	8	7
138	137	63	81	11	12	79	80	10	136	62	61
126	125	51	93	23	24	91	92	22	124	50	49
108	107	33	111	41	42	109	110	40	106	32	31
25	26	100	46	116	115	48	47	117	27	101	102
55	56	130	16	86	85	18	17	87	57	131	132
67	68	142	4	74	73	6	5	75	69	143	144

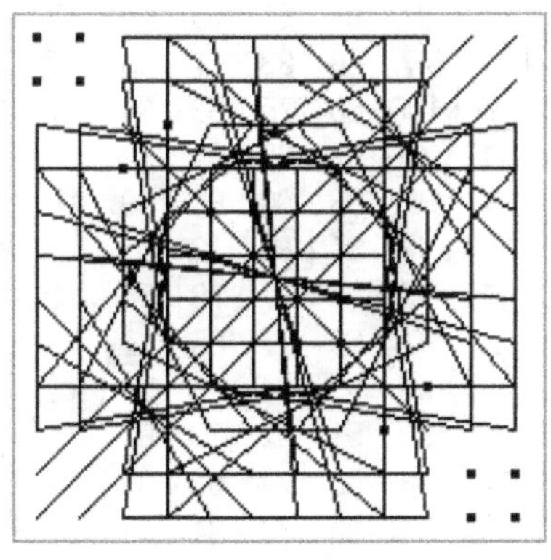

1	2	142	141	140	6	7	137	136	135	11	12
13	14	15	129	128	127	126	125	124	22	23	24
120	26	27	28	116	115	114	113	33	34	35	109
108	107	39	40	41	103	102	44	45	46	98	97
96	95	94	52	53	54	55	56	57	87	86	85
61	83	82	81	65	66	67	68	76	75	74	72
73	71	70	69	77	78	79	80	64	63	62	84
60	59	58	88	89	90	91	92	93	51	50	49
48	47	99	100	101	43	42	104	105	106	38	37
36	110	111	112	32	31	30	29	117	118	119	25
121	122	123	21	20	19	18	17	16	130	131	132
133	134	10	9	8	138	139	5	4	3	143	144

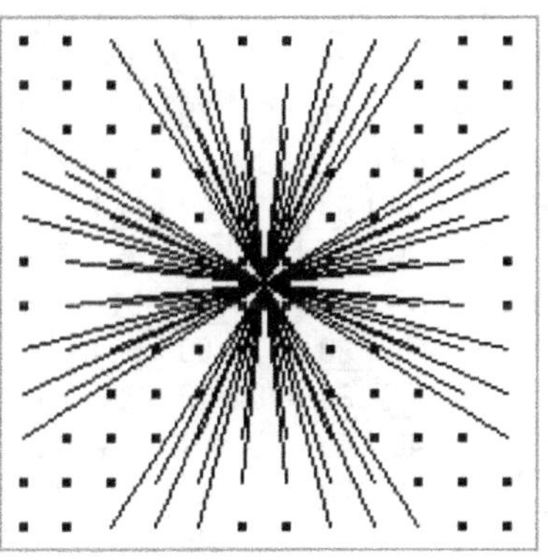

1	11	76	70	137	138	139	140	63	81	2	12
121	131	52	94	17	18	19	20	87	57	122	132
43	41	118	28	107	108	97	98	33	111	44	42
114	116	39	105	26	25	36	35	100	46	113	115
60	50	129	15	92	91	90	89	22	124	59	49
72	62	141	3	80	79	78	77	10	136	71	61
84	74	9	135	68	67	66	65	142	4	83	73
96	86	21	123	56	55	54	53	130	16	95	85
30	32	99	45	110	109	120	119	40	106	29	31
103	101	34	112	47	48	37	38	117	27	104	102
13	23	88	58	125	126	127	128	51	93	14	24
133	143	64	82	5	6	7	8	75	69	134	144

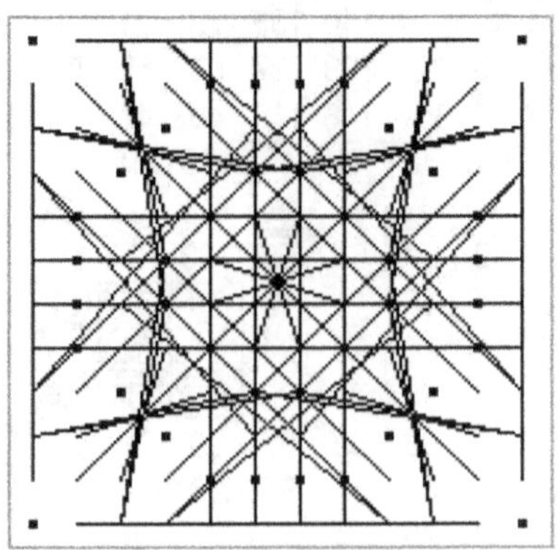

1	13	25	108	96	84	72	60	48	109	121	133
2	14	26	107	95	83	71	59	47	110	122	134
3	15	27	106	94	82	70	58	46	111	123	135
141	129	117	40	52	64	76	88	100	33	21	9
140	128	116	41	53	65	77	89	101	32	20	8
139	127	115	42	54	66	78	90	102	31	19	7
138	126	114	43	55	67	79	91	103	30	18	6
137	125	113	44	56	68	80	92	104	29	17	5
136	124	112	45	57	69	81	93	105	28	16	4
10	22	34	99	87	75	63	51	39	118	130	142
11	23	35	98	86	74	62	50	38	119	131	143
12	24	36	97	85	73	61	49	37	120	132	144

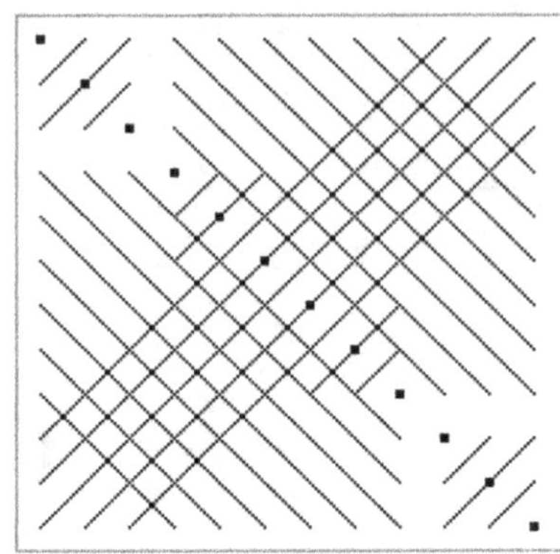

1	132	36	97	96	61	73	60	37	120	24	133
143	14	110	47	50	83	71	86	107	26	122	11
10	123	27	106	58	75	63	94	46	111	15	142
136	21	117	40	88	69	81	52	100	33	129	4
5	128	32	101	53	80	68	89	41	116	20	137
139	18	114	43	91	66	78	55	103	30	126	7
138	19	115	42	90	67	79	54	102	31	127	6
8	125	29	104	56	77	65	92	44	113	17	140
141	16	112	45	93	64	76	57	105	28	124	9
3	130	34	99	51	82	70	87	39	118	22	135
134	23	119	38	59	74	62	95	98	35	131	2
12	121	25	108	85	72	84	49	48	109	13	144

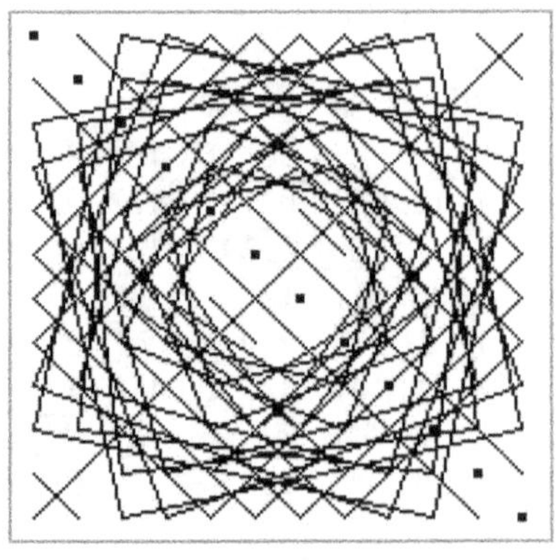

1	13	42	115	127	139	138	126	108	25	24	12
2	14	41	116	128	140	137	125	107	26	23	11
64	52	118	39	21	9	75	87	99	34	130	142
82	94	28	105	123	135	69	57	45	112	16	4
83	95	98	35	53	65	68	56	44	113	86	74
84	96	97	36	54	66	67	55	43	114	85	73
72	60	31	102	90	78	79	91	109	48	49	61
71	59	32	101	89	77	80	92	110	47	50	62
141	129	33	100	88	76	10	22	40	117	51	63
3	15	111	46	58	70	136	124	106	27	93	81
134	122	119	38	20	8	5	17	29	104	131	143
133	121	120	37	19	7	6	18	30	103	132	144

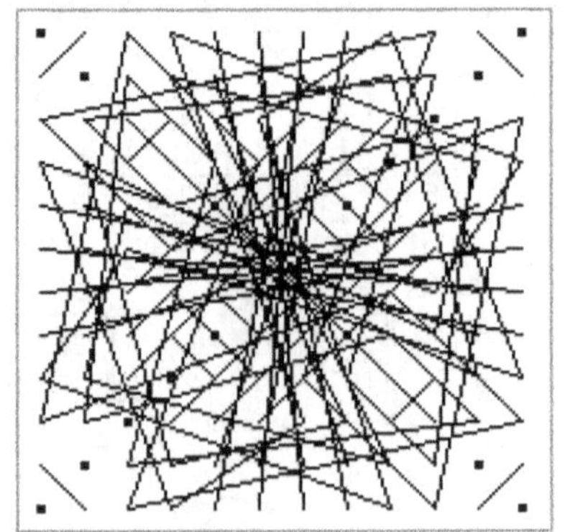

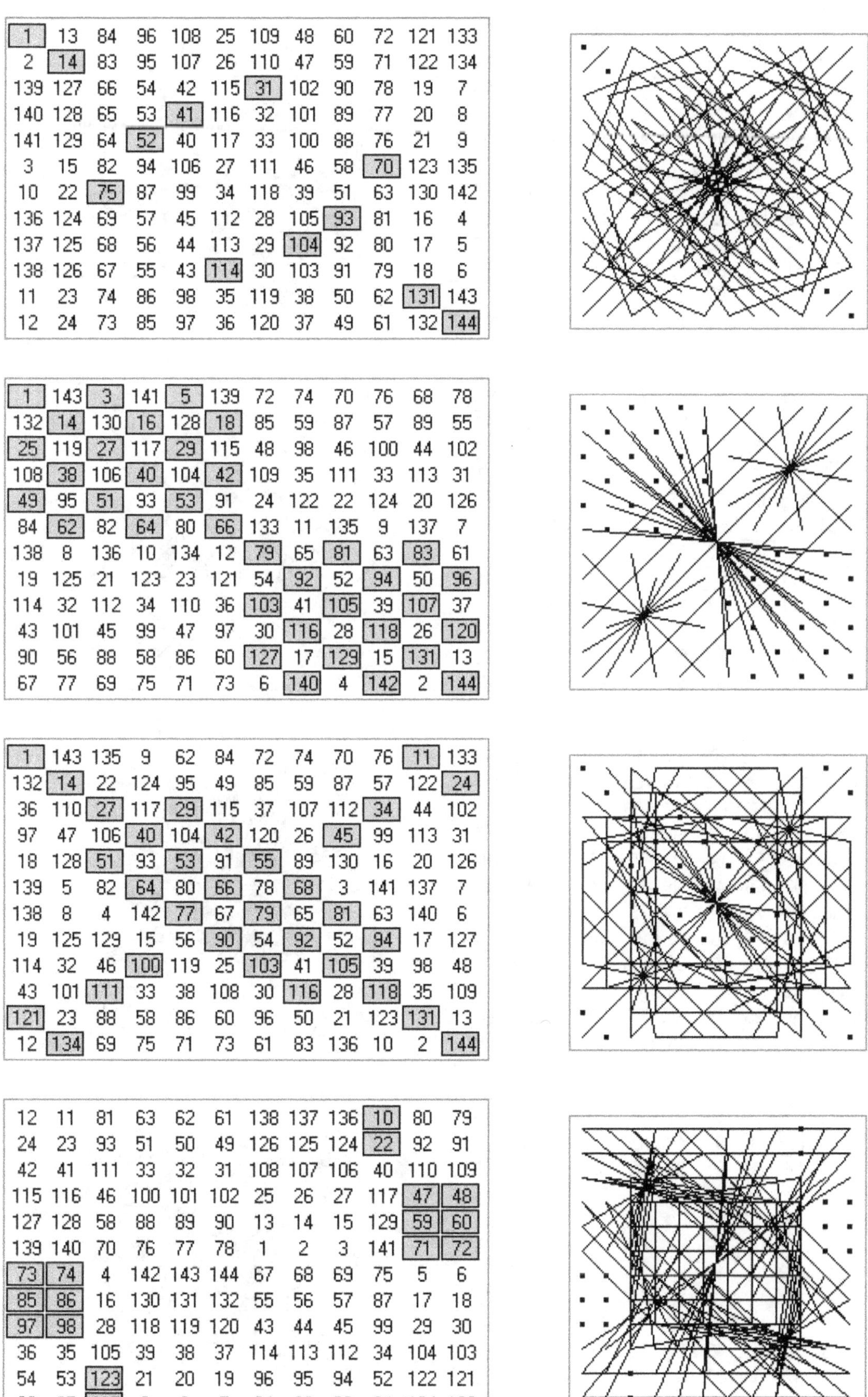

[1]	13	84	96	108	25	109	48	60	72	121	133
2	[14]	83	95	107	26	110	47	59	71	122	134
139	127	66	54	42	115	[31]	102	90	78	19	7
140	128	65	53	[41]	116	32	101	89	77	20	8
141	129	64	[52]	40	117	33	100	88	76	21	9
3	15	82	94	106	27	111	46	58	[70]	123	135
10	22	[75]	87	99	34	118	39	51	63	130	142
136	124	69	57	45	112	28	105	[93]	81	16	4
137	125	68	56	44	113	29	[104]	92	80	17	5
138	126	67	55	43	[114]	30	103	91	79	18	6
11	23	74	86	98	35	119	38	50	62	[131]	143
12	24	73	85	97	36	120	37	49	61	132	[144]

[1]	143	[3]	141	[5]	139	72	74	70	76	68	78
132	[14]	130	[16]	128	[18]	85	59	87	57	89	55
[25]	119	[27]	117	[29]	115	48	98	46	100	44	102
108	[38]	106	[40]	104	[42]	109	35	111	33	113	31
[49]	95	[51]	93	[53]	91	24	122	22	124	20	126
84	[62]	82	[64]	80	[66]	133	11	135	9	137	7
138	8	136	10	134	12	[79]	65	[81]	63	[83]	61
19	125	21	123	23	121	54	[92]	52	[94]	50	[96]
114	32	112	34	110	36	[103]	41	[105]	39	[107]	37
43	101	45	99	47	97	30	[116]	28	[118]	26	[120]
90	56	88	58	86	60	[127]	17	[129]	15	[131]	13
67	77	69	75	71	73	6	[140]	4	[142]	2	[144]

[1]	143	135	9	62	84	72	74	70	76	[11]	133
132	[14]	22	124	95	49	85	59	87	57	122	[24]
36	110	[27]	117	[29]	115	37	107	112	[34]	44	102
97	47	106	[40]	104	[42]	120	26	[45]	99	113	31
18	128	[51]	93	[53]	91	[55]	89	130	16	20	126
139	5	82	[64]	80	[66]	78	[68]	3	141	137	7
138	8	4	142	[77]	67	[79]	65	[81]	63	140	6
19	125	129	15	56	[90]	54	[92]	52	[94]	17	127
114	32	46	[100]	119	25	[103]	41	[105]	39	98	48
43	101	[111]	33	38	108	30	[116]	28	[118]	35	109
[121]	23	88	58	86	60	96	50	21	123	[131]	13
12	[134]	69	75	71	73	61	83	136	10	2	[144]

12	11	81	63	62	61	138	137	136	[10]	80	79
24	23	93	51	50	49	126	125	124	[22]	92	91
42	41	111	33	32	31	108	107	106	40	110	109
115	116	46	100	101	102	25	26	27	117	[47]	[48]
127	128	58	88	89	90	13	14	15	129	[59]	[60]
139	140	70	76	77	78	1	2	3	141	[71]	[72]
[73]	[74]	4	142	143	144	67	68	69	75	5	6
[85]	[86]	16	130	131	132	55	56	57	87	17	18
[97]	[98]	28	118	119	120	43	44	45	99	29	30
36	35	105	39	38	37	114	113	112	34	104	103
54	53	[123]	21	20	19	96	95	94	52	122	121
66	65	[135]	9	8	7	84	83	82	64	134	133

247

14	13	128	22	129	127	126	124	15	125	24	23
2	1	140	10	141	139	138	136	3	137	12	11
95	96	53	87	52	54	55	57	94	56	85	86
110	109	32	118	33	31	30	28	111	29	120	119
107	108	41	99	40	42	43	45	106	44	97	98
83	84	65	75	64	66	67	69	82	68	73	74
71	72	77	63	76	78	79	81	70	80	61	62
47	48	101	39	100	102	103	105	46	104	37	38
26	25	116	34	117	115	114	112	27	113	36	35
59	60	89	51	88	90	91	93	58	92	49	50
134	133	8	142	9	7	6	4	135	5	144	143
122	121	20	130	21	19	18	16	123	17	132	131

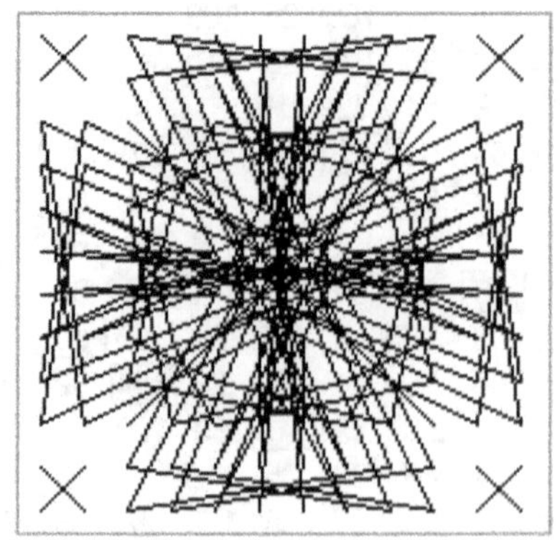

14	143	38	119	62	95	59	74	35	98	11	122
132	1	108	25	84	49	85	72	109	48	133	24
129	4	105	28	81	52	88	69	112	45	136	21
15	142	39	118	63	94	58	75	34	99	10	123
127	6	42	115	66	91	55	78	31	102	138	19
17	140	104	29	80	53	89	68	113	44	8	125
20	137	101	32	77	56	92	65	116	41	5	128
126	7	43	114	67	90	54	79	30	103	139	18
22	135	46	111	70	87	51	82	27	106	3	130
124	9	100	33	76	57	93	64	117	40	141	16
121	12	97	36	73	60	96	61	120	37	144	13
23	134	47	110	71	86	50	83	26	107	2	131

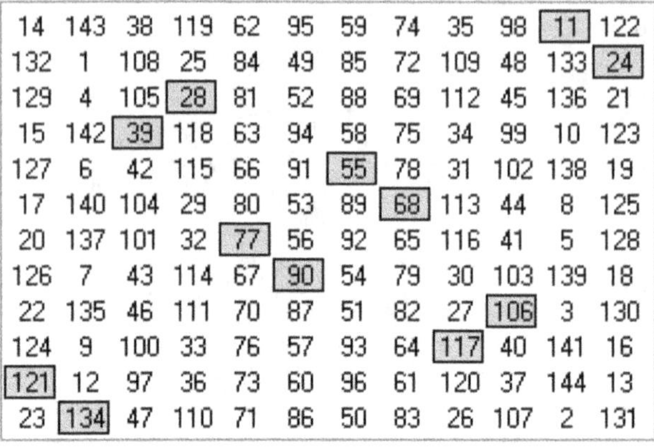

26	1	15	140	115	129	137	112	126	35	10	24
3	14	25	117	128	139	114	125	136	12	23	34
13	27	2	127	141	116	124	138	113	22	36	11
107	82	96	65	40	54	68	43	57	98	73	87
84	95	106	42	53	64	45	56	67	75	86	97
94	108	83	52	66	41	55	69	44	85	99	74
71	46	60	101	76	90	104	79	93	62	37	51
48	59	70	78	89	100	81	92	103	39	50	61
58	72	47	88	102	77	91	105	80	49	63	38
134	109	123	32	7	21	29	4	18	143	118	132
111	122	133	9	20	31	6	17	28	120	131	142
121	135	110	19	33	8	16	30	5	130	144	119

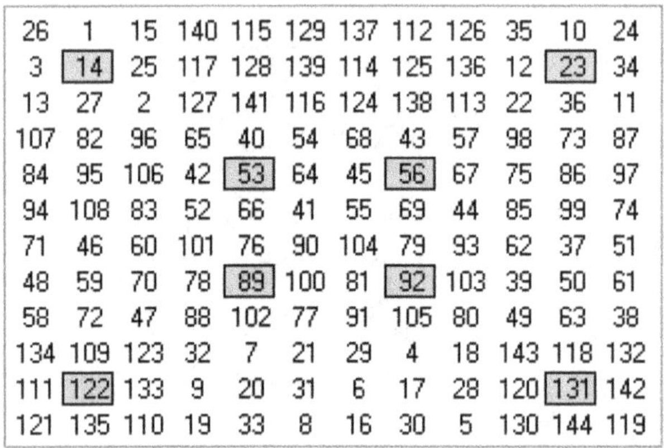

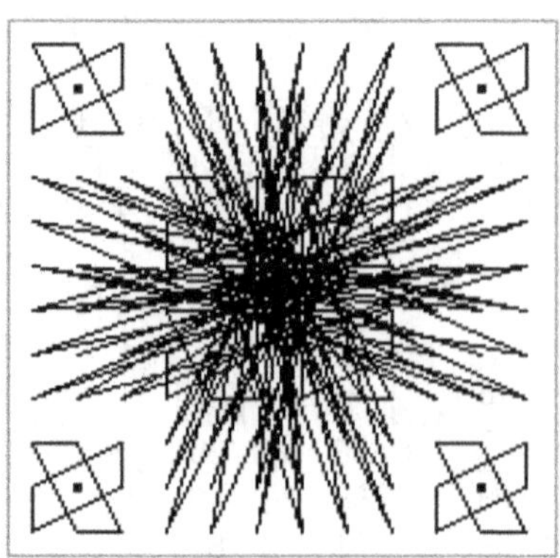

27	112	116	115	35	36	25	26	114	113	117	34
46	40	101	102	98	37	48	107	103	104	45	39
94	57	53	54	86	85	96	95	55	56	52	87
82	69	65	66	74	73	84	83	67	68	64	75
123	21	20	19	131	132	121	122	18	17	16	130
135	4	8	7	143	144	133	134	6	5	9	142
3	136	140	139	11	12	1	2	138	137	141	10
15	129	128	127	23	24	13	14	126	125	124	22
70	81	77	78	62	61	72	71	79	80	76	63
58	93	89	90	50	49	60	59	91	92	88	51
106	100	41	42	38	97	108	47	43	44	105	99
111	28	32	31	119	120	109	110	30	29	33	118

40	117	129	[4]	101	32	20	137	48	109	121	[12]
106	27	[15]	142	43	114	126	7	98	35	[23]	134
135	[26]	14	143	42	115	127	6	99	[34]	22	107
[37]	120	132	1	104	29	17	140	[45]	112	124	9
57	76	64	49	92	65	77	[56]	93	84	72	85
59	70	82	95	54	79	[67]	90	51	62	74	87
58	71	83	94	55	[78]	66	91	50	63	75	86
60	73	61	52	[89]	68	80	53	96	81	69	88
136	21	33	[100]	5	128	116	41	144	13	25	[108]
38	123	[111]	46	139	18	30	103	2	131	[119]	10
11	[122]	110	47	138	19	31	102	3	[130]	118	39
[133]	24	36	97	8	125	113	44	[141]	16	28	105

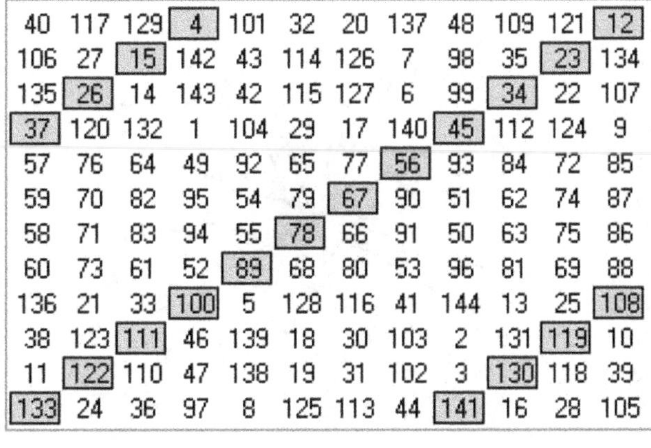
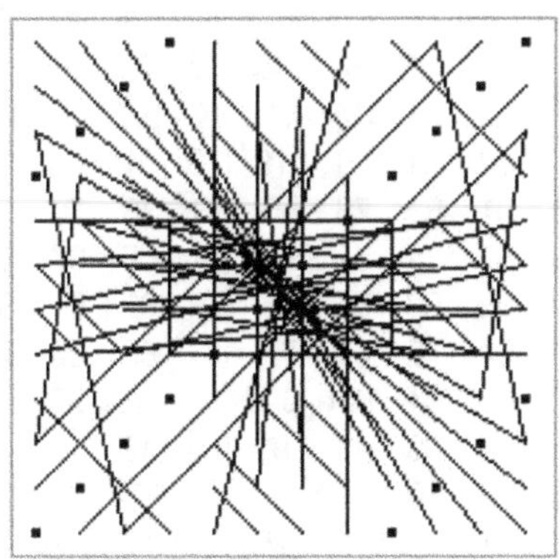

66	65	64	82	83	84	73	74	[9]	135	68	67
54	53	52	94	95	96	85	86	[21]	123	56	55
42	41	40	106	107	108	97	98	[33]	111	44	43
115	116	117	27	26	25	36	35	100	[46]	113	114
127	128	129	15	14	13	24	23	88	[58]	125	126
139	140	141	3	2	1	12	11	76	[70]	137	138
7	8	[75]	69	134	133	144	143	142	4	5	6
19	20	[87]	57	122	121	132	131	130	16	17	18
31	32	[99]	45	110	109	120	119	118	28	29	30
102	101	34	[112]	47	48	37	38	39	105	104	103
90	89	22	[124]	59	60	49	50	51	93	92	91
78	77	10	[136]	71	72	61	62	63	81	80	79

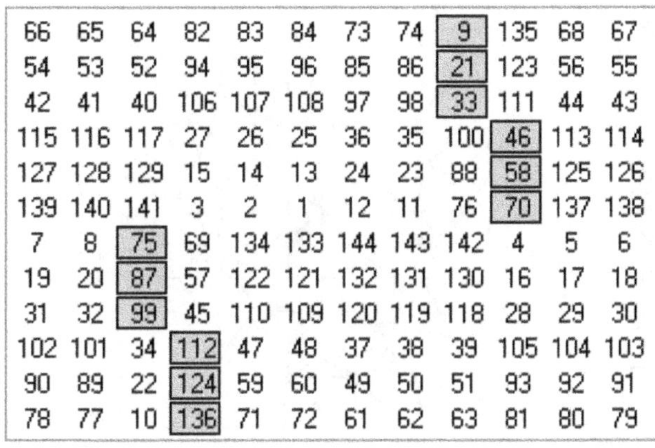
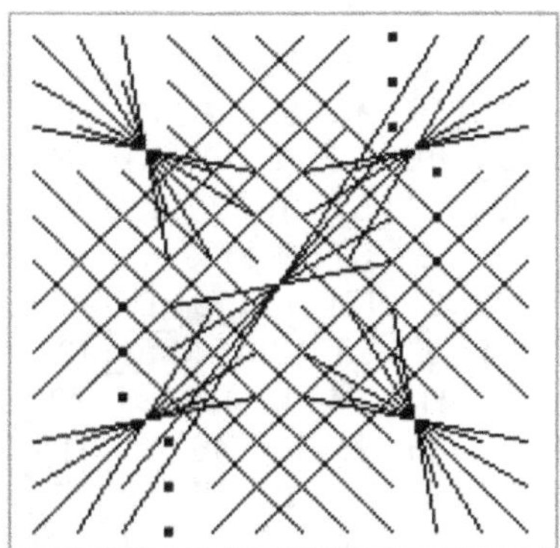

66	121	109	115	60	72	6	18	97	103	91	[12]
65	53	110	47	128	140	74	86	29	104	[23]	11
64	52	40	117	129	141	75	87	99	[34]	22	10
82	94	106	27	15	3	69	57	[45]	112	124	136
83	95	107	26	14	2	68	[56]	44	113	125	137
84	96	108	25	13	1	[67]	55	43	114	126	138
7	19	31	102	90	[78]	144	132	120	37	49	61
8	20	32	101	[89]	77	143	131	119	38	50	62
9	21	33	[100]	88	76	142	130	118	39	51	63
135	123	[111]	46	58	70	4	16	28	105	93	81
134	[122]	41	116	59	71	5	17	98	35	92	80
[133]	54	42	48	127	139	73	85	30	36	24	79

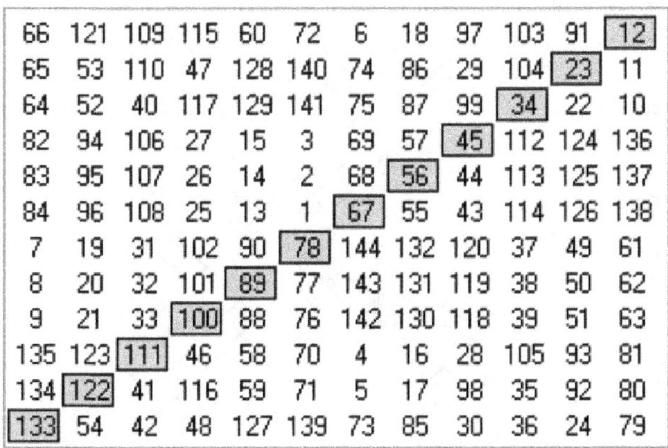
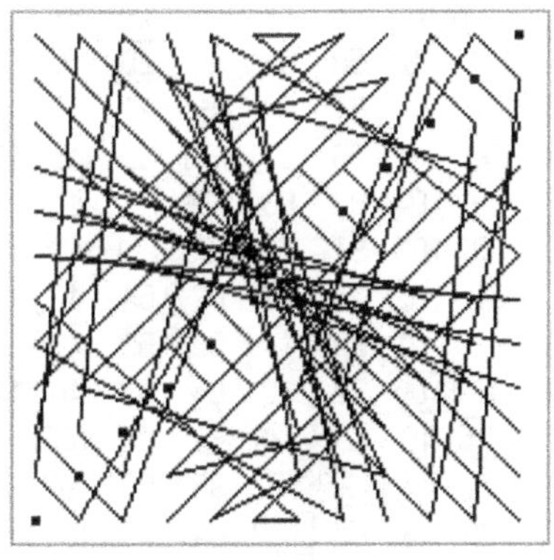

105	39	10	108	88	86	87	85	[9]	107	134	[12]
28	118	119	25	69	75	74	72	124	[22]	[23]	121
16	130	131	13	81	63	62	84	112	[34]	[35]	109
141	3	2	144	96	50	51	93	[45]	99	98	[48]
44	102	103	41	[53]	91	90	[56]	140	6	7	137
113	31	30	116	80	[66]	[67]	77	17	127	126	20
125	19	18	128	68	[78]	[79]	65	29	115	114	32
8	138	139	5	[89]	55	54	[92]	104	42	43	101
[97]	47	46	[100]	52	94	95	49	1	143	142	4
36	[110]	[111]	33	61	83	82	64	132	14	15	129
24	[122]	[123]	21	73	71	70	76	120	26	27	117
[133]	11	38	[136]	60	58	59	57	37	135	106	40

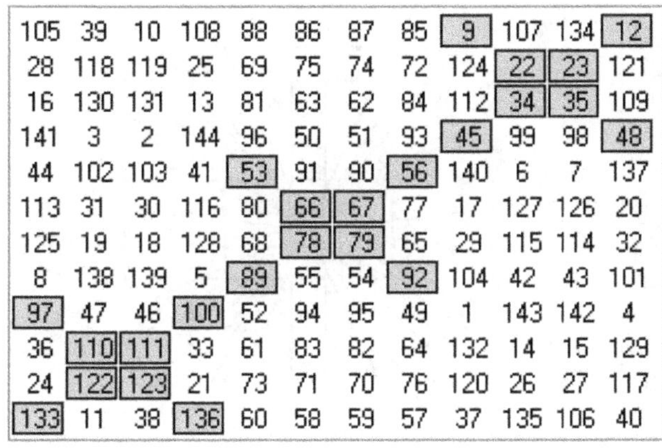
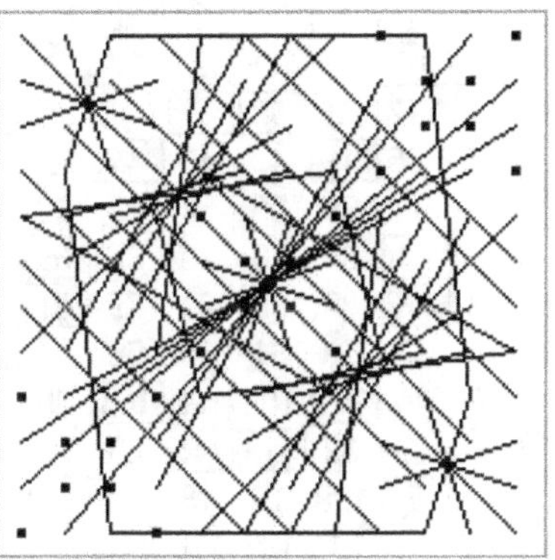

131	13	129	15	127	17	20	126	22	124	24	122
2	144	4	142	6	140	137	7	135	9	133	11
107	37	105	39	103	41	44	102	46	100	48	98
26	120	28	118	30	116	113	31	111	33	109	35
83	61	81	63	79	65	68	78	70	76	72	74
50	96	52	94	54	92	89	55	87	57	85	59
86	60	88	58	90	56	53	91	51	93	49	95
71	73	69	75	67	77	80	66	82	64	84	62
110	36	112	34	114	32	29	115	27	117	25	119
47	97	45	99	43	101	104	42	106	40	108	38
134	12	136	10	138	8	5	139	3	141	1	143
23	121	21	123	19	125	128	18	130	16	132	14

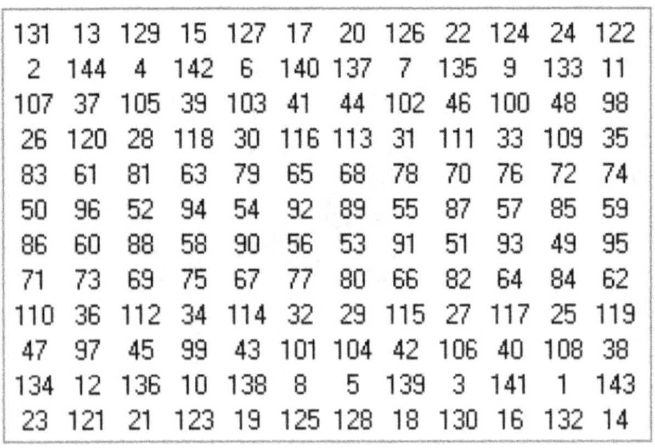

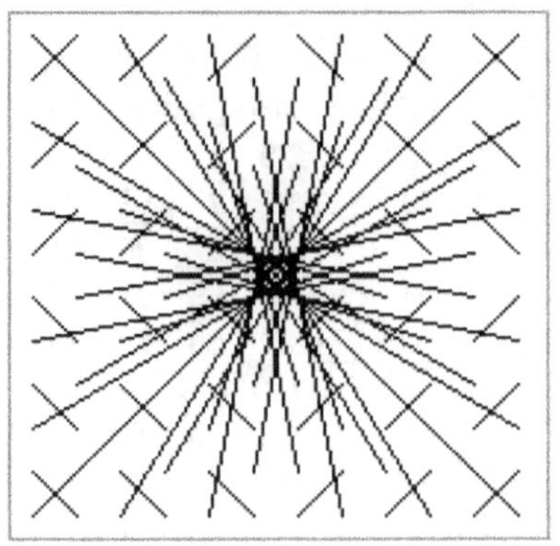

144	2	142	4	71	73	6	140	75	69	77	67
13	131	15	129	86	60	127	17	58	88	56	90
120	26	118	28	47	97	30	116	99	45	101	43
37	107	39	105	110	36	103	41	34	112	32	114
126	20	124	22	53	91	24	122	93	51	95	49
7	137	9	135	80	66	133	11	64	82	62	84
61	83	63	81	134	12	79	65	10	136	8	138
96	50	94	52	23	121	54	92	123	21	125	19
31	113	33	111	104	42	109	35	40	106	38	108
102	44	100	46	29	115	48	98	117	27	119	25
55	89	57	87	128	18	85	59	16	130	14	132
78	68	76	70	5	139	72	74	141	3	143	1

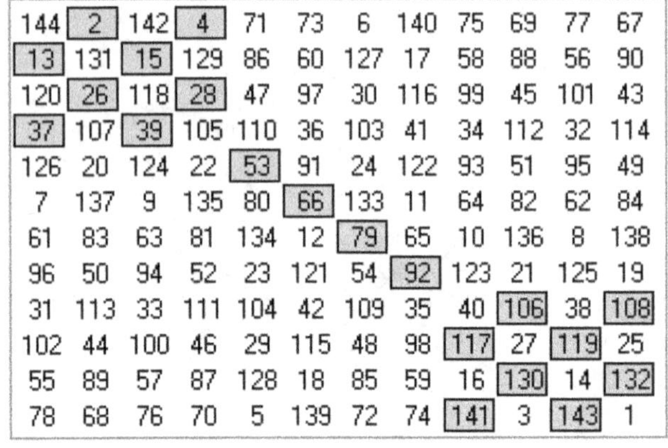

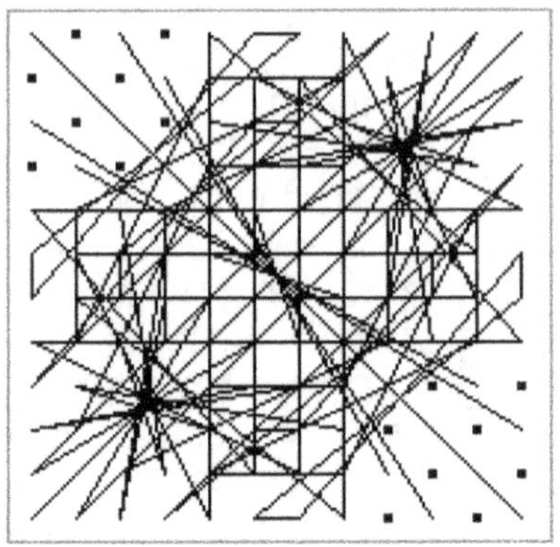

144	132	120	37	49	61	7	19	97	36	90	78
143	131	119	38	50	62	8	20	98	35	89	77
142	130	118	39	51	63	9	21	99	34	88	76
4	16	28	105	93	81	135	123	45	112	58	70
5	17	29	104	92	80	134	122	44	113	59	71
6	18	30	103	91	79	133	121	43	114	60	72
73	85	31	102	24	12	66	54	42	115	127	139
74	86	32	101	23	11	65	53	41	116	128	140
75	87	33	100	22	10	64	52	40	117	129	141
69	57	111	46	124	136	82	94	106	27	15	3
68	56	110	47	125	137	83	95	107	26	14	2
67	55	109	48	126	138	84	96	108	25	13	1

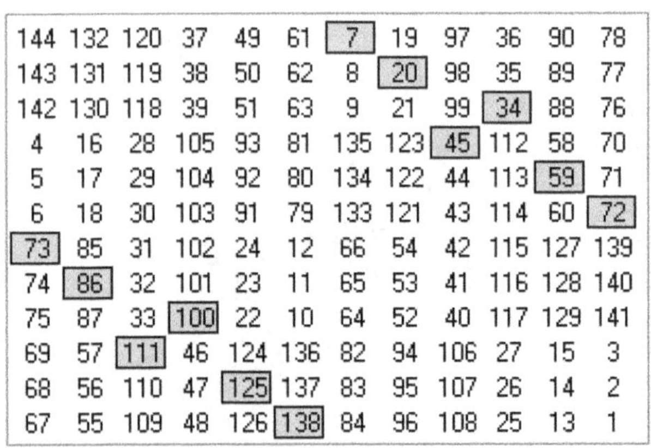

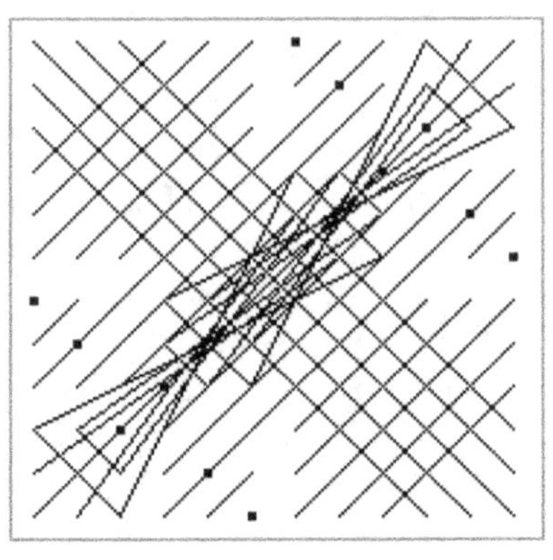

1	143	142	141	5	7	6	8	136	135	134	12
132	131	15	16	20	126	127	17	21	22	122	121
120	26	27	117	116	30	31	113	112	34	35	109
108	38	106	105	41	43	42	44	100	99	47	97
85	50	58	88	89	55	54	92	93	51	59	96
61	83	75	64	68	78	79	65	69	82	74	72
73	71	63	76	80	66	67	77	81	70	62	84
49	86	94	52	53	91	90	56	57	87	95	60
48	98	46	45	101	103	102	104	40	39	107	37
36	110	111	33	32	114	115	29	28	118	119	25
24	23	123	124	128	18	19	125	129	130	14	13
133	11	10	9	137	139	138	140	4	3	2	144

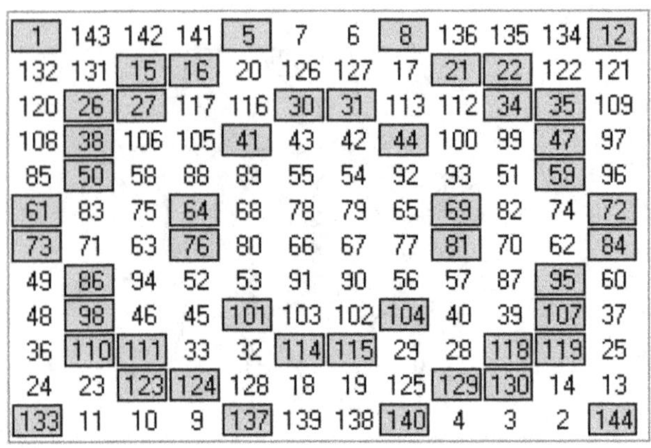

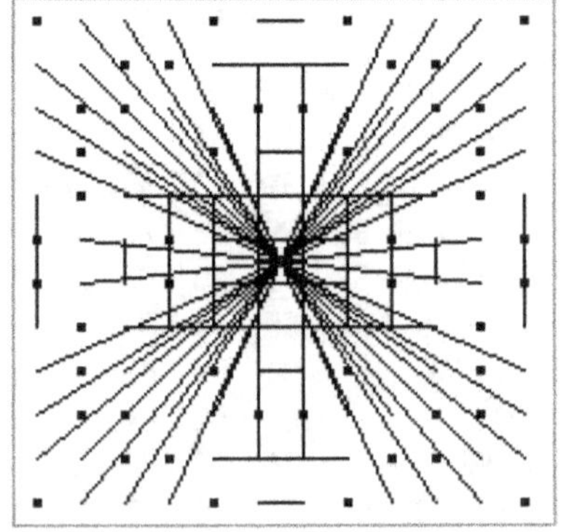

The following are regular 12th order magic squares. Odd numbers are highlighted on the left, and the lines are drawn in sequence on the right (from 1 to 2 to 3, etc.)

1	2	76	70	140	139	138	137	63	81	11	12
13	14	88	58	128	127	126	125	51	93	23	24
43	44	118	28	98	97	108	107	33	111	41	42
114	113	39	105	35	36	25	26	100	46	116	115
96	95	21	123	53	54	55	56	130	16	86	85
84	83	9	135	65	66	67	68	142	4	74	73
72	71	141	3	77	78	79	80	10	136	62	61
60	59	129	15	89	90	91	92	22	124	50	49
30	29	99	45	119	120	109	110	40	106	32	31
103	104	34	112	38	37	48	47	117	27	101	102
121	122	52	94	20	19	18	17	87	57	131	132
133	134	64	82	8	7	6	5	75	69	143	144

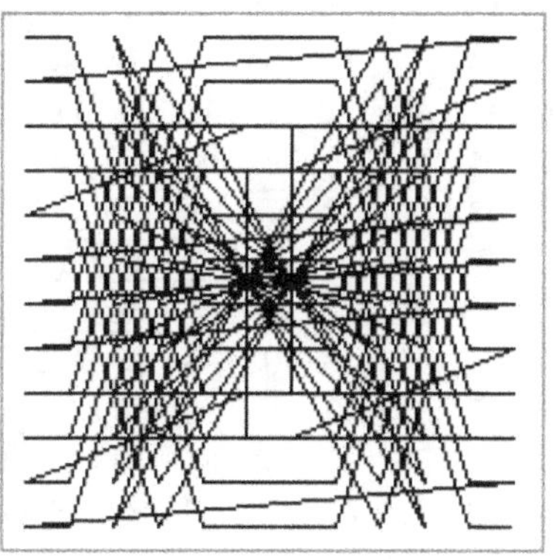

1	141	8	134	10	138	139	3	143	5	136	12
108	40	101	47	99	43	42	106	38	104	45	97
85	57	92	50	94	54	55	87	59	89	52	96
24	124	17	131	15	127	126	22	122	20	129	13
109	33	116	26	118	30	31	111	35	113	28	120
72	76	65	83	63	79	78	70	74	68	81	61
84	64	77	71	75	67	66	82	62	80	69	73
25	117	32	110	34	114	115	27	119	29	112	36
132	16	125	23	123	19	18	130	14	128	21	121
49	93	56	86	58	90	91	51	95	53	88	60
48	100	41	107	39	103	102	46	98	44	105	37
133	9	140	2	142	6	7	135	11	137	4	144

6	137	2	4	131	129	135	133	22	24	127	20
30	113	26	28	107	105	111	109	46	48	103	44
54	89	50	52	83	81	87	85	70	72	79	68
67	80	71	69	86	88	82	84	51	49	90	53
43	104	47	45	110	112	106	108	27	25	114	29
19	128	23	21	134	136	130	132	3	1	138	5
140	7	144	142	13	15	9	11	124	122	17	126
116	31	120	118	37	39	33	35	100	98	41	102
92	55	96	94	61	63	57	59	76	74	65	78
77	66	73	75	60	58	64	62	93	95	56	91
101	42	97	99	36	34	40	38	117	119	32	115
125	18	121	123	12	10	16	14	141	143	8	139

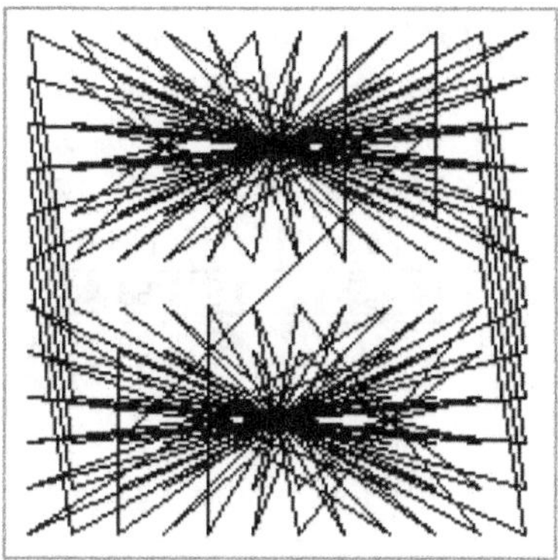

The following are regular magic squares with circular plots.

1	2	3	140	141	72	139	71	70	76	77	78
13	14	15	128	129	60	127	59	58	88	89	90
25	26	27	117	47	115	48	116	46	100	101	102
108	107	106	40	110	109	42	41	111	33	32	31
96	95	94	52	53	121	54	123	122	21	20	19
84	83	82	64	65	66	133	135	134	9	8	7
138	137	136	11	10	12	79	80	81	63	62	61
126	125	124	23	22	91	24	92	93	51	50	49
114	113	112	34	104	103	36	35	105	39	38	37
43	44	45	99	29	97	30	98	28	118	119	120
55	56	57	87	86	18	85	16	17	130	131	132
67	68	69	75	74	6	73	4	5	142	143	144

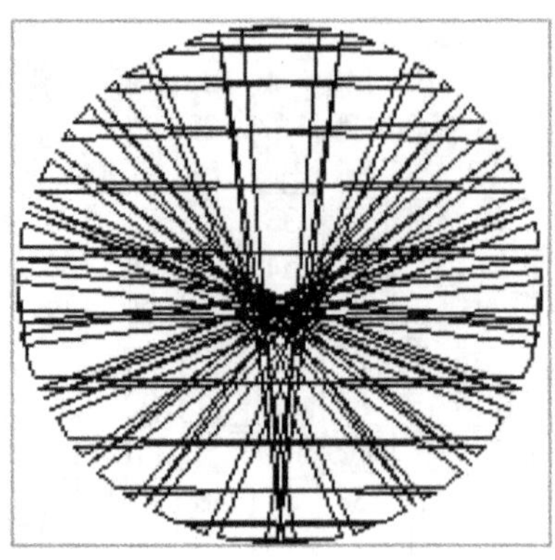

Circular Std Sequence

1	2	142	141	140	6	7	137	136	135	11	12
13	14	15	129	128	127	126	125	124	22	23	24
120	26	27	28	116	115	114	113	33	34	35	109
108	107	39	40	41	103	102	44	45	46	98	97
96	95	94	52	53	54	55	56	57	87	86	85
61	83	82	81	65	66	67	68	76	75	74	72
73	71	70	69	77	78	79	80	64	63	62	84
60	59	58	88	89	90	91	92	93	51	50	49
48	47	99	100	101	43	42	104	105	106	38	37
36	110	111	112	32	31	30	29	117	118	119	25
121	122	123	21	20	19	18	17	16	130	131	132
133	134	10	9	8	138	139	5	4	3	143	144

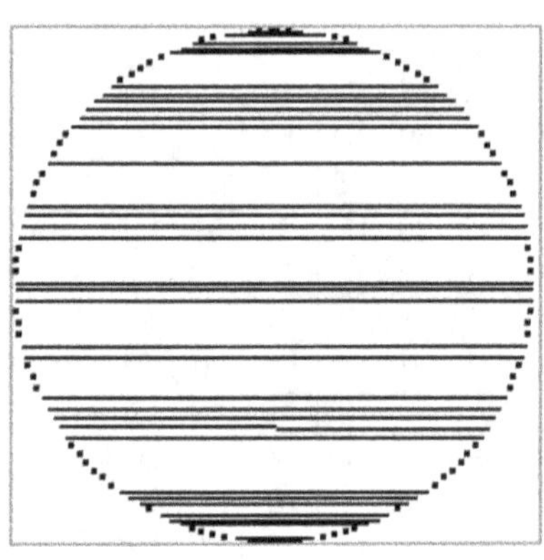

Circular Orig to Curr

1	2	3	141	140	139	138	137	136	10	11	12
108	107	106	40	41	42	43	44	45	99	98	97
96	95	94	52	53	54	55	56	57	87	86	85
13	14	15	129	128	127	126	125	124	22	23	24
25	26	27	117	116	115	114	113	112	34	35	36
84	83	82	64	65	66	67	68	69	75	74	73
72	71	70	76	77	78	79	80	81	63	62	61
109	110	111	33	32	31	30	29	28	118	119	120
121	122	123	21	20	19	18	17	16	130	131	132
60	59	58	88	89	90	91	92	93	51	50	49
48	47	46	100	101	102	103	104	105	39	38	37
133	134	135	9	8	7	6	5	4	142	143	144

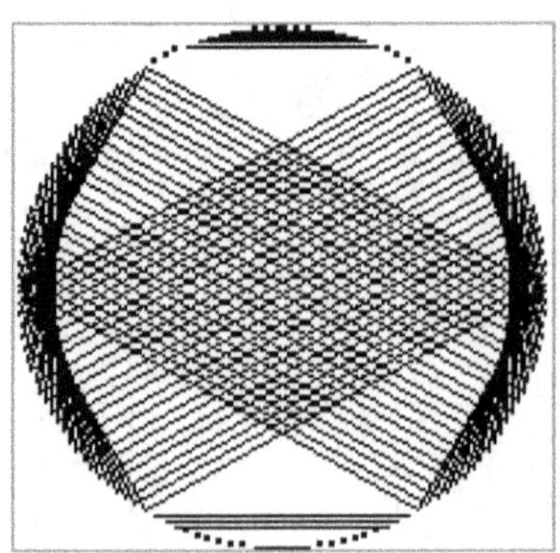

Circular Orig to Curr

Appendix J – Selected 13th Order Magic Squares

The following are regular 13th order magic squares. The boxed numbers in the left squares are in their original positions. The right square shows the movement of the numbers from their original to their current position.

1	98	26	110	38	122	50	134	62	146	74	158	86
100	15	112	27	124	52	136	64	148	76	160	88	3
17	114	29	126	41	138	53	150	78	162	90	5	102
116	31	128	43	140	55	152	67	164	79	7	104	19
33	130	45	142	57	154	69	166	81	9	93	21	105
119	47	131	59	156	71	168	83	11	95	23	107	35
49	133	61	145	73	157	85	13	97	25	109	37	121
135	63	147	75	159	87	2	99	14	111	39	123	51
65	149	77	161	89	4	101	16	113	28	125	40	137
151	66	163	91	6	103	18	115	30	127	42	139	54
68	165	80	8	92	20	117	32	129	44	141	56	153
167	82	10	94	22	106	34	118	46	143	58	155	70
84	12	96	24	108	36	120	48	132	60	144	72	169

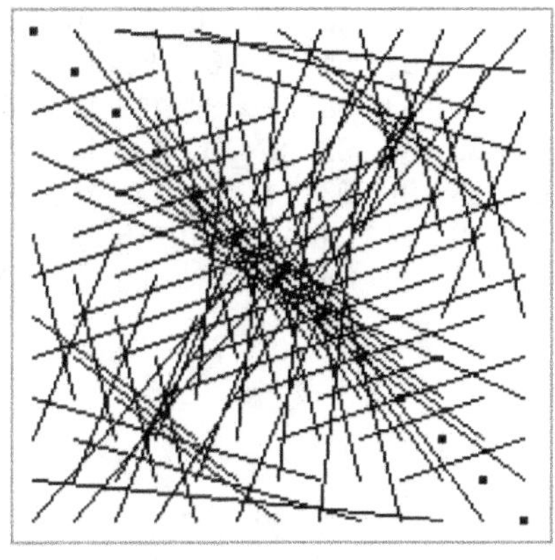

1	100	17	116	33	119	49	135	65	151	68	167	84
98	15	114	31	130	47	133	63	149	66	165	82	12
26	112	29	128	45	131	61	147	77	163	80	10	96
110	27	126	43	142	59	145	75	161	91	8	94	24
38	124	41	140	57	156	73	159	89	6	92	22	108
122	52	138	55	154	71	157	87	4	103	20	106	36
50	136	53	152	69	168	85	2	101	18	117	34	120
134	64	150	67	166	83	13	99	16	115	32	118	48
62	148	78	164	81	11	97	14	113	30	129	46	132
146	76	162	79	9	95	25	111	28	127	44	143	60
74	160	90	7	93	23	109	39	125	42	141	58	144
158	88	5	104	21	107	37	123	40	139	56	155	72
86	3	102	19	105	35	121	51	137	54	153	70	169

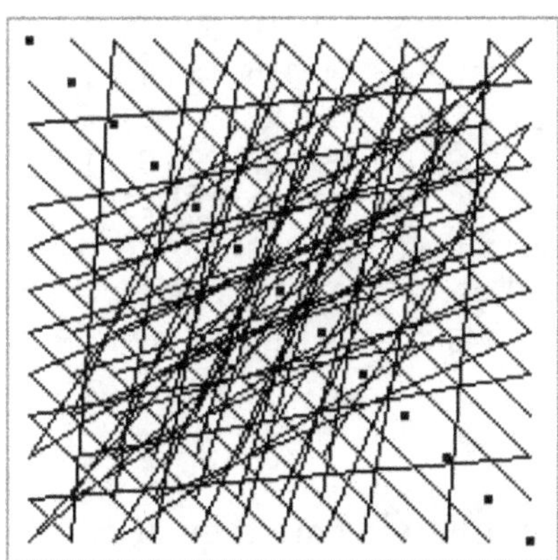

57	128	17	88	146	48	168	106	8	66	137	39	97
140	29	100	158	60	118	69	20	91	149	51	109	11
41	112	1	72	143	32	152	103	161	63	121	23	81
124	26	84	155	44	115	53	4	75	133	35	93	164
38	96	167	56	127	16	136	87	145	47	105	7	78
108	10	68	139	28	99	50	157	59	130	19	90	148
156	45	116	5	76	134	85	36	94	165	54	125	14
22	80	151	40	111	13	120	71	142	31	102	160	62
92	163	65	123	25	83	34	154	43	114	3	74	132
6	77	135	37	95	166	117	55	126	15	86	144	46
89	147	49	107	9	67	18	138	27	98	169	58	129
159	61	119	21	79	150	101	52	110	12	70	141	30
73	131	33	104	162	64	2	122	24	82	153	42	113

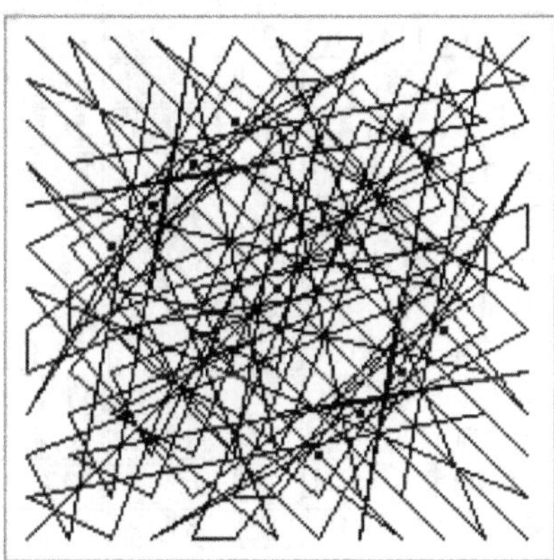

93	108	123	138	153	168	1	16	31	46	61	76	91
107	122	137	152	167	13	15	30	45	60	75	90	92
121	136	151	166	12	14	29	44	59	74	89	104	106
135	150	165	11	26	28	43	58	73	88	103	105	120
149	164	10	25	27	42	57	72	87	102	117	119	134
163	9	24	39	41	56	71	86	101	116	118	133	148
8	23	38	40	55	70	85	100	115	130	132	147	162
22	37	52	54	69	84	99	114	129	131	146	161	7
36	51	53	68	83	98	113	128	143	145	160	6	21
50	65	67	82	97	112	127	142	144	159	5	20	35
64	66	81	96	111	126	141	156	158	4	19	34	49
78	80	95	110	125	140	155	157	3	18	33	48	63
79	94	109	124	139	154	169	2	17	32	47	62	77

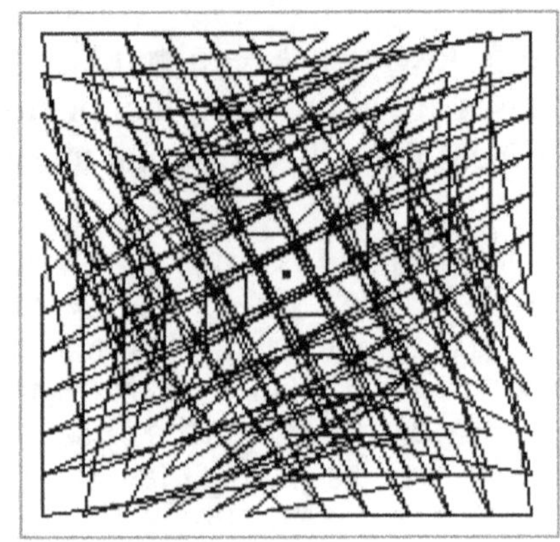

169	58	129	18	89	147	49	107	9	67	138	27	98
84	155	44	115	4	75	133	35	93	164	53	124	26
12	70	141	30	101	159	61	119	21	79	150	52	110
96	167	56	127	16	87	145	47	105	7	78	136	38
24	82	153	42	113	2	73	131	33	104	162	64	122
108	10	68	139	28	99	157	59	130	19	90	148	50
36	94	165	54	125	14	85	156	45	116	5	76	134
120	22	80	151	40	111	13	71	142	31	102	160	62
48	106	8	66	137	39	97	168	57	128	17	88	146
132	34	92	163	65	123	25	83	154	43	114	3	74
60	118	20	91	149	51	109	11	69	140	29	100	158
144	46	117	6	77	135	37	95	166	55	126	15	86
72	143	32	103	161	63	121	23	81	152	41	112	1

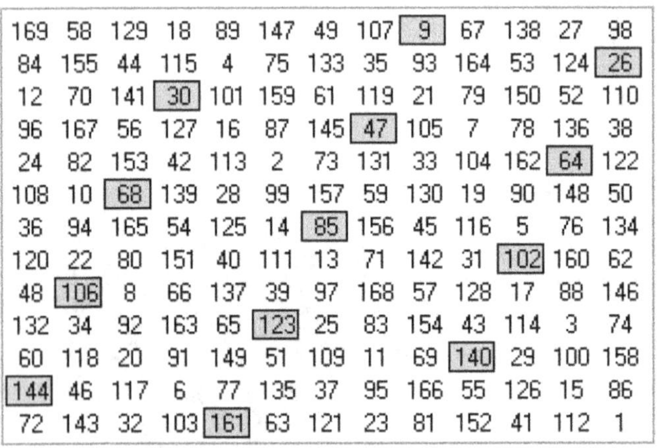
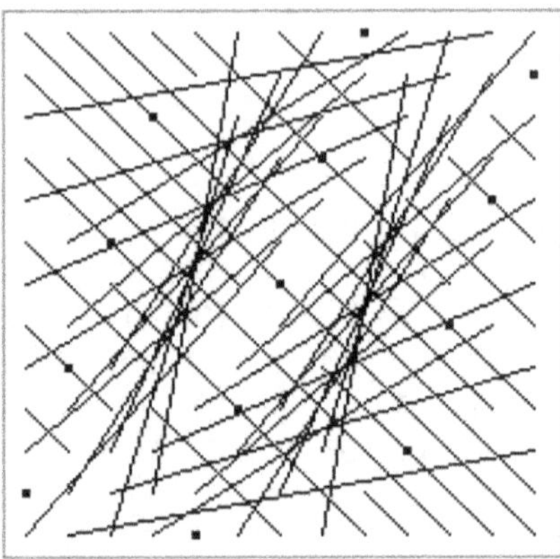

The following are regular 13th order magic squares. Odd numbers are highlighted on the left, and the lines are drawn in sequence on the right (from 1 to 2 to 3, etc.)

1	100	17	116	33	119	49	135	65	151	68	167	84
98	15	114	31	130	47	133	63	149	66	165	82	12
26	112	29	128	45	131	61	147	77	163	80	10	96
110	27	126	43	142	59	145	75	161	91	8	94	24
38	124	41	140	57	156	73	159	89	6	92	22	108
122	52	138	55	154	71	157	87	4	103	20	106	36
50	136	53	152	69	168	85	2	101	18	117	34	120
134	64	150	67	166	83	13	99	16	115	32	118	48
62	148	78	164	81	11	97	14	113	30	129	46	132
146	76	162	79	9	95	25	111	28	127	44	143	60
74	160	90	7	93	23	109	39	125	42	141	58	144
158	88	5	104	21	107	37	123	40	139	56	155	72
86	3	102	19	105	35	121	51	137	54	153	70	169

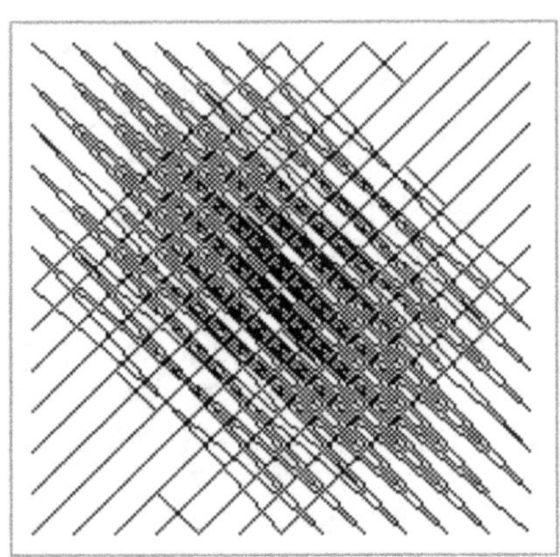

1	135	151	33	167	100	116	119	65	17	68	84	49
63	15	31	82	47	149	165	12	114	66	130	133	98
103	55	71	122	87	20	36	52	154	106	157	4	138
81	46	62	113	78	11	14	30	132	97	148	164	129
143	95	111	162	127	60	76	79	25	146	28	44	9
112	77	80	131	96	29	45	61	163	128	10	26	147
152	117	120	2	136	69	85	101	34	168	50	53	18
23	144	160	42	7	109	125	141	74	39	90	93	58
161	126	142	24	145	91	94	110	43	8	59	75	27
41	6	22	73	38	140	156	159	92	57	108	124	89
32	166	13	64	16	118	134	150	83	48	99	115	67
72	37	40	104	56	158	5	21	123	88	139	155	107
121	86	102	153	105	51	54	70	3	137	19	35	169

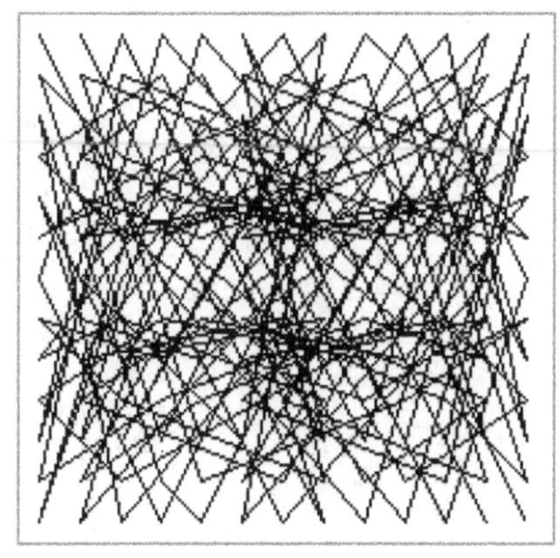

7	32	57	82	107	132	157	26	51	76	101	126	151
164	20	45	70	95	120	145	1	39	64	89	114	139
152	8	33	58	83	108	133	158	14	52	77	102	127
140	165	21	46	71	96	121	146	2	27	65	90	115
128	153	9	34	59	84	109	134	159	15	40	78	103
116	141	166	22	47	72	97	122	147	3	28	53	91
104	129	154	10	35	60	85	110	135	160	16	41	66
79	117	142	167	23	48	73	98	123	148	4	29	54
67	92	130	155	11	36	61	86	111	136	161	17	42
55	80	105	143	168	24	49	74	99	124	149	5	30
43	68	93	118	156	12	37	62	87	112	137	162	18
31	56	81	106	131	169	25	50	75	100	125	150	6
19	44	69	94	119	144	13	38	63	88	113	138	163

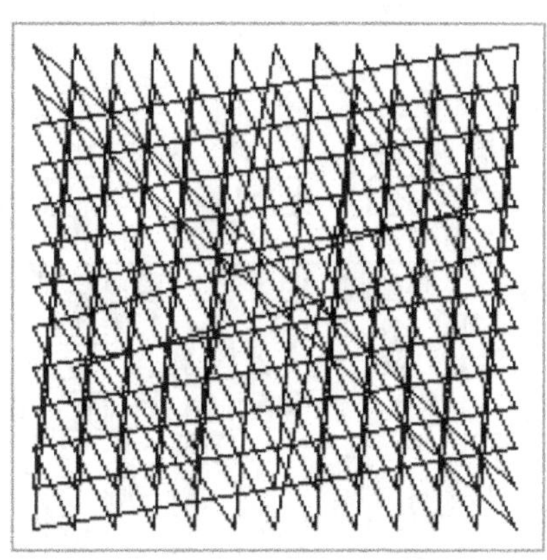

7	99	22	114	37	129	52	131	54	146	69	161	84
97	20	112	35	127	50	142	65	144	67	159	82	5
18	110	33	125	48	140	63	155	78	157	80	3	95
108	31	123	46	138	61	153	76	168	91	1	93	16
29	121	44	136	59	151	74	166	89	12	104	14	106
119	42	134	57	149	72	164	87	10	102	25	117	27
40	132	55	147	70	162	85	8	100	23	115	38	130
143	53	145	68	160	83	6	98	21	113	36	128	51
64	156	66	158	81	4	96	19	111	34	126	49	141
154	77	169	79	2	94	17	109	32	124	47	139	62
75	167	90	13	92	15	107	30	122	45	137	60	152
165	88	11	103	26	105	28	120	43	135	58	150	73
86	9	101	24	116	39	118	41	133	56	148	71	163

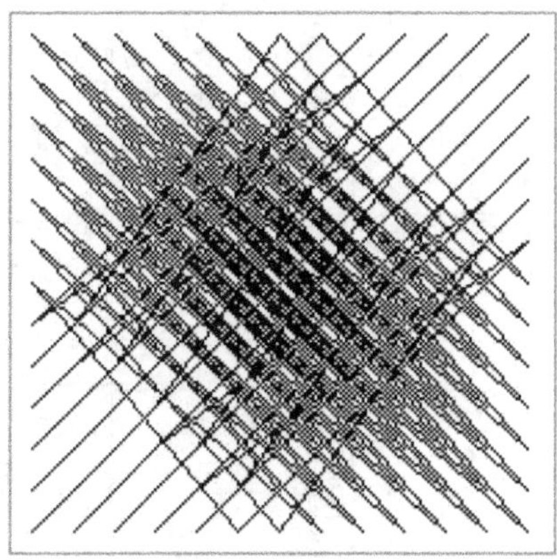

255

71	142	31	102	160	22	120	62	80	151	40	111	13
154	43	114	3	74	92	34	132	163	65	123	25	83
55	126	15	86	144	6	117	46	77	135	37	95	166
138	27	98	169	58	89	18	129	147	49	107	9	67
52	110	12	70	141	159	101	30	61	119	21	79	150
106	8	66	137	39	57	168	97	128	17	88	146	48
36	94	165	54	125	156	85	14	45	116	5	76	134
122	24	82	153	42	73	2	113	131	33	104	162	64
20	91	149	51	109	140	69	11	29	100	158	60	118
103	161	63	121	23	41	152	81	112	1	72	143	32
4	75	133	35	93	124	53	164	26	84	155	44	115
87	145	47	105	7	38	136	78	96	167	56	127	16
157	59	130	19	90	108	50	148	10	68	139	28	99

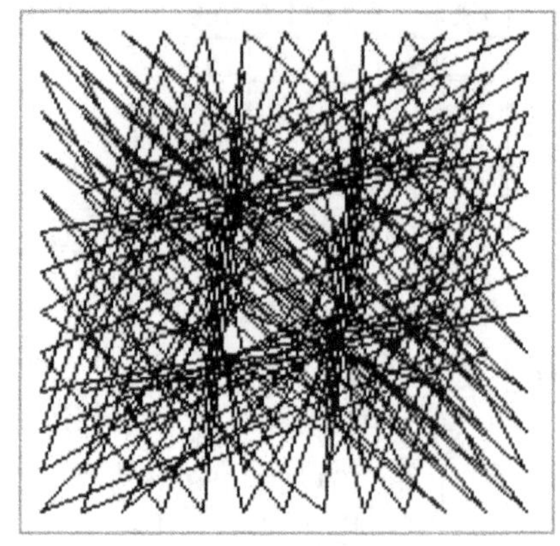

71	156	59	131	47	119	168	35	107	23	95	11	83
154	57	142	45	130	33	69	105	21	93	9	81	166
55	140	43	128	31	116	152	19	104	7	79	164	67
138	41	126	29	114	17	53	102	5	90	162	78	150
52	124	27	112	15	100	136	3	88	160	76	148	64
122	38	110	26	98	1	50	86	158	74	146	62	134
157	73	145	61	133	49	85	121	37	109	25	97	13
36	108	24	96	12	84	120	169	72	144	60	132	48
106	22	94	10	82	167	34	70	155	58	143	46	118
20	92	8	80	165	68	117	153	56	141	44	129	32
103	6	91	163	66	151	18	54	139	42	127	30	115
4	89	161	77	149	65	101	137	40	125	28	113	16
87	159	75	147	63	135	2	51	123	39	111	14	99

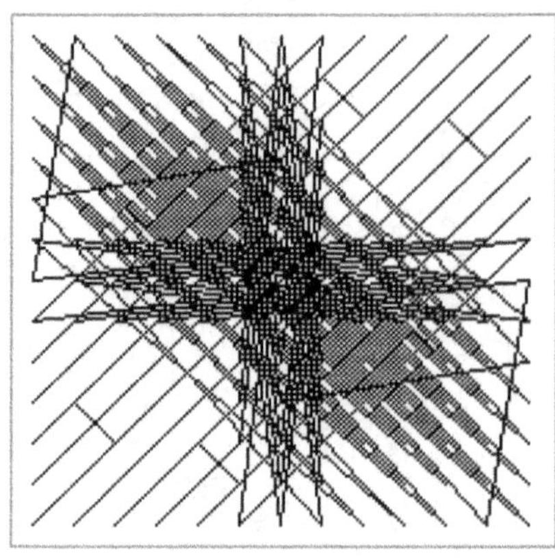

72	120	60	132	48	144	169	12	108	24	96	36	84
128	7	116	19	104	31	43	55	164	67	152	79	140
58	106	46	118	34	143	155	167	94	10	82	22	70
142	21	130	33	105	45	57	69	9	81	166	93	154
44	92	32	117	20	129	141	153	80	165	68	8	56
156	35	131	47	119	59	71	83	23	95	11	107	168
157	49	145	61	133	73	85	97	37	109	25	121	13
2	63	159	75	147	87	99	111	51	123	39	135	14
114	162	102	5	90	17	29	41	150	53	138	78	126
16	77	4	89	161	101	113	125	65	137	40	149	28
100	148	88	160	76	3	15	27	136	52	124	64	112
30	91	18	103	6	115	127	139	66	151	54	163	42
86	134	74	146	62	158	1	26	122	38	110	50	98

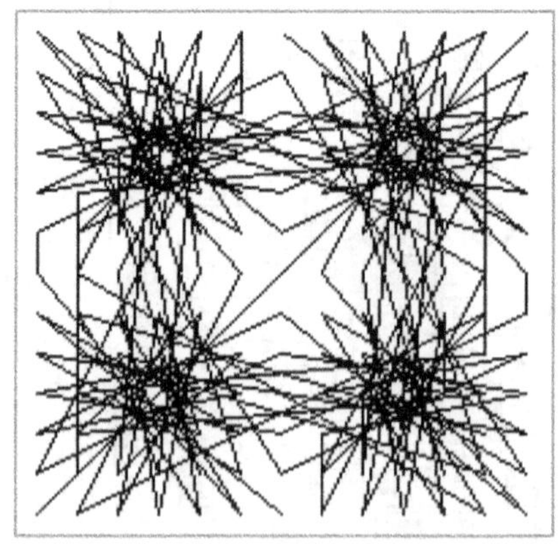

72	144	60	132	48	120	169	36	108	24	96	12	84
156	59	131	47	119	35	71	107	23	95	11	83	168
58	143	46	118	34	106	155	22	94	10	82	167	70
142	45	130	33	105	21	57	93	9	81	166	69	154
44	129	32	117	20	92	141	8	80	165	68	153	56
128	31	116	19	104	7	43	79	164	67	152	55	140
157	73	145	61	133	49	85	121	37	109	25	97	13
30	115	18	103	6	91	127	163	66	151	54	139	42
114	17	102	5	90	162	29	78	150	53	138	41	126
16	101	4	89	161	77	113	149	65	137	40	125	28
100	3	88	160	76	148	15	64	136	52	124	27	112
2	87	159	75	147	63	99	135	51	123	39	111	14
86	158	74	146	62	134	1	50	122	38	110	26	98

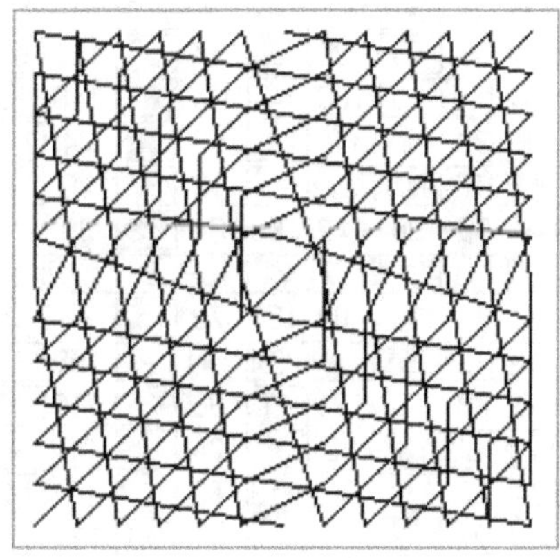

77	62	47	32	139	154	169	2	17	124	109	94	79
63	48	33	18	125	140	155	157	3	110	95	80	78
49	34	19	4	111	126	141	156	158	96	81	66	64
35	20	5	159	97	112	127	142	144	82	67	65	50
134	119	117	102	27	42	57	72	87	25	10	164	149
148	133	118	116	41	56	71	86	101	39	24	9	163
162	147	132	130	55	70	85	100	115	40	38	23	8
7	161	146	131	69	84	99	114	129	54	52	37	22
21	6	160	145	83	98	113	128	143	68	53	51	36
120	105	103	88	26	28	43	58	73	11	165	150	135
106	104	89	74	12	14	29	44	59	166	151	136	121
92	90	75	60	167	13	15	30	45	152	137	122	107
91	76	61	46	153	168	1	16	31	138	123	108	93

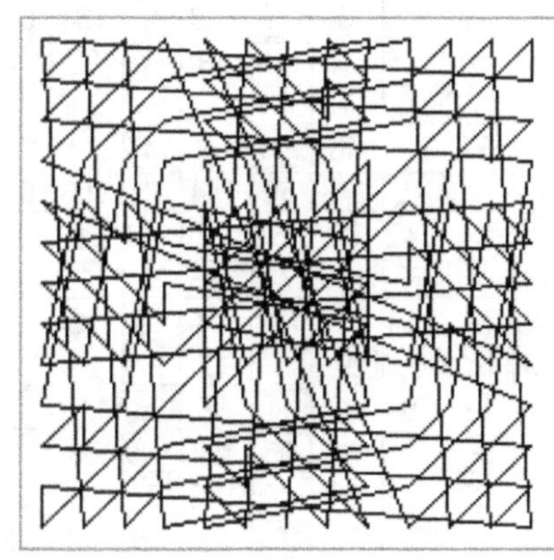

79	8	93	22	107	36	121	50	135	64	149	78	163
164	80	9	94	23	108	37	122	51	136	65	150	66
67	165	81	10	95	24	109	38	123	52	137	53	151
152	68	166	82	11	96	25	110	39	124	40	138	54
55	153	69	167	83	12	97	26	111	27	125	41	139
140	56	154	70	168	84	13	98	14	112	28	126	42
43	141	57	155	71	169	85	1	99	15	113	29	127
128	44	142	58	156	72	157	86	2	100	16	114	30
31	129	45	143	59	144	73	158	87	3	101	17	115
116	32	130	46	131	60	145	74	159	88	4	102	18
19	117	33	118	47	132	61	146	75	160	89	5	103
104	20	105	34	119	48	133	62	147	76	161	90	6
7	92	21	106	35	120	49	134	63	148	77	162	91

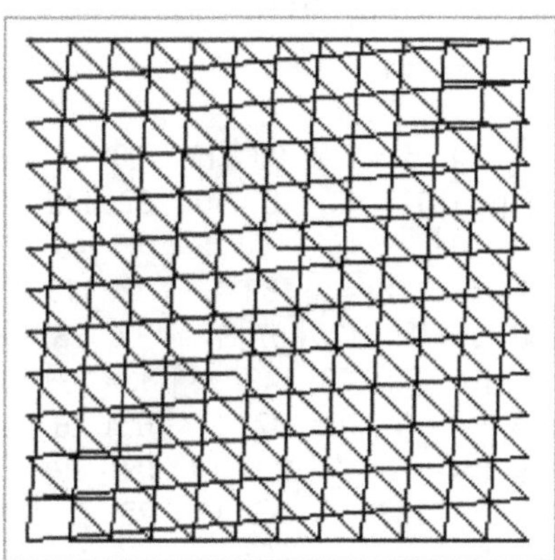

79	8	149	64	107	36	121	50	135	22	93	78	163
164	80	65	136	23	108	37	122	51	94	9	150	66
19	117	89	160	47	132	61	146	75	118	33	5	103
116	32	4	88	131	60	145	74	159	46	130	102	18
55	153	125	27	83	12	97	26	111	167	69	41	139
140	56	28	112	168	84	13	98	14	70	154	126	42
43	141	113	15	71	169	85	1	99	155	57	29	127
128	44	16	100	156	72	157	86	2	58	142	114	30
31	129	101	3	59	144	73	158	87	143	45	17	115
152	68	40	124	11	96	25	110	39	82	166	138	54
67	165	137	52	95	24	109	38	123	10	81	53	151
104	20	161	76	119	48	133	62	147	34	105	90	6
7	92	77	148	35	120	49	134	63	106	21	162	91

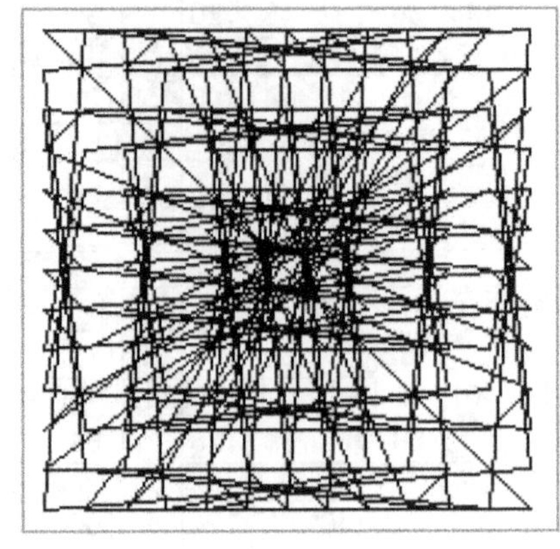

7	92	21	106	35	120	49	134	63	148	77	162	91
104	20	105	34	119	48	133	62	147	76	161	90	6
19	117	33	118	47	132	61	146	75	160	89	5	103
116	32	130	46	131	60	145	74	159	88	4	102	18
31	129	45	143	59	144	73	158	87	3	101	17	115
128	44	142	58	156	72	157	86	2	100	16	114	30
43	141	57	155	71	169	85	1	99	15	113	29	127
140	56	154	70	168	84	13	98	14	112	28	126	42
55	153	69	167	83	12	97	26	111	27	125	41	139
152	68	166	82	11	96	25	110	39	124	40	138	54
67	165	81	10	95	24	109	38	123	52	137	53	151
164	80	9	94	23	108	37	122	51	136	65	150	66
79	8	93	22	107	36	121	50	135	64	149	78	163

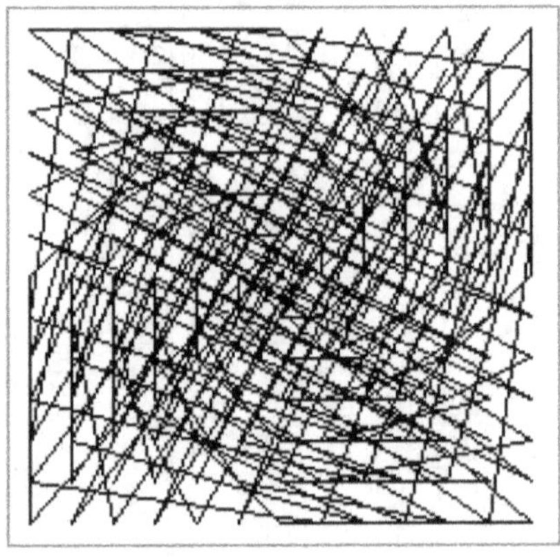

The following magic squares are 13th order pan-diagonal

1	111	52	149	90	18	128	56	166	94	35	132	73
87	15	125	53	163	104	32	142	70	11	108	49	146
160	101	29	139	67	8	105	46	156	84	25	122	63
77	5	115	43	153	81	22	119	60	157	98	39	136
150	91	19	129	57	167	95	36	133	74	2	112	40
54	164	92	33	143	71	12	109	50	147	88	16	126
140	68	9	106	47	144	85	26	123	64	161	102	30
44	154	82	23	120	61	158	99	27	137	78	6	116
130	58	168	96	37	134	75	3	113	41	151	79	20
34	131	72	13	110	51	148	89	17	127	55	165	93
107	48	145	86	14	124	65	162	103	31	141	69	10
24	121	62	159	100	28	138	66	7	117	45	155	83
97	38	135	76	4	114	42	152	80	21	118	59	169

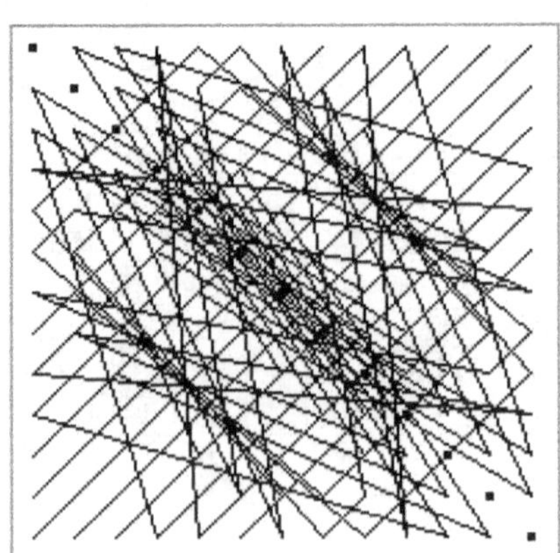

1	85	169	71	155	57	141	43	127	29	113	15	99
62	133	48	119	34	105	**20**	104	6	90	161	76	147
110	25	96	11	82	166	68	152	54	138	40	124	**39**
158	73	144	59	143	**45**	129	31	115	17	101	3	87
50	121	36	107	22	93	8	79	163	78	149	**64**	135
98	13	84	168	**70**	154	56	140	42	126	28	112	14
146	61	132	47	118	33	117	19	103	5	**89**	160	75
38	109	24	**95**	10	81	165	67	151	53	137	52	123
86	157	72	156	58	142	44	128	30	**114**	16	100	2
134	49	**120**	35	106	21	92	7	91	162	77	148	63
26	97	12	83	167	69	153	55	**139**	41	125	27	111
74	**145**	60	131	46	130	32	116	18	102	4	88	159
122	37	108	23	94	9	80	**164**	66	150	65	136	51

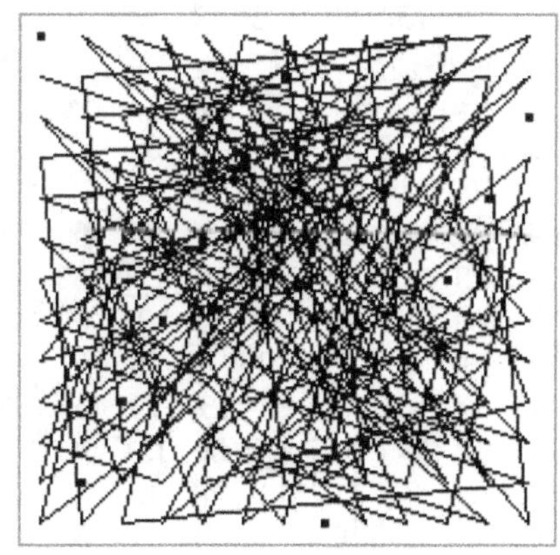

68	117	153	20	56	92	141	**8**	44	80	129	165	32
11	47	83	119	168	35	71	107	156	**23**	59	95	131
110	146	26	62	98	134	1	50	86	122	158	**38**	74
40	89	125	161	28	77	113	149	16	65	101	137	4
152	19	**55**	104	140	7	43	79	128	164	31	67	116
82	118	167	34	**70**	106	155	22	58	94	143	10	46
25	61	97	133	13	49	**85**	121	157	37	73	109	145
124	160	27	76	112	148	15	64	**100**	136	3	52	88
54	103	139	6	42	91	127	163	30	66	**115**	151	18
166	33	69	105	154	21	57	93	142	9	45	81	**130**
96	**132**	12	48	84	120	169	36	72	108	144	24	60
39	75	111	**147**	14	63	99	135	2	51	87	123	159
138	5	41	90	126	**162**	29	78	114	150	17	53	102

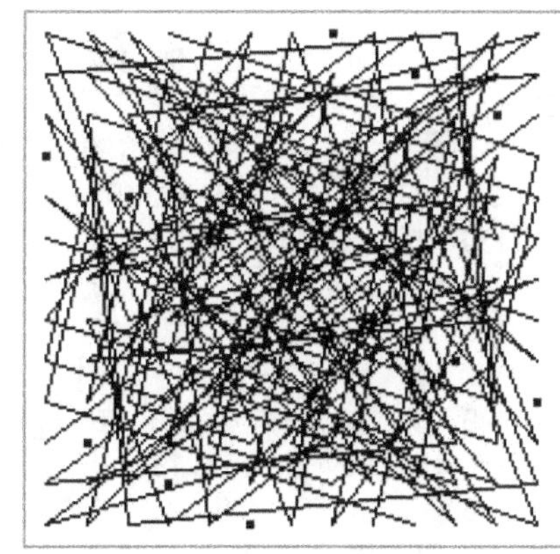

71	142	31	102	160	62	120	22	80	151	40	111	**13**
154	43	114	3	74	132	34	92	163	65	123	**25**	83
55	126	15	86	144	46	117	6	77	135	**37**	95	166
138	27	98	169	58	129	18	89	147	**49**	107	9	67
52	110	12	70	141	30	101	159	**61**	119	21	79	150
122	24	82	153	42	113	2	**73**	131	33	104	162	64
36	94	165	54	125	14	**85**	156	45	116	5	76	134
106	8	66	137	39	**97**	168	57	128	17	88	146	48
20	91	149	51	**109**	11	69	140	29	100	158	60	118
103	161	63	**121**	23	81	152	41	112	1	72	143	32
4	75	**133**	35	93	164	53	124	26	84	155	44	115
87	**145**	47	105	7	78	136	38	96	167	56	127	16
157	59	130	19	90	148	50	108	10	68	139	28	99

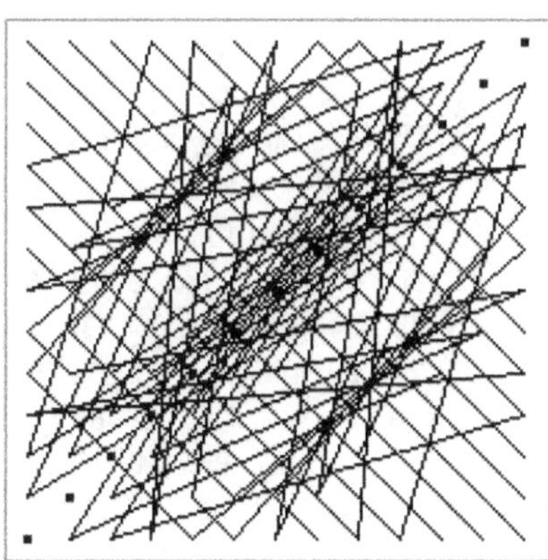

151	53	137	52	123	38	109	24	95	10	81	165	67
30	114	16	100	2	86	157	72	156	58	142	44	128
91	162	77	148	63	134	49	120	35	106	21	92	7
139	41	125	27	111	26	97	12	83	167	69	153	55
18	102	4	88	159	74	145	60	131	46	130	32	116
66	150	65	136	51	122	37	108	23	94	9	80	164
127	29	113	15	99	1	85	169	71	155	57	141	43
6	90	161	76	147	62	133	48	119	34	105	20	104
54	138	40	124	39	110	25	96	11	82	166	68	152
115	17	101	3	87	158	73	144	59	143	45	129	31
163	78	149	64	135	50	121	36	107	22	93	8	79
42	126	28	112	14	98	13	84	168	70	154	56	140
103	5	89	160	75	146	61	132	47	118	33	117	19

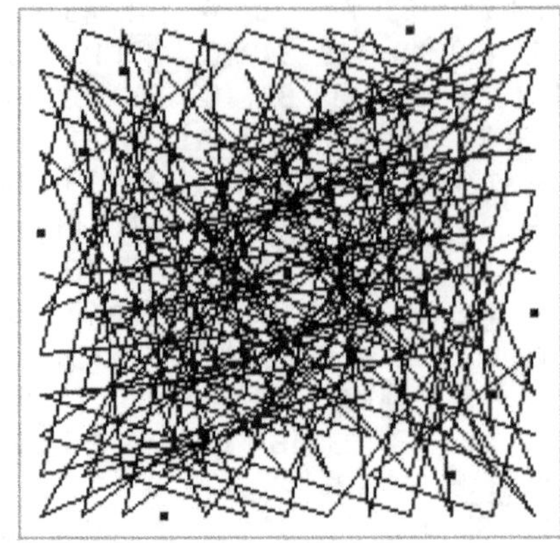

The following are regular magic squares with circular plots.

1	98	26	110	38	122	50	134	62	146	74	158	86
100	15	112	27	124	52	136	64	148	76	160	88	3
17	114	29	126	41	138	53	150	78	162	90	5	102
116	31	128	43	140	55	152	67	164	79	7	104	19
33	130	45	142	57	154	69	166	81	9	93	21	105
119	47	131	59	156	71	168	83	11	95	23	107	35
49	133	61	145	73	157	85	13	97	25	109	37	121
135	63	147	75	159	87	2	99	14	111	39	123	51
65	149	77	161	89	4	101	16	113	28	125	40	137
151	66	163	91	6	103	18	115	30	127	42	139	54
68	165	80	8	92	20	117	32	129	44	141	56	153
167	82	10	94	22	106	34	118	46	143	58	155	70
84	12	96	24	108	36	120	48	132	60	144	72	169

Circular Std Sequence

1	63	112	161	41	103	152	32	81	143	23	72	121
135	15	77	126	6	55	117	166	46	95	144	37	86
100	149	29	91	140	20	69	118	11	60	109	158	51
65	114	163	43	92	154	34	83	132	25	74	123	3
17	66	128	8	57	106	168	48	97	146	39	88	137
151	31	80	142	22	71	120	13	62	111	160	40	102
116	165	45	94	156	36	85	134	14	76	125	5	54
68	130	10	59	108	157	50	99	148	28	90	139	19
33	82	131	24	73	122	2	64	113	162	42	104	153
167	47	96	145	38	87	136	16	78	127	7	56	105
119	12	61	110	159	52	101	150	30	79	141	21	70
84	133	26	75	124	4	53	115	164	44	93	155	35
49	98	147	27	89	138	18	67	129	9	58	107	169

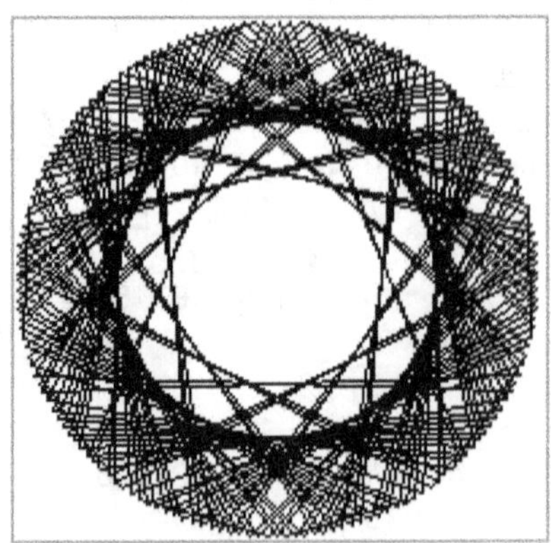

Circular Std Sequence

Appendix K – Selected 14th Order Magic Squares

The following are regular 14th order magic squares. The left side highlights those numbers in their original positions while the right side connects the lines from a numbers original position to its current position.

1	3	133	135	41	43	146	147	84	82	188	186	96	94
2	4	134	136	42	44	145	148	83	81	187	185	95	93
125	127	33	35	165	167	74	75	180	178	88	86	24	22
126	128	34	36	166	168	73	76	179	177	87	85	23	21
53	55	157	159	65	67	171	170	108	106	16	14	120	118
54	56	158	160	66	68	172	169	107	105	15	13	119	117
150	152	60	59	191	189	97	98	5	6	140	138	46	48
151	149	57	58	190	192	99	100	8	7	137	139	47	45
80	78	184	182	92	90	25	28	129	131	37	39	141	143
79	77	183	181	91	89	27	26	130	132	38	40	142	144
176	174	112	110	20	18	124	121	29	31	161	163	69	71
175	173	111	109	19	17	123	122	30	32	162	164	70	72
104	102	12	10	116	114	52	49	153	155	61	63	193	195
103	101	11	9	115	113	51	50	154	156	62	64	194	196

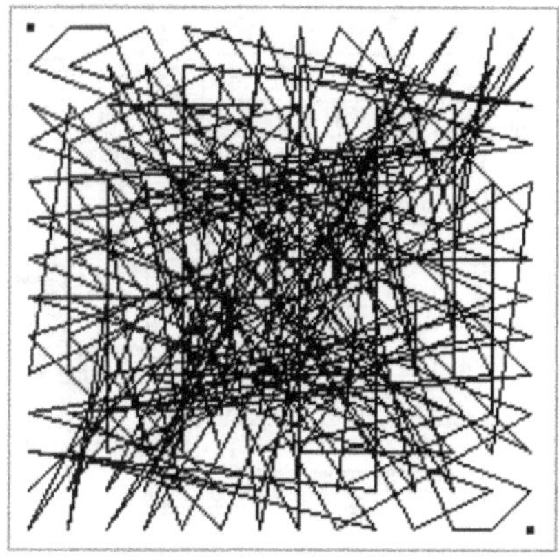

1	188	82	3	133	43	145	148	84	186	96	41	135	94
53	16	106	55	157	67	169	172	108	14	120	65	159	118
176	161	31	174	112	18	123	122	29	163	69	20	110	71
175	162	32	173	111	17	124	121	30	164	70	19	109	72
2	187	81	4	134	44	146	147	83	185	95	42	136	93
54	15	105	56	158	68	171	170	107	13	119	66	160	117
150	137	7	152	58	191	99	100	5	139	45	189	60	47
151	140	6	149	59	190	97	98	8	138	48	192	57	46
80	37	131	78	184	90	27	26	129	39	141	92	182	143
104	61	155	102	12	114	51	50	153	63	193	116	10	195
125	88	178	127	33	167	73	76	180	86	24	165	35	22
126	87	177	128	34	168	74	75	179	85	23	166	36	21
79	38	132	77	183	89	28	25	130	40	142	91	181	144
103	62	156	101	11	113	52	49	154	64	194	115	9	196

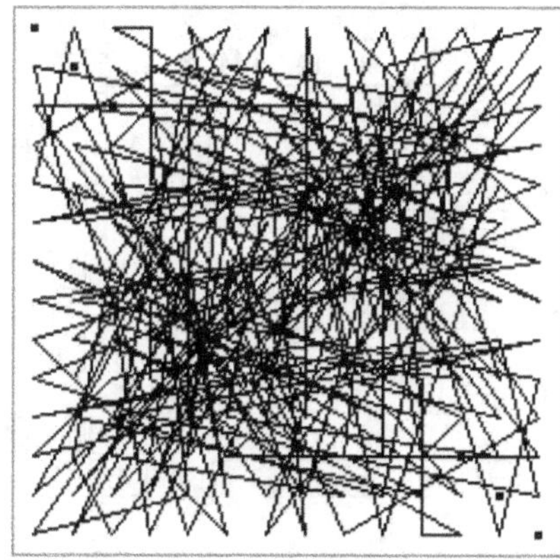

87	85	123	121	159	157	194	195	6	8	42	44	78	80
88	86	124	122	160	158	196	193	5	7	41	43	77	79
83	81	91	89	127	125	163	162	170	172	10	12	46	48
84	82	92	90	128	126	164	161	169	171	9	11	45	47
51	49	59	57	95	93	131	130	166	168	174	176	14	16
52	50	60	58	96	94	129	132	165	167	173	175	13	15
18	17	55	53	63	64	99	100	136	135	141	143	177	178
19	20	54	56	62	61	97	98	133	134	144	142	180	179
182	184	22	24	30	32	65	68	103	101	139	137	147	145
181	183	21	23	29	31	67	66	104	102	140	138	148	146
150	152	186	188	26	28	33	36	71	69	107	105	115	113
149	151	185	187	25	27	34	35	72	70	108	106	116	114
118	120	154	156	190	192	4	1	39	37	75	73	111	109
117	119	153	155	189	191	3	2	40	38	76	74	112	110

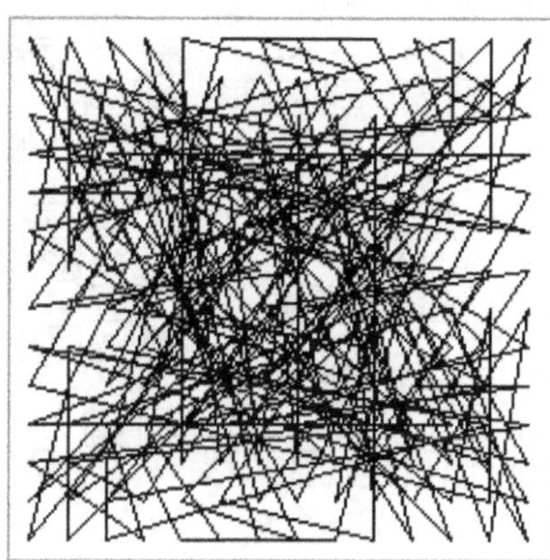

The highlighted numbers on the left side are odd numbers. The right side shows lines drawn in sequence (from 1 to 2 to 3, etc.) The first three magic squares are almost identical to each other. The difference can be found in the numbers that were swapped in the center rows and center columns.

1	3	133	135	41	43	146	147	84	82	188	186	96	94
2	4	134	136	42	44	145	148	83	81	187	185	95	93
125	127	33	35	165	167	74	75	180	178	88	86	24	22
126	128	34	36	166	168	73	76	179	177	87	85	23	21
53	55	157	159	65	67	171	170	108	106	16	14	120	118
54	56	158	160	66	68	172	169	107	105	15	13	119	117
150	152	60	59	191	189	97	98	5	6	140	138	46	48
151	149	57	58	190	192	99	100	8	7	137	139	47	45
80	78	184	182	92	90	25	28	129	131	37	39	141	143
79	77	183	181	91	89	27	26	130	132	38	40	142	144
176	174	112	110	20	18	124	121	29	31	161	163	69	71
175	173	111	109	19	17	123	122	30	32	162	164	70	72
104	102	12	10	116	114	52	49	153	155	61	63	193	195
103	101	11	9	115	113	51	50	154	156	62	64	194	196

1	3	133	135	41	43	146	147	84	82	188	186	96	94
2	4	134	136	42	44	148	145	83	81	187	185	95	93
129	131	37	39	141	143	79	78	184	182	92	90	28	26
130	132	38	40	142	144	80	77	183	181	91	89	27	25
33	35	165	167	73	75	179	178	88	86	24	22	128	126
34	36	166	168	74	76	177	180	87	85	23	21	127	125
163	164	70	72	174	173	109	110	17	18	124	122	32	31
162	161	71	69	175	176	111	112	20	19	121	123	29	30
68	66	172	170	108	106	13	16	117	119	53	55	157	159
67	65	171	169	107	105	15	14	118	120	54	56	158	160
196	194	104	102	12	10	113	116	49	51	153	155	61	63
195	193	103	101	11	9	115	114	50	52	154	156	62	64
100	98	8	6	140	138	48	45	149	151	57	59	189	191
99	97	7	5	139	137	46	47	150	152	58	60	190	192

1	3	133	135	41	43	148	145	84	82	188	186	96	94
2	4	134	136	42	44	147	146	83	81	187	185	95	93
125	127	33	35	165	167	74	75	180	178	88	86	24	22
126	128	34	36	166	168	73	76	179	177	87	85	23	21
53	55	157	159	65	67	172	169	108	106	16	14	120	118
54	56	158	160	66	68	171	170	107	105	15	13	119	117
151	152	58	57	191	189	97	99	6	5	140	139	48	47
150	149	59	60	190	192	98	100	7	8	137	138	45	46
80	78	184	182	92	90	27	26	129	131	37	39	141	143
79	77	183	181	91	89	28	25	130	132	38	40	142	144
176	174	112	110	20	18	121	124	29	31	161	163	69	71
175	173	111	109	19	17	122	123	30	32	162	164	70	72
104	102	12	10	116	114	51	49	154	155	61	63	193	195
103	101	11	9	115	113	50	52	153	156	62	64	194	196

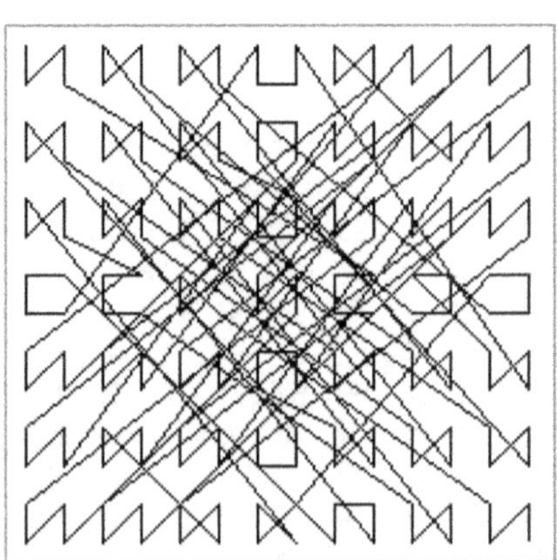

1	191	38	155	74	119	110	117	112	153	76	189	40	4
3	192	37	156	73	120	109	118	111	154	75	190	39	2
36	27	70	187	106	151	114	149	116	185	108	25	72	33
34	28	69	188	105	152	113	150	115	186	107	26	71	35
68	31	102	23	138	183	146	181	148	21	140	29	104	65
66	32	101	24	137	104	145	182	147	22	139	30	103	67
98	61	134	56	142	20	179	19	180	54	144	62	133	97
100	64	135	53	143	17	178	18	177	55	141	63	136	99
130	94	167	58	175	50	15	52	13	60	173	96	165	131
132	93	168	57	176	49	16	51	14	59	174	95	166	129
162	126	171	90	11	82	47	84	45	92	9	128	169	163
161	125	172	89	12	81	48	83	46	91	10	127	170	164
195	158	7	122	43	86	79	88	77	124	41	160	5	194
193	157	8	121	44	85	80	87	78	123	42	159	6	196

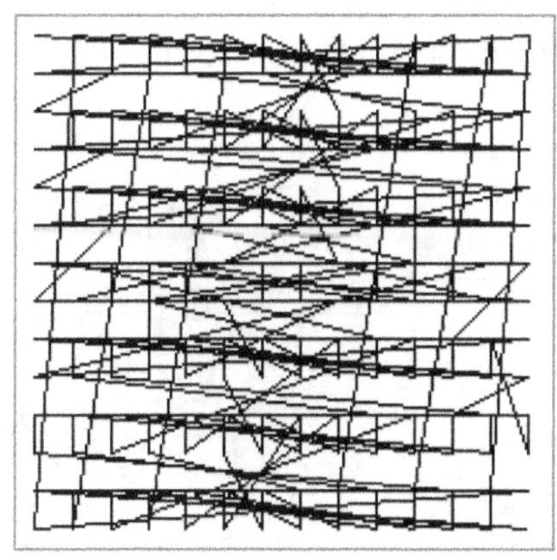

1	96	133	188	41	84	146	147	43	82	135	186	3	94
104	193	12	61	116	153	52	49	114	155	10	63	102	195
125	24	33	88	165	180	74	75	167	178	35	86	127	22
176	69	112	161	20	29	124	121	18	31	110	163	174	71
53	120	157	16	65	108	171	170	67	106	159	14	55	118
80	141	184	37	92	129	25	28	90	131	182	39	78	143
150	46	60	140	191	5	97	98	189	6	59	138	152	48
151	47	57	137	190	8	99	100	192	7	58	139	149	45
54	119	158	15	66	107	172	169	68	105	160	13	56	117
79	142	183	38	91	130	27	26	89	132	181	40	77	144
126	23	34	87	166	179	73	76	168	177	36	85	128	21
175	70	111	162	19	30	123	122	17	32	109	164	173	72
2	95	134	187	42	83	145	148	44	81	136	185	4	93
103	194	11	62	115	154	51	50	113	156	9	64	101	196

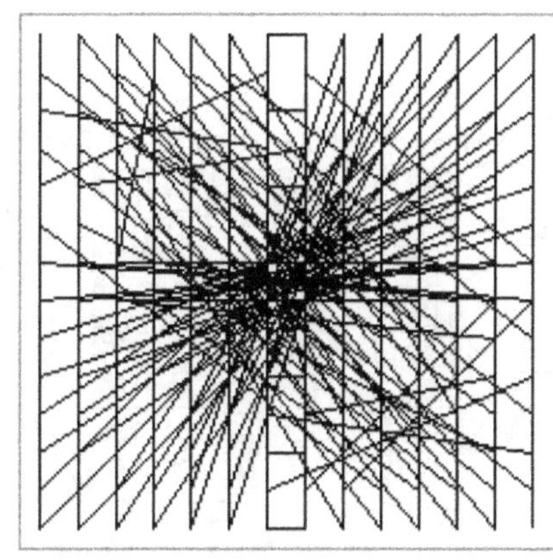

1	109	73	111	192	37	156	75	190	39	120	154	118	4
130	16	176	14	93	168	57	174	95	166	49	59	51	131
161	47	11	45	126	171	90	9	128	169	82	92	84	164
162	48	12	46	125	172	89	10	127	170	81	91	83	163
98	179	142	180	61	134	56	144	62	133	20	54	19	97
132	15	175	13	94	167	58	173	96	165	50	60	52	129
2	110	74	112	191	38	155	76	189	40	119	153	117	3
195	80	44	78	157	8	121	42	159	6	85	123	87	194
68	145	137	147	32	101	24	139	30	103	184	22	182	65
100	178	143	177	64	135	53	141	63	136	17	55	18	99
35	114	106	116	27	70	187	108	25	72	151	185	149	34
36	113	105	115	28	69	188	107	26	71	152	186	150	33
66	146	138	148	31	102	23	140	29	104	183	21	181	67
193	79	43	77	158	7	122	41	160	5	86	124	88	196

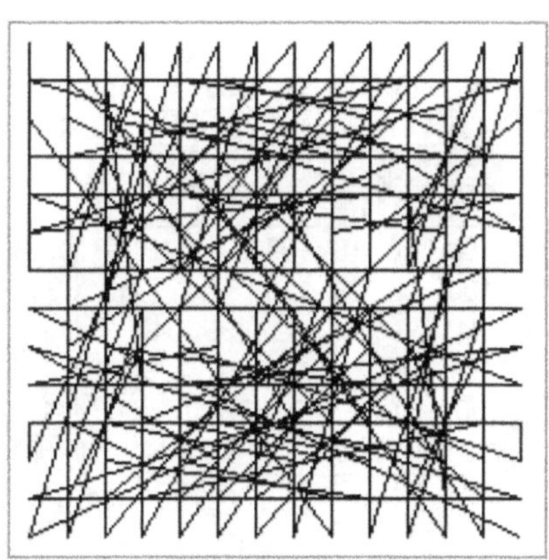

263

The next three magic squares share the characteristic that they have the numbers 97, 98, 99, and 100 in their four corners.

97	62	64	54	20	18	141	56	179	177	143	136	134	98
65	32	30	24	182	184	137	22	147	145	139	101	103	68
67	31	29	23	181	183	138	21	148	146	140	102	104	66
36	28	26	188	150	152	105	186	115	113	107	69	71	33
2	191	189	155	117	119	74	153	112	110	76	38	40	3
4	192	190	156	118	120	73	154	111	109	75	37	39	1
163	125	127	89	83	81	12	91	46	48	10	172	170	162
35	27	25	187	149	151	106	185	116	114	108	70	72	34
193	158	160	122	88	86	43	124	77	79	41	7	5	196
194	157	159	121	87	85	44	123	78	80	42	8	6	195
164	126	128	90	84	82	11	92	45	47	9	171	169	161
131	93	95	57	51	49	176	59	14	16	174	168	166	130
129	94	96	58	52	50	175	60	13	15	173	167	165	132
99	63	61	55	17	19	144	53	178	180	142	133	135	100

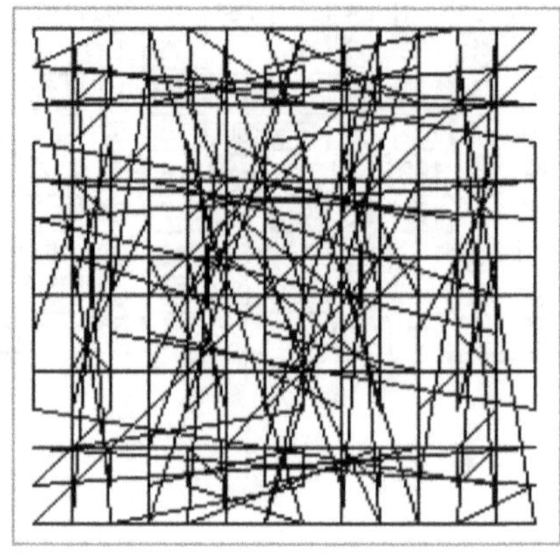

97	171	172	73	74	148	145	52	51	122	121	28	27	98
190	68	67	168	167	43	44	113	114	17	18	89	90	191
189	66	65	166	165	42	41	115	116	19	20	91	92	192
60	160	159	36	35	136	135	9	10	109	110	181	182	57
59	158	157	34	33	134	133	11	12	111	112	183	184	58
152	56	55	128	127	4	3	101	102	173	174	77	78	149
151	54	53	126	125	2	1	103	104	175	176	79	80	150
47	117	118	21	22	93	94	196	195	72	71	144	143	46
48	119	120	23	24	95	96	194	193	70	69	142	141	45
138	13	14	85	86	185	186	64	63	164	163	40	39	139
137	15	16	87	88	187	188	62	61	162	161	38	37	140
7	105	106	177	178	81	82	156	155	32	31	132	131	6
5	107	108	179	180	83	84	154	153	30	29	130	129	8
99	170	169	76	75	146	147	49	50	123	124	25	26	100

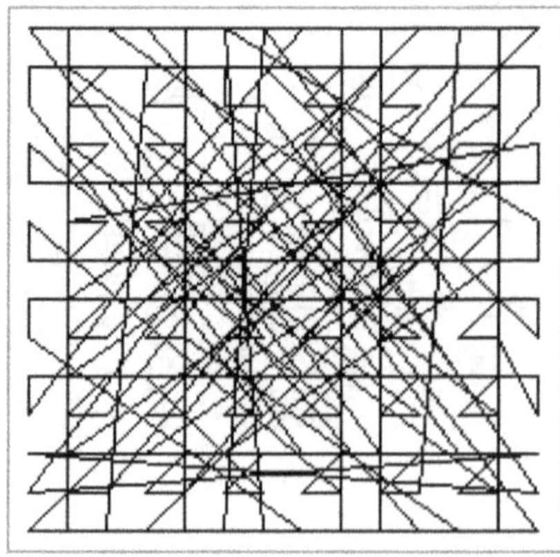

97	5	46	140	6	138	48	150	60	191	59	152	189	98
25	129	141	37	131	39	143	80	184	92	182	78	90	28
52	153	193	61	155	63	195	104	12	116	10	102	114	49
124	29	69	161	31	163	71	176	112	20	110	174	18	121
171	108	120	16	106	14	118	53	157	65	159	55	67	170
74	180	24	88	178	86	22	125	33	165	35	127	167	75
146	84	96	188	82	186	94	1	133	41	135	3	43	147
51	154	194	62	156	64	196	103	11	115	9	101	113	50
123	30	70	162	32	164	72	175	111	19	109	173	17	122
27	130	142	38	132	40	144	79	183	91	181	77	89	26
73	179	23	87	177	85	21	126	34	166	36	128	168	76
145	83	95	187	81	185	93	2	134	42	136	4	44	148
172	107	119	15	105	13	117	54	158	66	160	56	68	169
99	8	47	137	7	139	45	151	57	190	58	149	192	100

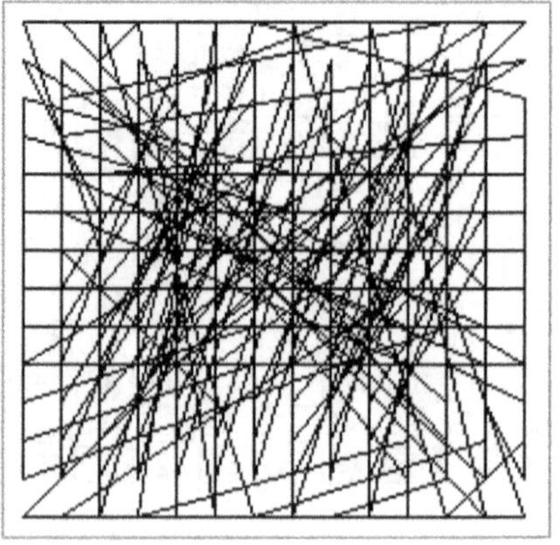

These last three magic squares show odd numbers highlighted on the left, and the right
shows lines connecting sequence pairs.

1	3	133	135	41	43	148	145	84	82	188	186	96	94
2	4	134	136	42	44	147	146	83	81	187	185	95	93
129	131	37	39	141	143	77	80	184	182	92	90	28	26
130	132	38	40	142	144	79	78	183	181	91	89	27	25
33	35	165	167	73	75	177	180	88	86	24	22	128	126
34	36	166	168	74	76	179	178	87	85	23	21	127	125
164	162	70	72	175	173	109	110	17	18	124	123	32	30
161	163	71	69	174	176	111	112	20	19	121	122	29	31
68	66	172	170	108	106	14	15	117	119	53	55	157	159
67	65	171	169	107	105	13	16	118	120	54	56	158	160
196	194	104	102	12	10	114	115	49	51	153	155	61	63
195	193	103	101	11	9	116	113	50	52	154	156	62	64
100	98	8	6	140	138	47	46	149	151	57	59	189	191
99	97	7	5	139	137	48	45	150	152	58	60	190	192

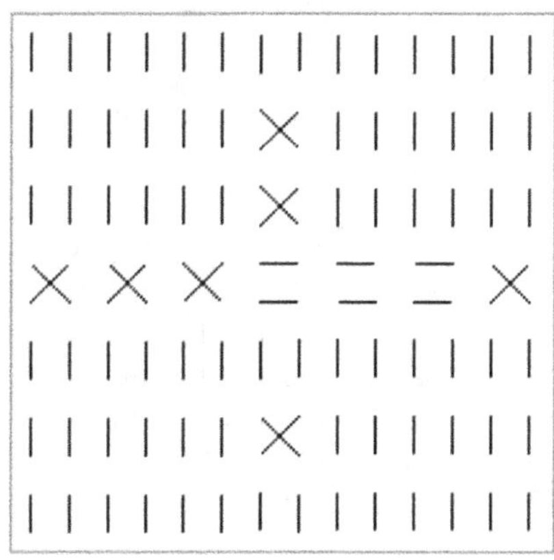

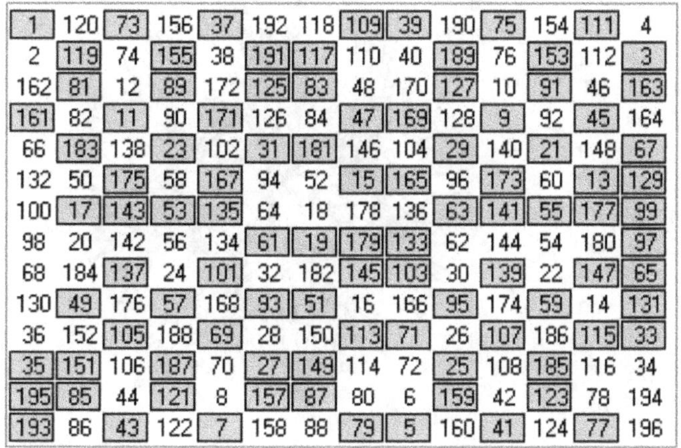

1	120	73	156	37	192	118	109	39	190	75	154	111	4
2	119	74	155	38	191	117	110	40	189	76	153	112	3
162	81	12	89	172	125	83	48	170	127	10	91	46	163
161	82	11	90	171	126	84	47	169	128	9	92	45	164
66	183	138	23	102	31	181	146	104	29	140	21	148	67
132	50	175	58	167	94	52	15	165	96	173	60	13	129
100	17	143	53	135	64	18	178	136	63	141	55	177	99
98	20	142	56	134	61	19	179	133	62	144	54	180	97
68	184	137	24	101	32	182	145	103	30	139	22	147	65
130	49	176	57	168	93	51	16	166	95	174	59	14	131
36	152	105	188	69	28	150	113	71	26	107	186	115	33
35	151	106	187	70	27	149	114	72	25	108	185	116	34
195	85	44	121	8	157	87	80	6	159	42	123	78	194
193	86	43	122	7	158	88	79	5	160	41	124	77	196

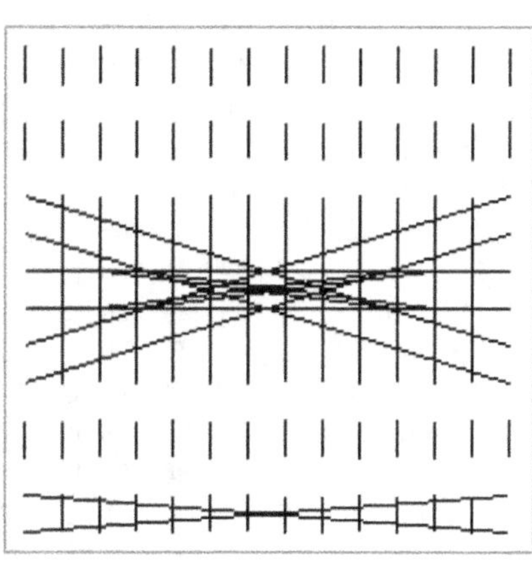

1	191	189	76	74	119	117	110	112	153	155	38	40	4
3	192	190	75	73	120	118	109	111	154	156	37	39	2
161	125	127	10	12	81	83	48	46	91	89	172	170	164
162	126	128	9	11	82	84	47	45	92	90	171	169	163
68	31	29	140	138	183	181	146	148	21	23	102	104	65
66	32	30	139	137	184	182	145	147	22	24	101	103	67
100	64	63	141	143	17	18	178	177	55	53	135	136	99
98	61	62	144	142	20	19	179	180	54	56	134	133	97
130	94	96	173	175	50	52	15	13	60	58	167	165	131
132	93	95	174	176	49	51	16	14	59	57	168	166	129
34	28	26	107	105	152	150	113	115	186	188	69	71	35
36	27	25	108	106	151	149	114	116	185	187	70	72	33
195	158	160	41	43	86	88	79	77	124	122	7	5	194
193	157	159	42	44	85	87	80	78	123	121	8	6	196

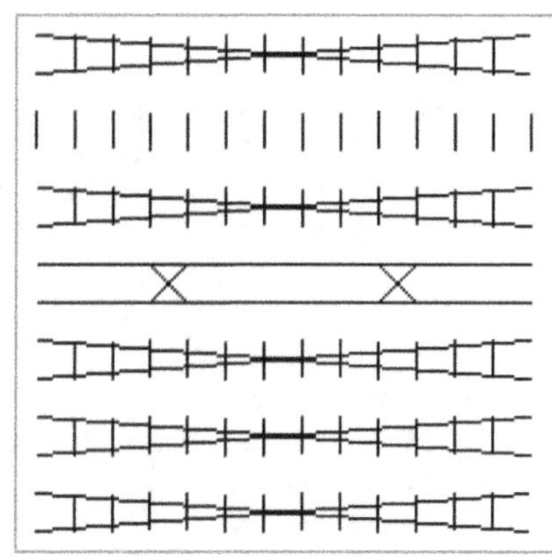

The following are regular magic squares with circular plots.

117	112	153	76	189	40	4	1	191	38	155	74	119	110
118	111	154	75	190	39	2	3	192	37	156	73	120	109
149	116	185	108	25	72	33	36	27	70	187	106	151	114
150	115	186	107	26	71	35	34	28	69	188	105	152	113
181	148	21	140	29	104	65	68	31	102	23	138	183	146
182	147	22	139	30	103	67	66	32	101	24	137	184	145
19	180	54	144	62	133	97	98	61	134	56	142	20	179
18	177	55	141	63	136	99	100	64	135	53	143	17	178
52	13	60	173	96	165	131	130	94	167	58	175	50	15
51	14	59	174	95	166	129	132	93	168	57	176	49	16
84	45	92	9	128	169	163	162	126	171	90	11	82	47
83	46	91	10	127	170	164	161	125	172	89	12	81	48
88	77	124	41	160	5	194	195	158	7	122	43	86	79
87	78	123	42	159	6	196	193	157	8	121	44	85	80

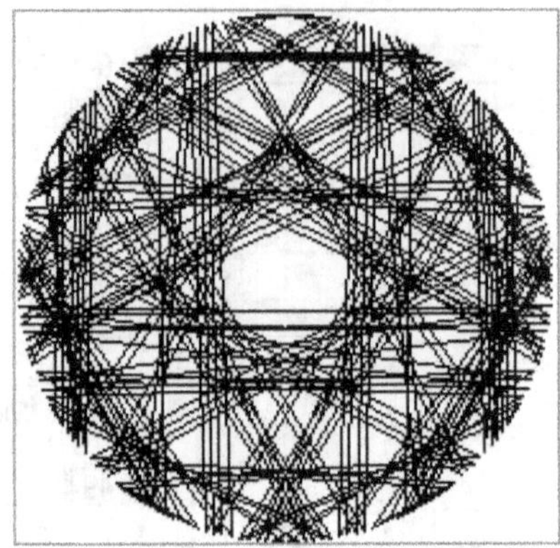

Circular Std Sequence

97	171	172	73	74	148	145	52	51	122	121	28	27	98
190	68	67	168	167	43	44	113	114	17	18	89	90	191
189	66	65	166	165	42	41	115	116	19	20	91	92	192
60	160	159	36	35	136	135	9	10	109	110	181	182	57
59	158	157	34	33	134	133	11	12	111	112	183	184	58
152	56	55	128	127	4	3	101	102	173	174	77	78	149
151	54	53	126	125	2	1	103	104	175	176	79	80	150
47	117	118	21	22	93	94	196	195	72	71	144	143	46
48	119	120	23	24	95	96	194	193	70	69	142	141	45
138	13	14	85	86	185	186	64	63	164	163	40	39	139
137	15	16	87	88	187	188	62	61	162	161	38	37	140
7	105	106	177	178	81	82	156	155	32	31	132	131	6
5	107	108	179	180	83	84	154	153	30	29	130	129	8
99	170	169	76	75	146	147	49	50	123	124	25	26	100

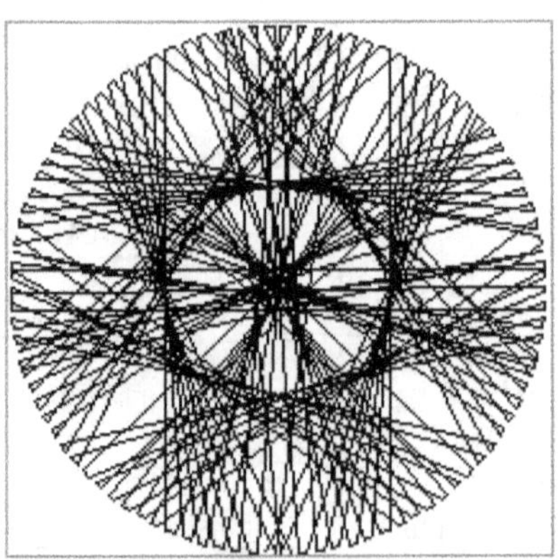

Circular Std Sequence

87	85	123	121	159	157	194	195	6	8	42	44	78	80
88	86	124	122	160	158	196	193	5	7	41	43	77	79
83	81	91	89	127	125	163	162	170	172	10	12	46	48
84	82	92	90	128	126	164	161	169	171	9	11	45	47
51	49	59	57	95	93	131	130	166	168	174	176	14	16
52	50	60	58	96	94	129	132	165	167	173	175	13	15
18	17	55	53	63	64	99	100	136	135	141	143	177	178
19	20	54	56	62	61	97	98	133	134	144	142	180	179
182	184	22	24	30	32	65	68	103	101	139	137	147	145
181	183	21	23	29	31	67	66	104	102	140	138	148	146
150	152	186	188	26	28	33	36	71	69	107	105	115	113
149	151	185	187	25	27	34	35	72	70	108	106	116	114
118	120	154	156	190	192	4	1	39	37	75	73	111	109
117	119	153	155	189	191	3	2	40	38	76	74	112	110

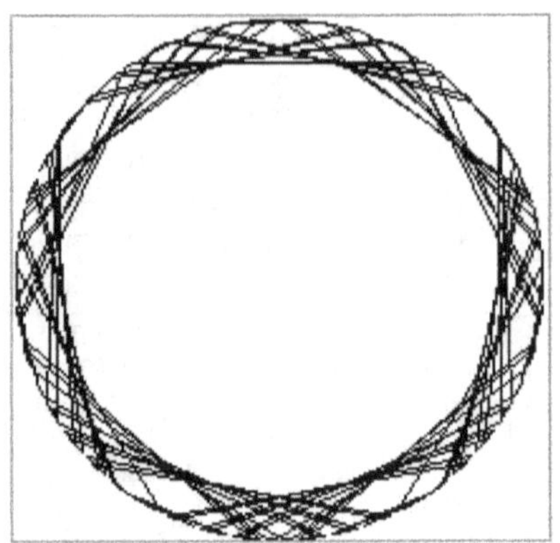

Circular Std Sequence

Appendix L – Selected 15th Order Magic Squares

The following are regular 15th order magic squares. The boxed numbers in the left squares are numbers in their original positions. The right square shows the movement of the numbers from their original to their current position.

104	87	70	53	36	19	2	225	208	191	174	157	140	123	106
88	71	54	37	[20]	3	211	209	192	175	158	141	124	107	105
72	55	38	21	4	212	210	193	176	159	142	125	108	91	89
56	39	22	5	213	196	194	177	160	143	126	109	92	90	73
40	23	6	214	197	195	178	161	144	127	110	93	76	[74]	57
24	7	215	198	181	179	162	145	128	111	94	77	75	58	41
8	216	199	182	180	163	146	129	112	95	78	61	59	42	25
217	200	183	166	164	147	130	[113]	96	79	62	60	43	26	9
201	184	167	165	148	131	114	97	80	63	46	44	27	10	218
185	168	151	149	132	115	98	81	64	47	45	28	11	219	202
169	[152]	150	133	116	99	82	65	48	31	29	12	220	203	186
153	136	134	117	100	83	66	49	32	30	13	221	204	187	170
137	135	118	101	84	67	50	33	16	14	222	205	188	171	154
121	119	102	85	68	51	34	17	15	223	[206]	189	172	155	138
120	103	86	69	52	35	18	1	224	207	190	173	156	139	122

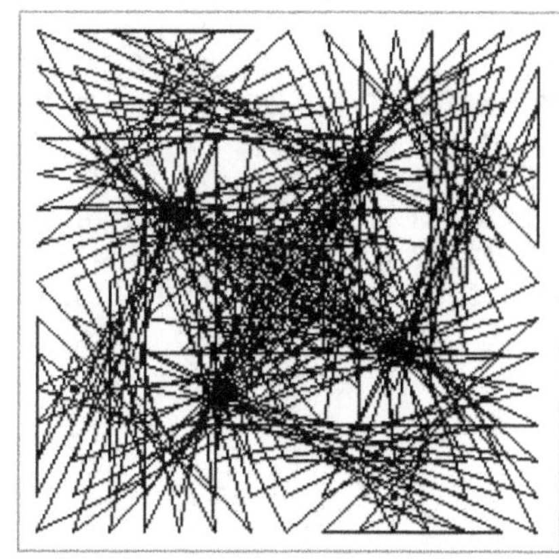

8	121	24	137	40	153	56	169	72	185	88	201	104	217	120
135	23	136	39	152	55	168	71	184	87	200	103	216	119	7
22	150	38	151	54	167	70	183	86	199	102	215	118	6	134
149	37	165	53	166	69	182	85	198	101	214	117	5	133	21
36	164	52	180	68	181	84	197	100	213	116	4	132	20	148
163	51	179	67	195	83	196	99	212	115	3	131	19	147	35
50	178	66	194	82	210	98	211	114	2	130	18	146	34	162
177	65	193	81	209	97	225	[113]	1	129	17	145	33	161	49
64	192	80	208	96	224	112	15	128	16	144	32	160	48	176
191	79	207	95	223	111	14	127	30	143	31	159	47	175	63
78	206	94	222	110	13	126	29	142	45	158	46	174	62	190
205	93	221	109	12	125	28	141	44	157	60	173	61	189	77
92	220	108	11	124	27	140	43	156	59	172	75	188	76	204
219	107	10	123	26	139	42	155	58	171	74	187	90	203	91
106	9	122	25	138	41	154	57	170	73	186	89	202	105	218

8	217	201	185	169	153	137	121	120	104	88	72	56	40	24
39	23	7	216	200	184	168	152	136	135	119	103	87	71	55
70	54	38	22	6	215	199	183	167	151	150	134	118	102	86
101	85	69	53	37	21	5	214	198	182	166	165	149	133	117
132	116	100	84	68	52	36	20	4	213	197	181	180	164	148
163	147	131	115	99	83	67	51	35	19	3	212	196	195	179
194	178	162	146	130	114	98	82	66	50	34	18	2	211	210
225	209	193	177	161	145	129	[113]	97	81	65	49	33	17	1
16	15	224	208	192	176	160	144	128	112	96	80	64	48	32
47	31	30	14	223	207	191	175	159	143	127	111	95	79	63
78	62	46	45	29	13	222	206	190	174	158	142	126	110	94
109	93	77	61	60	44	28	12	221	205	189	173	157	141	125
140	124	108	92	76	75	59	43	27	11	220	204	188	172	156
171	155	139	123	107	91	90	74	58	42	26	10	219	203	187
202	186	170	154	138	122	106	105	89	73	57	41	25	9	218

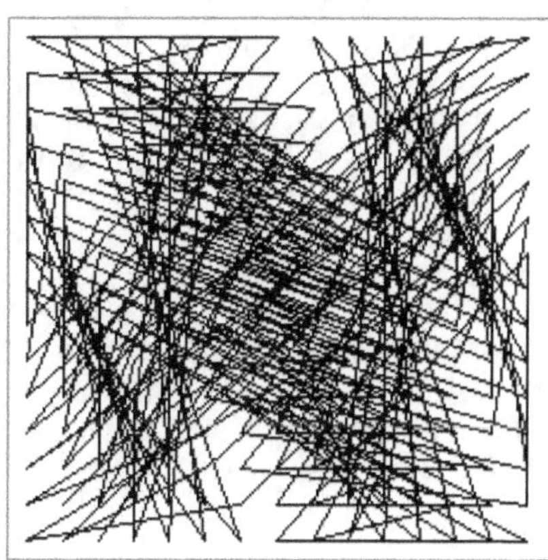

66	50	34	18	2	211	210	82	194	178	162	146	130	114	98
35	19	3	212	196	195	179	51	163	147	131	115	99	83	67
4	213	197	181	180	164	148	20	132	116	100	84	68	52	36
198	182	166	165	149	133	117	214	101	85	69	53	37	21	5
167	151	150	134	118	102	86	183	70	54	38	22	6	215	199
136	135	119	103	87	71	55	152	39	23	7	216	200	184	168
120	104	88	72	56	40	24	121	8	217	201	185	169	153	137
97	81	65	49	33	17	1	113	225	209	193	177	161	145	129
89	73	57	41	25	9	218	105	202	186	170	154	138	122	106
58	42	26	10	219	203	187	74	171	155	139	123	107	91	90
27	11	220	204	188	172	156	43	140	124	108	92	76	75	59
221	205	189	173	157	141	125	12	109	93	77	61	60	44	28
190	174	158	142	126	110	94	206	78	62	46	45	29	13	222
159	143	127	111	95	79	63	175	47	31	30	14	223	207	191
128	112	96	80	64	48	32	144	16	15	224	208	192	176	160

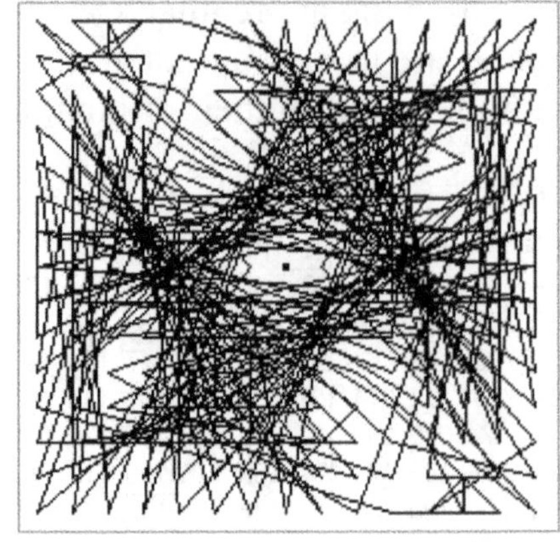

98	196	84	182	70	168	56	225	154	42	140	28	126	14	112
210	83	181	69	167	55	153	97	41	139	27	125	13	111	224
82	195	68	166	54	152	40	209	138	26	124	12	110	223	96
194	67	180	53	151	39	137	81	25	123	11	109	222	95	208
66	179	52	165	38	136	24	193	122	10	108	221	94	207	80
178	51	164	37	150	23	121	65	9	107	220	93	206	79	192
50	163	36	149	22	135	8	177	106	219	92	205	78	191	64
211	99	197	85	183	71	169	113	57	155	43	141	29	127	15
162	35	148	21	134	7	120	49	218	91	204	77	190	63	176
34	147	20	133	6	119	217	161	105	203	76	189	62	175	48
146	19	132	5	118	216	104	33	202	90	188	61	174	47	160
18	131	4	117	215	103	201	145	89	187	75	173	46	159	32
130	3	116	214	102	200	88	17	186	74	172	60	158	31	144
2	115	213	101	199	87	185	129	73	171	59	157	45	143	16
114	212	100	198	86	184	72	1	170	58	156	44	142	30	128

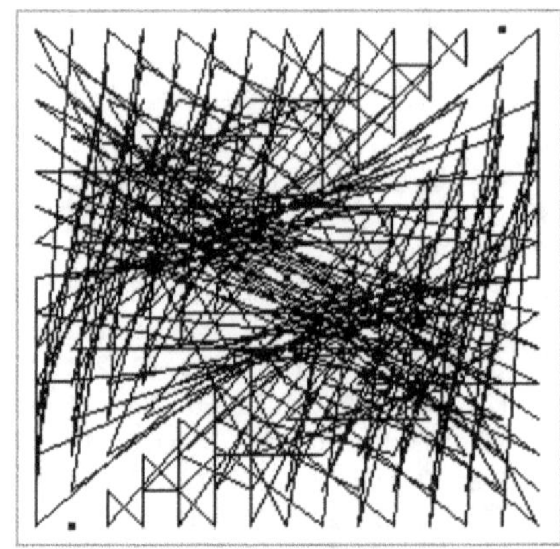

118	216	104	19	132	5	160	33	146	61	174	47	202	90	188
6	119	217	147	20	133	48	161	34	189	62	175	105	203	76
134	7	120	35	148	21	176	49	162	77	190	63	218	91	204
199	87	185	115	213	101	16	129	2	157	45	143	73	171	59
102	200	88	3	116	214	144	17	130	60	158	31	186	74	172
215	103	201	131	4	117	32	145	18	173	46	159	89	187	75
70	168	56	196	84	182	112	225	98	28	126	14	154	42	140
183	71	169	99	197	85	15	113	211	141	29	127	57	155	43
86	184	72	212	100	198	128	1	114	44	142	30	170	58	156
151	39	137	67	180	53	208	81	194	109	222	95	25	123	11
54	152	40	195	68	166	96	209	82	12	110	223	138	26	124
167	55	153	83	181	69	224	97	210	125	13	111	41	139	27
22	135	8	163	36	149	64	177	50	205	78	191	106	219	92
150	23	121	51	164	37	192	65	178	93	206	79	9	107	220
38	136	24	179	52	165	80	193	66	221	94	207	122	10	108

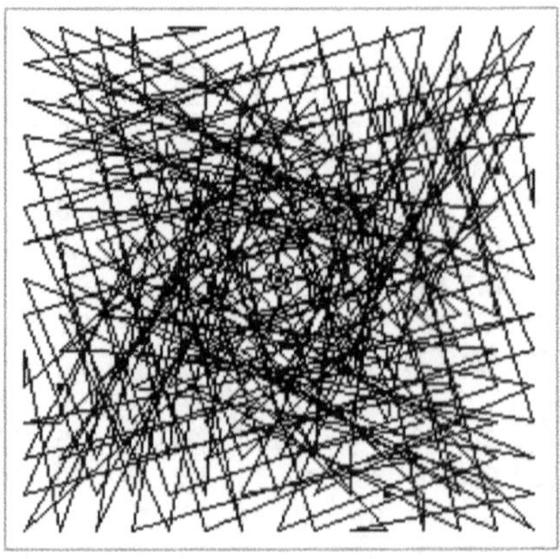

These are regular 15th order magic squares. Odd numbers are highlighted in the square on the left, and the lines are drawn in sequence on the right (from 1 to 2 to 3, etc.)

8	1	6	179	172	177	80	73	78	206	199	204	107	100	105
3	5	7	174	176	178	75	77	79	201	203	205	102	104	106
4	9	2	175	180	173	76	81	74	202	207	200	103	108	101
161	154	159	62	55	60	188	181	186	134	127	132	35	28	33
156	158	160	57	59	61	183	185	187	129	131	133	30	32	34
157	162	155	58	63	56	184	189	182	130	135	128	31	36	29
89	82	87	215	208	213	116	109	114	17	10	15	143	136	141
84	86	88	210	212	214	111	113	115	12	14	16	138	140	142
85	90	83	211	216	209	112	117	110	13	18	11	139	144	137
197	190	195	98	91	96	44	37	42	170	163	168	71	64	69
192	194	196	93	95	97	39	41	43	165	167	169	66	68	70
193	198	191	94	99	92	40	45	38	166	171	164	67	72	65
125	118	123	26	19	24	152	145	150	53	46	51	224	217	222
120	122	124	21	23	25	147	149	151	48	50	52	219	221	223
121	126	119	22	27	20	148	153	146	49	54	47	220	225	218

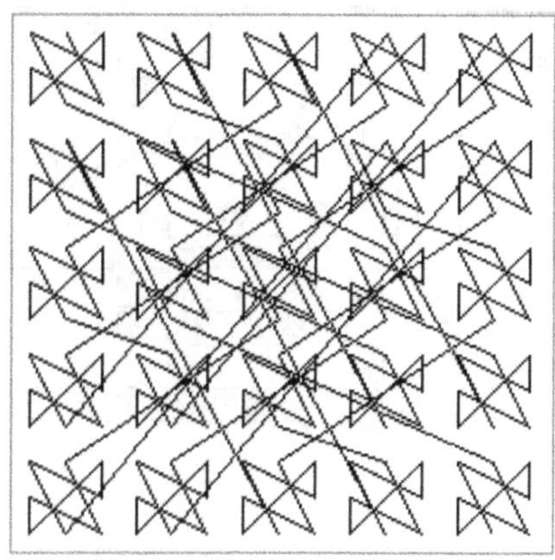

44	33	30	47	36	211	222	205	208	219	94	83	80	97	86
31	50	42	39	28	203	214	217	225	206	81	100	92	89	78
27	41	38	35	49	224	210	213	216	202	77	91	88	85	99
48	37	34	26	45	220	201	209	212	223	98	87	84	76	95
40	29	46	43	32	207	218	221	204	215	90	79	96	93	82
165	154	171	168	157	107	118	121	104	115	65	54	71	68	57
173	162	159	151	170	120	101	109	112	123	73	62	59	51	70
152	166	163	160	174	124	110	113	116	102	52	66	63	60	74
156	175	167	164	153	103	114	117	125	106	56	75	67	64	53
169	158	155	172	161	111	122	105	108	119	69	58	55	72	61
144	133	130	147	136	11	22	5	8	19	194	183	180	197	186
131	150	142	139	128	3	14	17	25	6	181	200	192	189	178
127	141	138	135	149	24	10	13	16	2	177	191	188	185	199
148	137	134	126	145	20	1	9	12	23	198	187	184	176	195
140	129	146	143	132	7	18	21	4	15	190	179	196	193	182

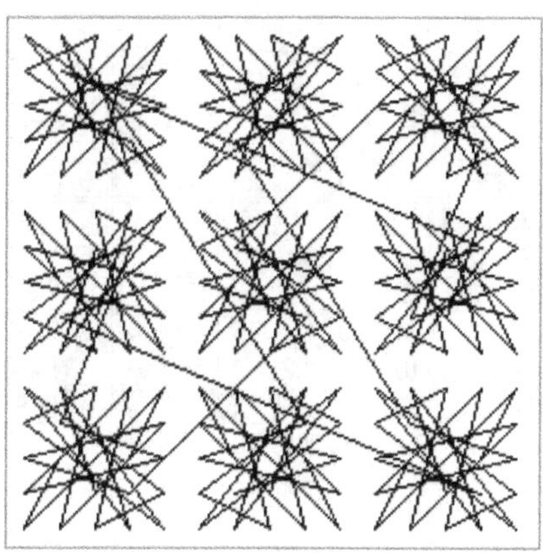

8	135	22	205	36	163	50	177	64	191	78	149	92	219	106
121	23	150	93	164	51	178	65	192	79	206	37	220	107	9
24	136	38	221	52	179	66	193	80	207	94	165	108	10	122
201	103	215	173	4	131	18	145	32	159	46	117	75	187	89
40	152	54	12	68	195	82	209	96	223	110	166	124	26	138
153	55	167	125	181	83	210	97	224	111	13	69	27	139	41
56	168	70	28	84	196	98	225	112	14	126	182	140	42	154
169	71	183	141	197	99	211	113	15	127	29	85	43	155	57
72	184	86	44	100	212	114	1	128	30	142	198	156	58	170
185	87	199	157	213	115	2	129	16	143	45	101	59	171	73
88	200	102	60	116	3	130	17	144	31	158	214	172	74	186
137	39	151	109	180	67	194	81	208	95	222	53	11	123	25
104	216	118	61	132	19	146	33	160	47	174	5	188	90	202
217	119	6	189	20	147	34	161	48	175	62	133	76	203	105
120	7	134	77	148	35	162	49	176	63	190	21	204	91	218

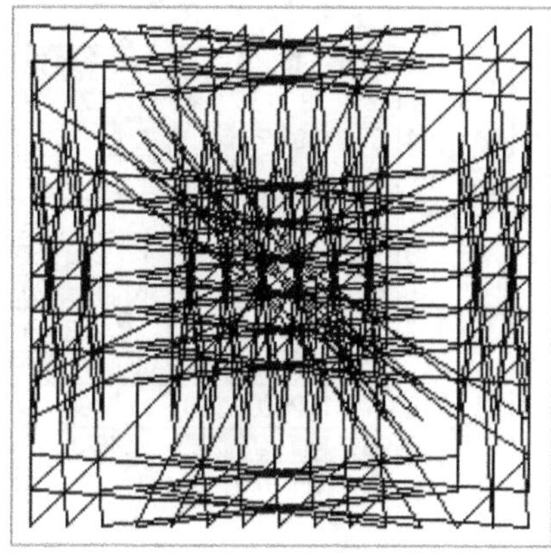

269

14	111	223	95	207	79	191	127	63	175	47	159	31	143	30
112	224	96	208	80	192	64	15	176	48	160	32	144	16	128
28	125	12	109	221	93	205	141	77	189	61	173	60	157	44
126	13	110	222	94	206	78	29	190	62	174	46	158	45	142
42	139	26	123	10	107	219	155	91	203	90	187	74	171	58
140	27	124	11	108	220	92	43	204	76	188	75	172	59	156
56	153	40	137	24	121	8	169	120	217	104	201	88	185	72
225	97	209	81	193	65	177	113	49	161	33	145	17	129	1
154	41	138	25	122	9	106	57	218	105	202	89	186	73	170
70	167	54	151	38	150	22	183	134	6	118	215	102	199	86
168	55	152	39	136	23	135	71	7	119	216	103	200	87	184
84	181	68	180	52	164	36	197	148	20	132	4	116	213	100
182	69	166	53	165	37	149	85	21	133	5	117	214	101	198
98	210	82	194	66	178	50	211	162	34	146	18	130	2	114
196	83	195	67	179	51	163	99	35	147	19	131	3	115	212

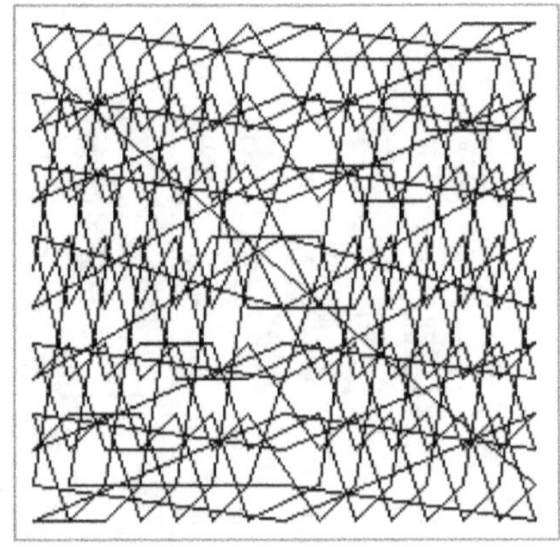

98	50	82	66	194	34	2	211	210	178	18	146	130	162	114
56	8	40	24	137	217	185	169	153	121	201	104	88	120	72
84	36	68	52	180	20	213	197	181	164	4	132	116	148	100
70	22	54	38	151	6	199	183	167	150	215	118	102	134	86
182	149	166	165	53	133	101	85	69	37	117	5	214	21	198
42	219	26	10	123	203	171	155	139	107	187	90	74	91	58
14	191	223	207	95	175	143	127	111	79	159	47	31	63	30
225	177	209	193	81	161	129	113	97	65	145	33	17	49	1
196	163	195	179	67	147	115	99	83	51	131	19	3	35	212
168	135	152	136	39	119	87	71	55	23	103	216	200	7	184
28	205	12	221	109	189	157	141	125	93	173	61	60	77	44
140	92	124	108	11	76	59	43	27	220	75	188	172	204	156
126	78	110	94	222	62	45	29	13	206	46	174	158	190	142
154	106	138	122	25	105	73	57	41	9	89	202	186	218	170
112	64	96	80	208	48	16	15	224	192	32	160	144	176	128

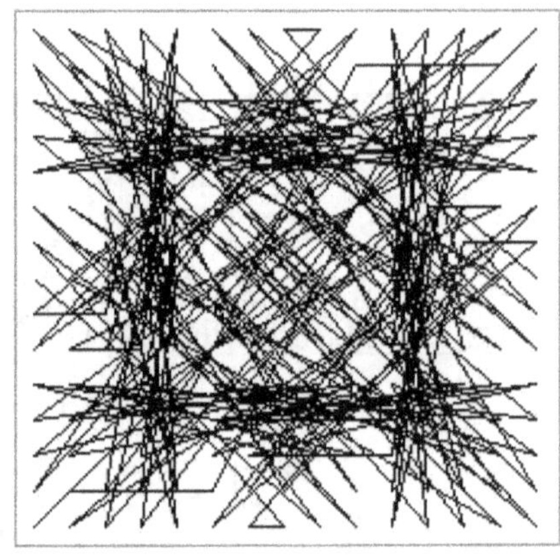

98	196	84	182	70	168	56	225	154	42	140	28	126	14	112
210	83	181	69	167	55	153	97	41	139	27	125	13	111	224
82	195	68	166	54	152	40	209	138	26	124	12	110	223	96
194	67	180	53	151	39	137	81	25	123	11	109	222	95	208
66	179	52	165	38	136	24	193	122	10	108	221	94	207	80
178	51	164	37	150	23	121	65	9	107	220	93	206	79	192
50	163	36	149	22	135	8	177	106	219	92	205	78	191	64
211	99	197	85	183	71	169	113	57	155	43	141	29	127	15
162	35	148	21	134	7	120	49	218	91	204	77	190	63	176
34	147	20	133	6	119	217	161	105	203	76	189	62	175	48
146	19	132	5	118	216	104	33	202	90	188	61	174	47	160
18	131	4	117	215	103	201	145	89	187	75	173	46	159	32
130	3	116	214	102	200	88	17	186	74	172	60	158	31	144
2	115	213	101	199	87	185	129	73	171	59	157	45	143	16
114	212	100	198	86	184	72	1	170	58	156	44	142	30	128

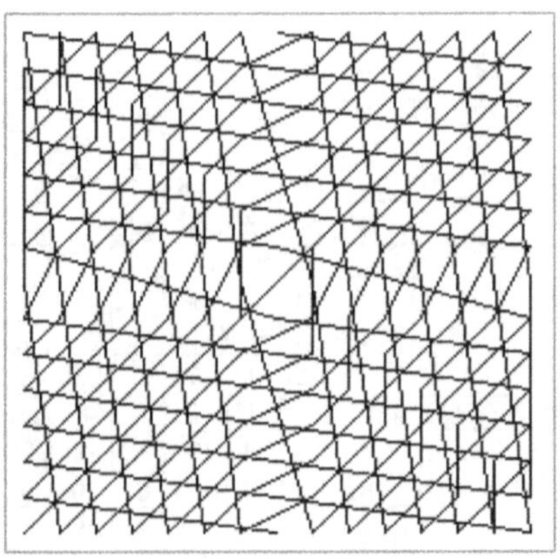

98	202	81	185	64	168	47	219	151	45	149	28	132	11	115
204	83	187	66	170	49	153	100	32	136	30	134	13	117	221
85	189	68	172	51	155	34	206	138	17	121	15	119	223	102
191	70	174	53	157	36	140	87	19	123	2	106	225	104	208
72	176	55	159	38	142	21	193	125	4	108	212	91	210	89
178	57	161	40	144	23	127	74	6	110	214	93	197	76	195
59	163	42	146	25	129	8	180	112	216	95	199	78	182	61
217	96	200	79	183	62	166	113	60	164	43	147	26	130	9
165	44	148	27	131	10	114	46	218	97	201	80	184	63	167
31	150	29	133	12	116	220	152	99	203	82	186	65	169	48
137	16	135	14	118	222	101	33	205	84	188	67	171	50	154
18	122	1	120	224	103	207	139	86	190	69	173	52	156	35
124	3	107	211	105	209	88	20	192	71	175	54	158	37	141
5	109	213	92	196	90	194	126	73	177	56	160	39	143	22
111	215	94	198	77	181	75	7	179	58	162	41	145	24	128

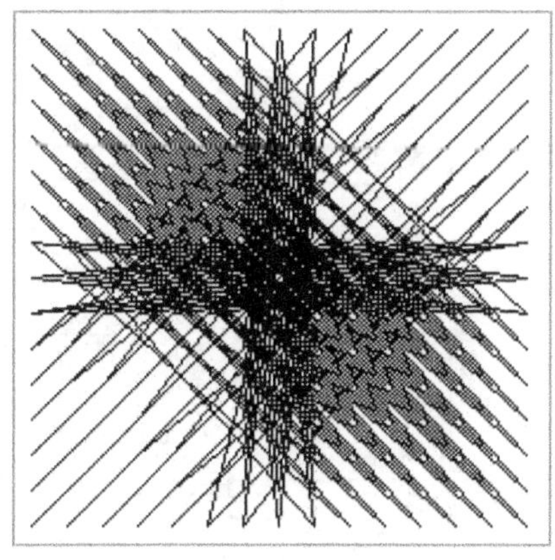

111	215	94	198	77	181	75	7	179	58	162	41	145	24	128
5	109	213	92	196	90	194	126	73	177	56	160	39	143	22
124	3	107	211	105	209	88	20	192	71	175	54	158	37	141
18	122	1	120	224	103	207	139	86	190	69	173	52	156	35
72	176	55	159	38	142	21	193	125	4	108	212	91	210	89
178	57	161	40	144	23	127	74	6	110	214	93	197	76	195
59	163	42	146	25	129	8	180	112	216	95	199	78	182	61
217	96	200	79	183	62	166	113	60	164	43	147	26	130	9
165	44	148	27	131	10	114	46	218	97	201	80	184	63	167
31	150	29	133	12	116	220	152	99	203	82	186	65	169	48
137	16	135	14	118	222	101	33	205	84	188	67	171	50	154
191	70	174	53	157	36	140	87	19	123	2	106	225	104	208
85	189	68	172	51	155	34	206	138	17	121	15	119	223	102
204	83	187	66	170	49	153	100	32	136	30	134	13	117	221
98	202	81	185	64	168	47	219	151	45	149	28	132	11	115

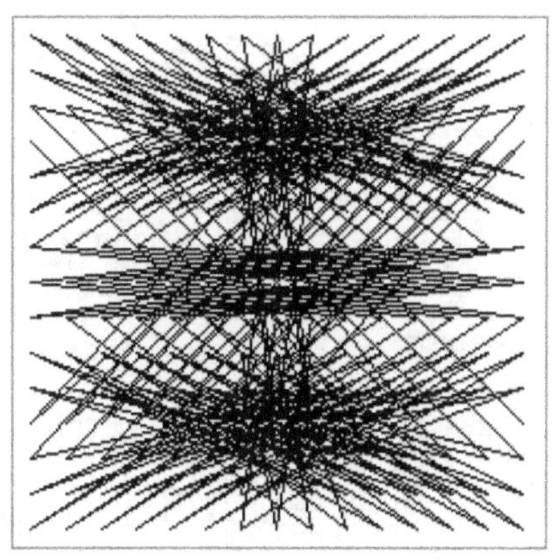

114	92	172	194	201	223	5	121	27	34	56	63	85	165	143
122	115	195	202	224	6	28	144	35	57	64	86	93	173	151
52	45	110	132	139	161	168	74	190	197	219	1	23	103	81
44	22	102	109	131	138	160	51	167	189	196	218	15	80	73
21	14	79	101	108	130	137	43	159	166	188	210	217	72	50
13	216	71	78	100	107	129	20	136	158	180	187	209	49	42
215	208	48	70	77	99	106	12	128	150	157	179	186	41	19
91	84	164	171	193	200	222	113	4	26	33	55	62	142	135
207	185	40	47	69	76	98	214	120	127	149	156	178	18	11
184	177	17	39	46	68	90	206	97	119	126	148	155	10	213
176	154	9	16	38	60	67	183	89	96	118	125	147	212	205
153	146	211	8	30	37	59	175	66	88	95	117	124	204	182
145	123	203	225	7	29	36	152	58	65	87	94	116	181	174
75	53	133	140	162	169	191	82	198	220	2	24	31	111	104
83	61	141	163	170	192	199	105	221	3	25	32	54	134	112

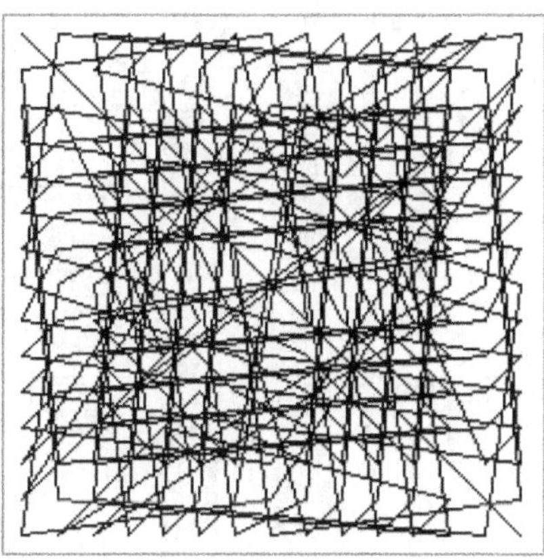

114	122	145	153	21	13	215	91	207	184	176	44	52	75	83
92	115	123	146	14	216	208	84	185	177	154	22	45	53	61
85	93	116	124	217	209	186	62	178	155	147	15	23	31	54
63	86	94	117	210	187	179	55	156	148	125	218	1	24	32
201	224	7	30	108	100	77	193	69	46	38	131	139	162	170
223	6	29	37	130	107	99	200	76	68	60	138	161	169	192
5	28	36	59	137	129	106	222	98	90	67	160	168	191	199
121	144	152	175	43	20	12	113	214	206	183	51	74	82	105
27	35	58	66	159	136	128	4	120	97	89	167	190	198	221
34	57	65	88	166	158	150	26	127	119	96	189	197	220	3
56	64	87	95	188	180	157	33	149	126	118	196	219	2	25
194	202	225	8	101	78	70	171	47	39	16	109	132	140	163
172	195	203	211	79	71	48	164	40	17	9	102	110	133	141
165	173	181	204	72	49	41	142	18	10	212	80	103	111	134
143	151	174	182	50	42	19	135	11	213	205	73	81	104	112

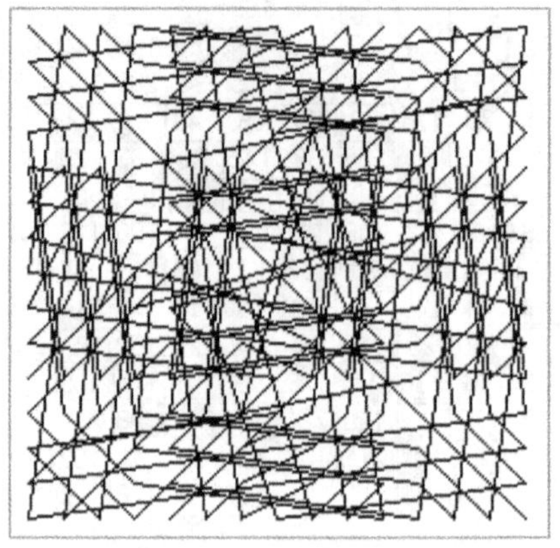

115	87	157	199	213	185	2	129	16	73	45	59	101	171	143
147	119	189	6	20	217	34	161	48	105	62	76	133	203	175
67	39	109	151	180	137	194	81	208	25	222	11	53	123	95
19	216	61	118	132	104	146	33	160	202	174	188	5	90	47
3	200	60	102	116	88	130	17	144	186	158	172	214	74	31
35	7	77	134	148	120	162	49	176	218	190	204	21	91	63
212	184	44	86	100	72	114	1	128	170	142	156	198	58	30
99	71	141	183	197	169	211	113	15	57	29	43	85	155	127
196	168	28	70	84	56	98	225	112	154	126	140	182	42	14
163	135	205	22	36	8	50	177	64	106	78	92	149	219	191
195	152	12	54	68	40	82	209	96	138	110	124	166	26	223
179	136	221	38	52	24	66	193	80	122	94	108	165	10	207
131	103	173	215	4	201	18	145	32	89	46	75	117	187	159
51	23	93	150	164	121	178	65	192	9	206	220	37	107	79
83	55	125	167	181	153	210	97	224	41	13	27	69	139	111

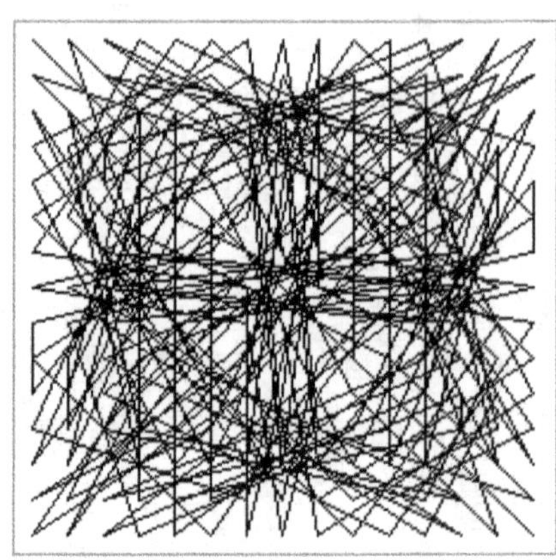

120	127	149	156	178	185	207	214	11	18	40	47	69	76	98
97	119	126	148	155	177	184	206	213	10	17	39	46	68	90
89	96	118	125	147	154	176	183	205	212	9	16	38	60	67
66	88	95	117	124	146	153	175	182	204	211	8	30	37	59
58	65	87	94	116	123	145	152	174	181	203	225	7	29	36
35	57	64	86	93	115	122	144	151	173	195	202	224	6	28
27	34	56	63	85	92	114	121	143	165	172	194	201	223	5
4	26	33	55	62	84	91	113	135	142	164	171	193	200	222
221	3	25	32	54	61	83	105	112	134	141	163	170	192	199
198	220	2	24	31	53	75	82	104	111	133	140	162	169	191
190	197	219	1	23	45	52	74	81	103	110	132	139	161	168
167	189	196	218	15	22	44	51	73	80	102	109	131	138	160
159	166	188	210	217	14	21	43	50	72	79	101	108	130	137
136	158	180	187	209	216	13	20	42	49	71	78	100	107	129
128	150	157	179	186	208	215	12	19	41	48	70	77	99	106

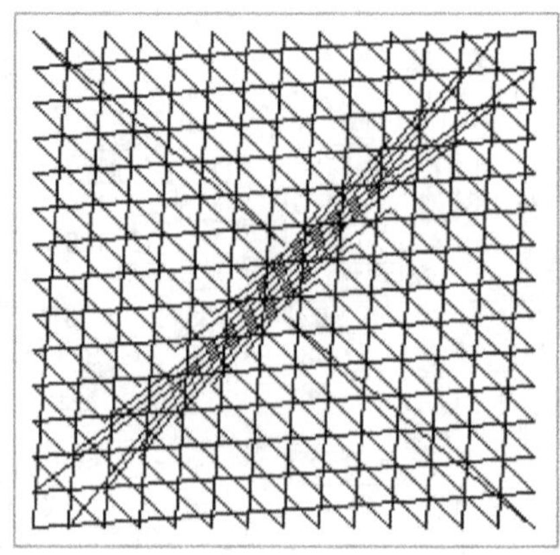

The following are regular magic squares with circular plots.

120	7	134	21	148	35	162	49	176	63	190	77	204	91	218
217	119	6	133	20	147	34	161	48	175	62	189	76	203	105
104	216	118	5	132	19	146	33	160	47	174	61	188	90	202
201	103	215	117	4	131	18	145	32	159	46	173	75	187	89
88	200	102	214	116	3	130	17	144	31	158	60	172	74	186
185	87	199	101	213	115	2	129	16	143	45	157	59	171	73
72	184	86	198	100	212	114	1	128	30	142	44	156	58	170
169	71	183	85	197	99	211	113	15	127	29	141	43	155	57
56	168	70	182	84	196	98	225	112	14	126	28	140	42	154
153	55	167	69	181	83	210	97	224	111	13	125	27	139	41
40	152	54	166	68	195	82	209	96	223	110	12	124	26	138
137	39	151	53	180	67	194	81	208	95	222	109	11	123	25
24	136	38	165	52	179	66	193	80	207	94	221	108	10	122
121	23	150	37	164	51	178	65	192	79	206	93	220	107	9
8	135	22	149	36	163	50	177	64	191	78	205	92	219	106

Circular Std Sequence

8	121	24	137	40	153	56	169	72	185	88	201	104	217	120
135	23	136	39	152	55	168	71	184	87	200	103	216	119	7
22	150	38	151	54	167	70	183	86	199	102	215	118	6	134
149	37	165	53	166	69	182	85	198	101	214	117	5	133	21
36	164	52	180	68	181	84	197	100	213	116	4	132	20	148
163	51	179	67	195	83	196	99	212	115	3	131	19	147	35
50	178	66	194	82	210	98	211	114	2	130	18	146	34	162
177	65	193	81	209	97	225	113	1	129	17	145	33	161	49
64	192	80	208	96	224	112	15	128	16	144	32	160	48	176
191	79	207	95	223	111	14	127	30	143	31	159	47	175	63
78	206	94	222	110	13	126	29	142	45	158	46	174	62	190
205	93	221	109	12	125	28	141	44	157	60	173	61	189	77
92	220	108	11	124	27	140	43	156	59	172	75	188	76	204
219	107	10	123	26	139	42	155	58	171	74	187	90	203	91
106	9	122	25	138	41	154	57	170	73	186	89	202	105	218

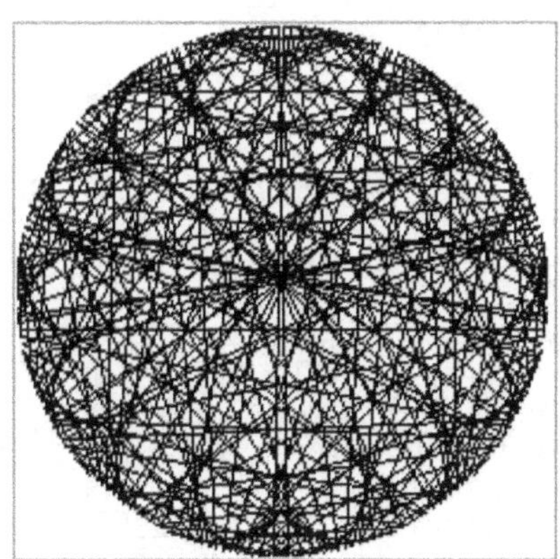

Circular Orig to Curr

54	175	141	107	3	209	105	71	37	158	124	20	211	192	88
35	156	122	103	224	190	86	52	18	139	120	1	207	173	69
16	137	118	84	205	171	67	33	14	135	101	222	188	154	50
12	133	99	65	186	152	48	29	220	116	82	203	169	150	31
218	114	80	46	167	148	44	10	201	97	63	184	165	131	27
134	30	221	187	83	49	170	136	117	13	204	100	66	32	153
115	11	202	168	64	45	151	132	98	219	185	81	47	28	149
96	217	183	164	60	26	147	113	79	200	166	62	43	9	130
77	198	179	145	41	7	128	94	75	181	162	58	24	215	111
73	194	160	126	22	213	109	90	56	177	143	39	5	196	92
199	95	61	42	163	129	25	216	182	78	59	180	146	112	8
195	76	57	23	144	110	6	197	178	74	40	161	127	93	214
176	72	38	4	125	91	212	193	159	55	21	142	108	89	210
157	53	19	225	106	87	208	174	140	36	2	123	104	70	191
138	34	15	206	102	68	189	155	121	17	223	119	85	51	172

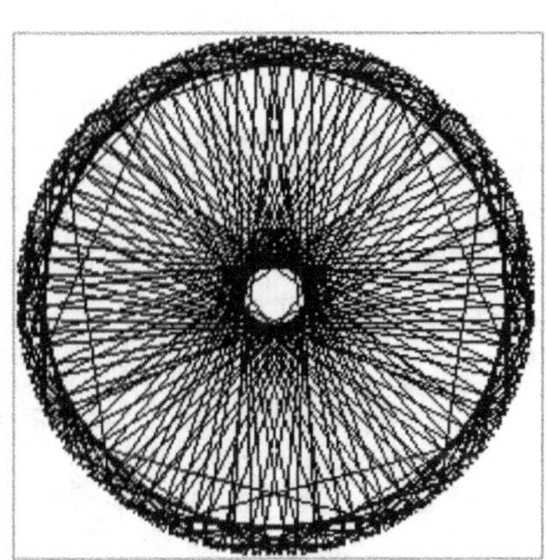

Circular Std Sequence

Appendix M – Selected 16th Order Magic Squares

The following are regular 16th order magic squares. The shaded numbers on the left are numbers in their original positions. The right squares show the movement of numbers from their original to their current position.

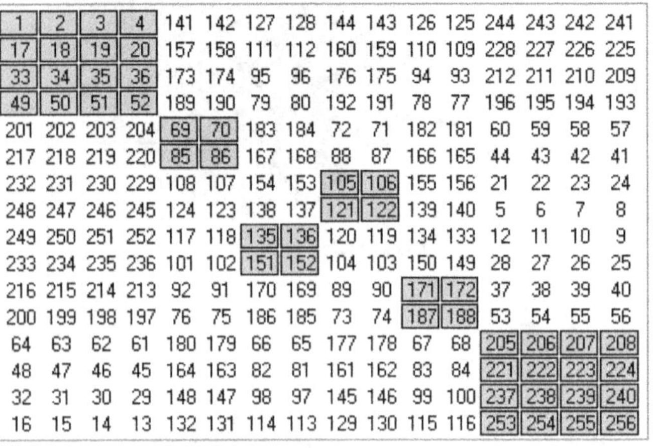

1	2	3	4	141	142	127	128	144	143	126	125	244	243	242	241
17	18	19	20	157	158	111	112	160	159	110	109	228	227	226	225
33	34	35	36	173	174	95	96	176	175	94	93	212	211	210	209
49	50	51	52	189	190	79	80	192	191	78	77	196	195	194	193
201	202	203	204	69	70	183	184	72	71	182	181	60	59	58	57
217	218	219	220	85	86	167	168	88	87	166	165	44	43	42	41
232	231	230	229	108	107	154	153	105	106	155	156	21	22	23	24
248	247	246	245	124	123	138	137	121	122	139	140	5	6	7	8
249	250	251	252	117	118	135	136	120	119	134	133	12	11	10	9
233	234	235	236	101	102	151	152	104	103	150	149	28	27	26	25
216	215	214	213	92	91	170	169	89	90	171	172	37	38	39	40
200	199	198	197	76	75	186	185	73	74	187	188	53	54	55	56
64	63	62	61	180	179	66	65	177	178	67	68	205	206	207	208
48	47	46	45	164	163	82	81	161	162	83	84	221	222	223	224
32	31	30	29	148	147	98	97	145	146	99	100	237	238	239	240
16	15	14	13	132	131	114	113	129	130	115	116	253	254	255	256

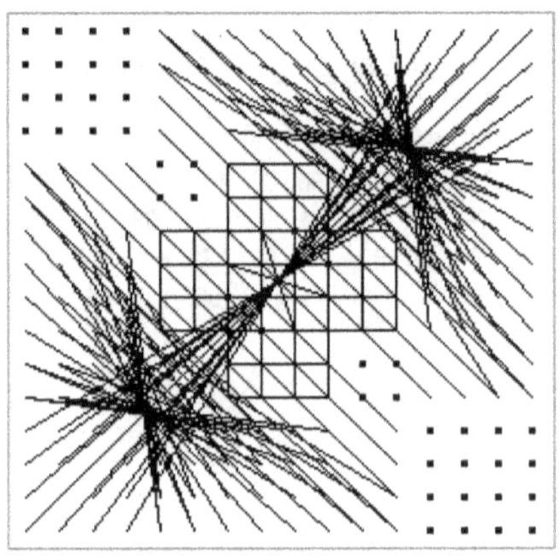

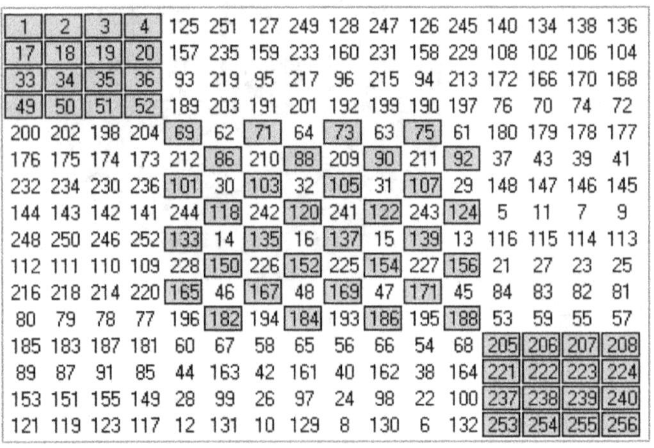

1	2	3	4	125	251	127	249	128	247	126	245	140	134	138	136
17	18	19	20	157	235	159	233	160	231	158	229	108	102	106	104
33	34	35	36	93	219	95	217	96	215	94	213	172	166	170	168
49	50	51	52	189	203	191	201	192	199	190	197	76	70	74	72
200	202	198	204	69	62	71	64	73	63	75	61	180	179	178	177
176	175	174	173	212	86	210	88	209	90	211	92	37	43	39	41
232	234	230	236	101	30	103	32	105	31	107	29	148	147	146	145
144	143	142	141	244	118	242	120	241	122	243	124	5	11	7	9
248	250	246	252	133	14	135	16	137	15	139	13	116	115	114	113
112	111	110	109	228	150	226	152	225	154	227	156	21	27	23	25
216	218	214	220	165	46	167	48	169	47	171	45	84	83	82	81
80	79	78	77	196	182	194	184	193	186	195	188	53	59	55	57
185	183	187	181	60	67	58	65	56	66	54	68	205	206	207	208
89	87	91	85	44	163	42	161	40	162	38	164	221	222	223	224
153	151	155	149	28	99	26	97	24	98	22	100	237	238	239	240
121	119	123	117	12	131	10	129	8	130	6	132	253	254	255	256

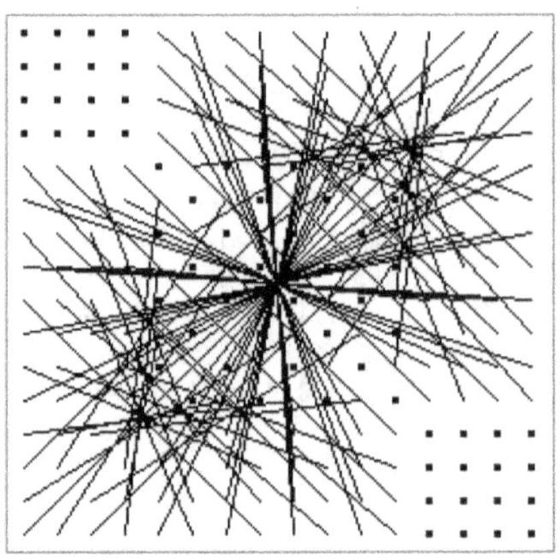

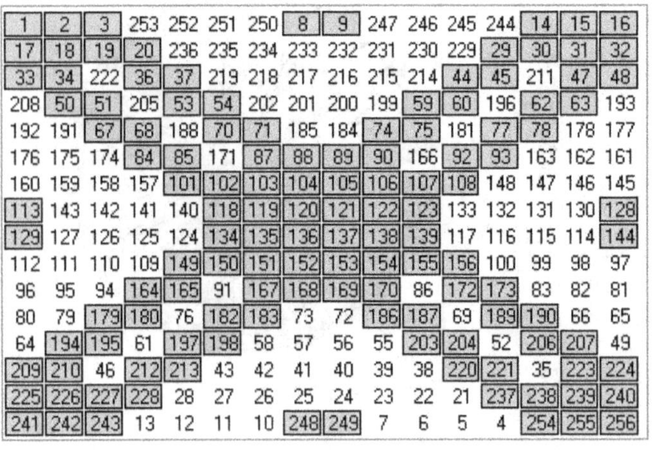

1	2	3	253	252	251	250	8	9	247	246	245	244	14	15	16
17	18	19	20	236	235	234	233	232	231	230	229	29	30	31	32
33	34	222	36	37	219	218	217	216	215	214	44	45	211	47	48
208	50	51	205	53	54	202	201	200	199	59	60	196	62	63	193
192	191	67	68	188	70	71	185	184	74	75	181	77	78	178	177
176	175	174	84	85	171	87	88	89	90	166	92	93	163	162	161
160	159	158	157	101	102	103	104	105	106	107	108	148	147	146	145
113	143	142	141	140	118	119	120	121	122	123	133	132	131	130	128
129	127	126	125	124	134	135	136	137	138	139	117	116	115	114	144
112	111	110	109	149	150	151	152	153	154	155	156	100	99	98	197
96	95	94	164	165	91	167	168	169	170	86	172	173	83	82	81
80	79	179	180	76	182	183	73	72	186	187	69	189	190	66	65
64	194	195	61	197	198	58	57	56	55	203	204	52	206	207	49
209	210	46	212	213	43	42	41	40	39	38	220	221	35	223	224
225	226	227	228	28	27	26	25	24	23	22	21	237	238	239	240
241	242	243	13	12	11	10	248	249	7	6	5	4	254	255	256

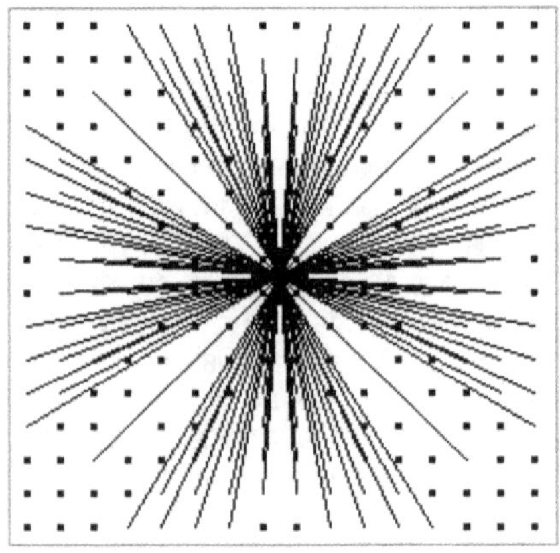

1	2	126	125	133	134	250	249	248	247	139	140	116	115	15	16
17	18	110	109	149	150	234	233	232	231	155	156	100	99	31	32
176	175	35	36	220	219	87	88	89	90	214	213	45	46	162	161
192	191	51	52	204	203	71	72	73	74	198	197	61	62	178	177
193	194	190	189	69	70	58	57	56	55	75	76	180	179	207	208
209	210	174	173	85	86	42	41	40	39	91	92	164	163	223	224
160	159	19	20	236	235	103	104	105	106	230	229	29	30	146	145
144	143	3	4	252	251	119	120	121	122	246	245	13	14	130	129
128	127	243	244	12	11	135	136	137	138	6	5	253	254	114	113
112	111	227	228	28	27	151	152	153	154	22	21	237	238	98	97
33	34	94	93	165	166	218	217	216	215	171	172	84	83	47	48
49	50	78	77	181	182	202	201	200	199	187	188	68	67	63	64
80	79	195	196	60	59	183	184	185	186	54	53	205	206	66	65
96	95	211	212	44	43	167	168	169	170	38	37	221	222	82	81
225	226	158	157	101	102	26	25	24	23	107	108	148	147	239	240
241	242	142	141	117	118	10	9	8	7	123	124	132	131	255	256

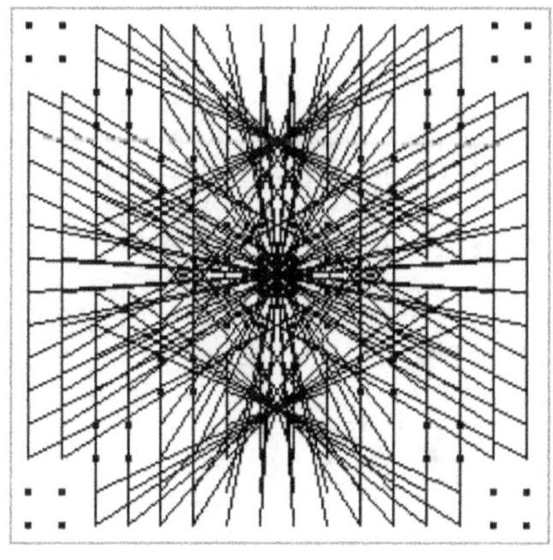

1	2	134	133	12	11	250	249	248	247	115	116	253	254	15	16
17	18	150	149	28	27	234	233	232	231	99	100	237	238	31	32
89	90	35	36	173	174	175	176	216	215	94	93	212	211	34	33
73	74	51	52	189	190	191	192	200	199	78	77	196	195	50	49
177	178	203	204	69	70	71	72	64	63	182	181	60	59	202	201
161	162	219	220	85	86	87	88	48	47	166	165	44	43	218	217
160	159	235	236	101	102	103	104	105	106	30	29	148	147	146	145
144	143	251	252	117	118	119	120	121	122	14	13	132	131	130	129
128	127	126	125	244	243	135	136	137	138	139	140	5	6	114	113
112	111	110	109	228	227	151	152	153	154	155	156	21	22	98	97
40	39	214	213	92	91	210	209	169	170	171	172	37	38	95	96
56	55	198	197	76	75	194	193	185	186	187	188	53	54	79	80
208	207	62	61	180	179	58	57	65	66	67	68	205	206	183	184
224	223	46	45	164	163	42	41	81	82	83	84	221	222	167	168
225	226	19	20	157	158	26	25	24	23	230	229	108	107	239	240
241	242	3	4	141	142	10	9	8	7	246	245	124	123	255	256

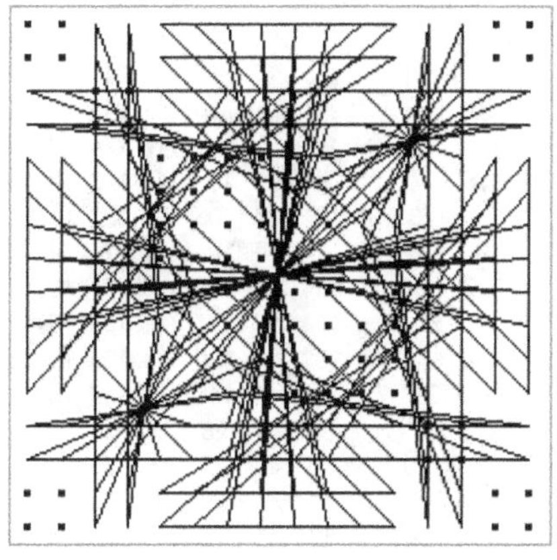

1	17	33	49	192	176	160	144	128	112	96	80	193	209	225	241
2	18	34	50	191	175	159	143	127	111	95	79	194	210	226	242
3	19	35	51	190	174	158	142	126	110	94	78	195	211	227	243
4	20	36	52	189	173	157	141	125	109	93	77	196	212	228	244
252	236	220	204	69	85	101	117	133	149	165	181	60	44	28	12
251	235	219	203	70	86	102	118	134	150	166	182	59	43	27	11
250	234	218	202	71	87	103	119	135	151	167	183	58	42	26	10
249	233	217	201	72	88	104	120	136	152	168	184	57	41	25	9
248	232	216	200	73	89	105	121	137	153	169	185	56	40	24	8
247	231	215	199	74	90	106	122	138	154	170	186	55	39	23	7
246	230	214	198	75	91	107	123	139	155	171	187	54	38	22	6
245	229	213	197	76	92	108	124	140	156	172	188	53	37	21	5
13	29	45	61	180	164	148	132	116	100	84	68	205	221	237	253
14	30	46	62	179	163	147	131	115	99	83	67	206	222	238	254
15	31	47	63	178	162	146	130	114	98	82	66	207	223	239	255
16	32	48	64	177	161	145	129	113	97	81	65	208	224	240	256

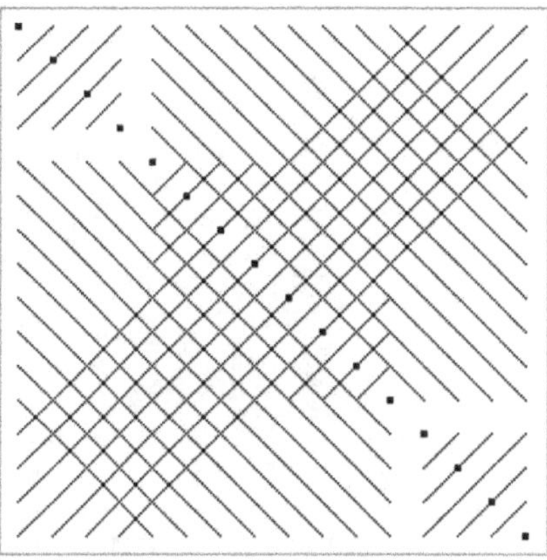

1	17	89	73	177	161	160	144	128	112	40	56	208	224	225	241
2	18	90	74	178	162	159	143	127	111	39	55	207	223	226	242
134	150	35	51	203	219	235	251	126	110	214	198	62	46	19	3
133	149	36	52	204	220	236	252	125	109	213	197	61	45	20	4
12	28	173	189	69	85	101	117	244	228	92	76	180	164	157	141
11	27	174	190	70	86	102	118	243	227	91	75	179	163	158	142
250	234	175	191	71	87	103	119	135	151	210	194	58	42	26	10
249	233	176	192	72	88	104	120	136	152	209	193	57	41	25	9
248	232	216	200	64	48	105	121	137	153	169	185	65	81	24	8
247	231	215	199	63	47	106	122	138	154	170	186	66	82	23	7
115	99	94	78	182	166	30	14	139	155	171	187	67	83	230	246
116	100	93	77	181	165	29	13	140	156	172	188	68	84	229	245
253	237	212	196	60	44	148	132	5	21	37	53	205	221	108	124
254	238	211	195	59	43	147	131	6	22	38	54	206	222	107	123
15	31	34	50	202	218	146	130	114	98	95	79	183	167	239	255
16	32	33	49	201	217	145	129	113	97	96	80	184	168	240	256

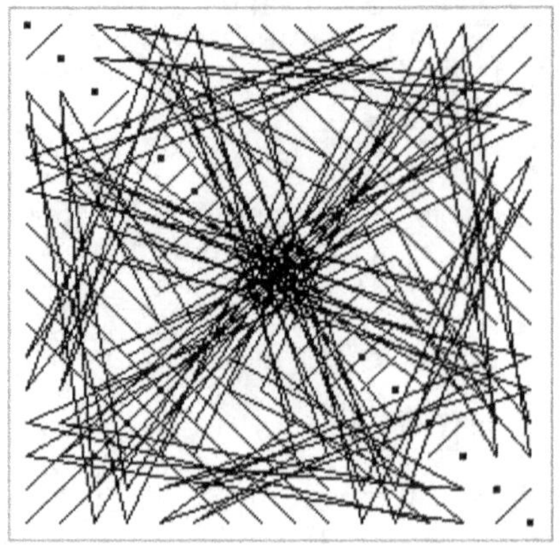

1	17	33	49	201	217	233	249	248	232	216	200	64	48	32	16
2	18	34	50	202	218	234	250	247	231	215	199	63	47	31	15
3	19	35	51	203	219	235	251	246	230	214	198	62	46	30	14
4	20	36	52	204	220	236	252	245	229	213	197	61	45	29	13
141	157	173	189	69	85	101	117	124	108	92	76	180	164	148	132
142	158	174	190	70	86	102	118	123	107	91	75	179	163	147	131
143	159	175	191	71	87	103	119	122	106	90	74	178	162	146	130
144	160	176	192	72	88	104	120	121	105	89	73	177	161	145	129
128	112	96	80	184	168	152	136	137	153	169	185	65	81	97	113
127	111	95	79	183	167	151	135	138	154	170	186	66	82	98	114
126	110	94	78	182	166	150	134	139	155	171	187	67	83	99	115
125	109	93	77	181	165	149	133	140	156	172	188	68	84	100	116
244	228	212	196	60	44	28	12	5	21	37	53	205	221	237	253
243	227	211	195	59	43	27	11	6	22	38	54	206	222	238	254
242	226	210	194	58	42	26	10	7	23	39	55	207	223	239	255
241	225	209	193	57	41	25	9	8	24	40	56	208	224	240	256

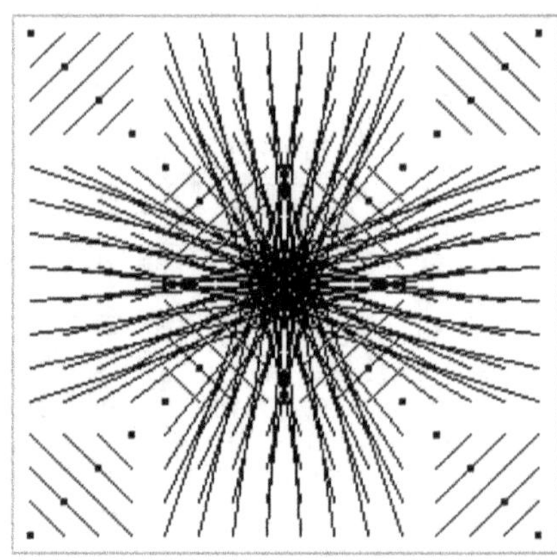

1	2	243	244	125	126	127	128	144	143	142	141	4	3	242	241
17	18	227	228	109	110	111	112	160	159	158	157	20	19	226	225
48	47	222	221	84	83	82	81	161	162	163	164	45	46	223	224
64	63	206	205	68	67	66	65	177	178	179	180	61	62	207	208
200	199	54	53	188	187	186	185	73	74	75	76	197	198	55	56
216	215	38	37	172	171	170	169	89	90	91	92	213	214	39	40
232	231	22	21	156	155	154	153	105	106	107	108	229	230	23	24
248	247	6	5	140	139	138	137	121	122	123	124	245	246	7	8
249	250	11	12	133	134	135	136	120	119	118	117	252	251	10	9
233	234	27	28	149	150	151	152	104	103	102	101	236	235	26	25
217	218	43	44	165	166	167	168	88	87	86	85	220	219	42	41
201	202	59	60	181	182	183	184	72	71	70	69	204	203	58	57
49	50	195	196	77	78	79	80	192	191	190	189	52	51	194	193
33	34	211	212	93	94	95	96	176	175	174	173	36	35	210	209
32	31	238	237	100	99	98	97	145	146	147	148	29	30	239	240
16	15	254	253	116	115	114	113	129	130	131	132	13	14	255	256

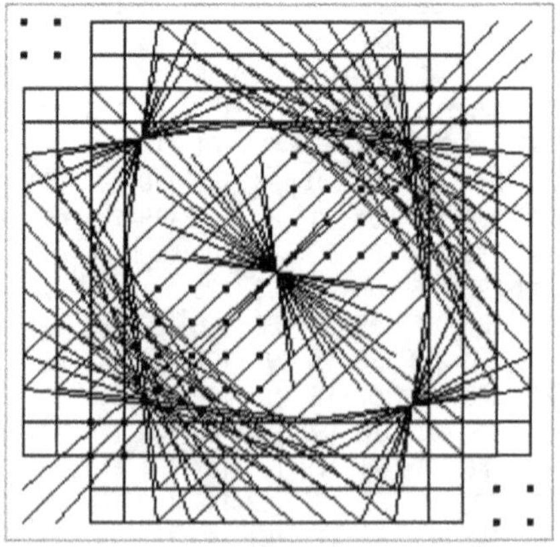

1	51	50	4	201	251	250	204	197	247	246	200	13	63	62	16
36	18	19	33	236	218	219	233	232	214	215	229	48	30	31	45
20	34	35	17	220	234	235	217	216	230	231	213	32	46	47	29
49	3	2	52	249	203	202	252	245	199	198	248	61	15	14	64
141	191	190	144	69	119	118	72	73	123	122	76	129	179	178	132
176	158	159	173	104	86	87	101	108	90	91	105	164	146	147	161
160	174	175	157	88	102	103	85	92	106	107	89	148	162	163	145
189	143	142	192	117	71	70	120	121	75	74	124	177	131	130	180
77	127	126	80	133	183	182	136	137	187	186	140	65	115	114	68
112	94	95	109	168	150	151	165	172	154	155	169	100	82	83	97
96	110	111	93	152	166	167	149	156	170	171	153	84	98	99	81
125	79	78	128	181	135	134	184	185	139	138	188	113	67	66	116
193	243	242	196	9	59	58	12	5	55	54	8	205	255	254	208
228	210	211	225	44	26	27	41	40	22	23	37	240	222	223	237
212	226	227	209	28	42	43	25	24	38	39	21	224	238	239	221
241	195	194	244	57	11	10	60	53	7	6	56	253	207	206	256

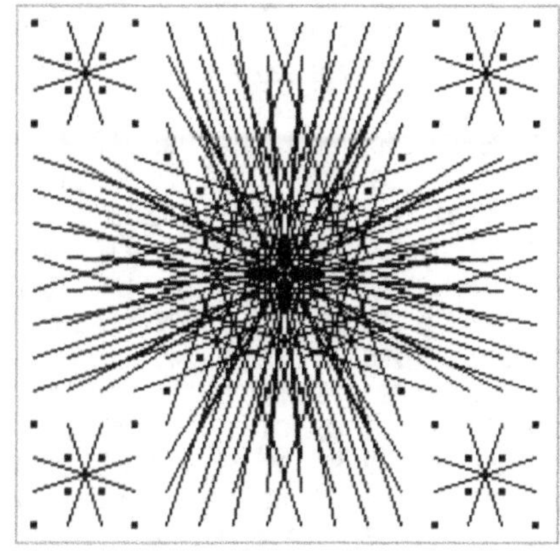

16	15	14	13	116	115	114	113	129	130	131	132	253	254	255	256
32	31	30	29	100	99	98	97	145	146	147	148	237	238	239	240
48	47	46	45	84	83	82	81	161	162	163	164	221	222	223	224
64	63	62	61	68	67	66	65	177	178	179	180	205	206	207	208
201	202	203	204	181	182	183	184	72	71	70	69	60	59	58	57
217	218	219	220	165	166	167	168	88	87	86	85	44	43	42	41
233	234	235	236	149	150	151	152	104	103	102	101	28	27	26	25
249	250	251	252	133	134	135	136	120	119	118	117	12	11	10	9
248	247	246	245	140	139	138	137	121	122	123	124	5	6	7	8
232	231	230	229	156	155	154	153	105	106	107	108	21	22	23	24
216	215	214	213	172	171	170	169	89	90	91	92	37	38	39	40
200	199	198	197	188	187	186	185	73	74	75	76	53	54	55	56
49	50	51	52	77	78	79	80	192	191	190	189	196	195	194	193
33	34	35	36	93	94	95	96	176	175	174	173	212	211	210	209
17	18	19	20	109	110	111	112	160	159	158	157	228	227	226	225
1	2	3	4	125	126	127	128	144	143	142	141	244	243	242	241

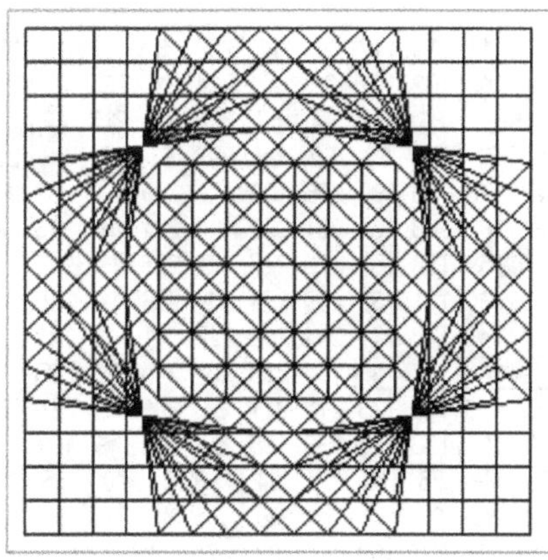

49	33	17	1	249	233	217	201	200	216	232	248	16	32	48	64
50	34	18	2	250	234	218	202	199	215	231	247	15	31	47	63
51	35	19	3	251	235	219	203	198	214	230	246	14	30	46	62
52	36	20	4	252	236	220	204	197	213	229	245	13	29	45	61
192	176	160	144	120	104	88	72	73	89	105	121	129	145	161	177
191	175	159	143	119	103	87	71	74	90	106	122	130	146	162	178
190	174	158	142	118	102	86	70	75	91	107	123	131	147	163	179
189	173	157	141	117	101	85	69	76	92	108	124	132	148	164	180
77	93	109	125	133	149	165	181	188	172	156	140	116	100	84	68
78	94	110	126	134	150	166	182	187	171	155	139	115	99	83	67
79	95	111	127	135	151	167	183	186	170	154	138	114	98	82	66
80	96	112	128	136	152	168	184	185	169	153	137	113	97	81	65
196	212	228	244	12	28	44	60	53	37	21	5	253	237	221	205
195	211	227	243	11	27	43	59	54	38	22	6	254	238	222	206
194	210	226	242	10	26	42	58	55	39	23	7	255	239	223	207
193	209	225	241	9	25	41	57	56	40	24	8	256	240	224	208

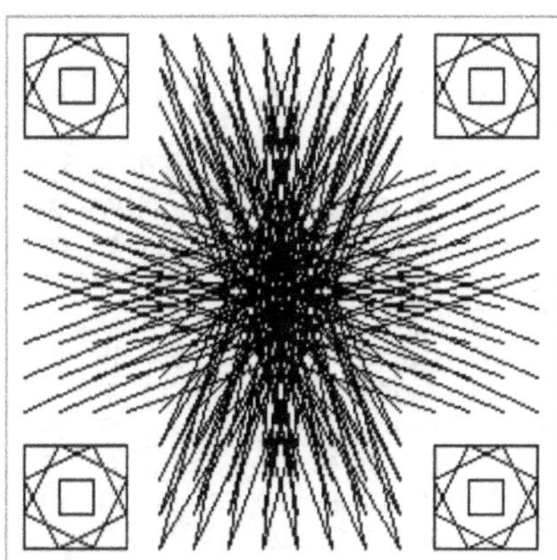

```
 52  51  50  49 197 198 199 200 141 142 143 144 124 123 122 121
 36  35  34  33 213 214 215 216 173 174 175 176  92  91  90  89
 20  19  18  17 229 230 231 232 157 158 159 160 108 107 106 105
  4   3   2   1 245 246 247 248 189 190 191 192  76  75  74  73
 77  78  79  80 188 187 186 185 244 243 242 241   5   6   7   8
 93  94  95  96 172 171 170 169 212 211 210 209  37  38  39  40
109 110 111 112 156 155 154 153 228 227 226 225  21  22  23  24
125 126 127 128 140 139 138 137 196 195 194 193  53  54  55  56
201 202 203 204  64  63  62  61 120 119 118 117 129 130 131 132
233 234 235 236  32  31  30  29 104 103 102 101 145 146 147 148
217 218 219 220  48  47  46  45  88  87  86  85 161 162 163 164
249 250 251 252  16  15  14  13  72  71  70  69 177 178 179 180
184 183 182 181  65  66  67  68   9  10  11  12 256 255 254 253
152 151 150 149  97  98  99 100  25  26  27  28 240 239 238 237
168 167 166 165  81  82  83  84  41  42  43  44 224 223 222 221
136 135 134 133 113 114 115 116  57  58  59  60 208 207 206 205
```

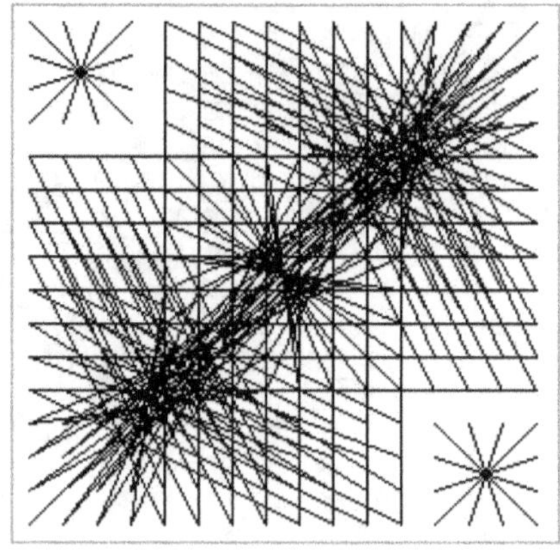

```
120 119 118 117  12  11 250   9 249 [10] 251 252 133 134 135 136
104 103 102 101  28  27 234  25 233 [26] 235 236 149 150 151 152
 88  87  86  85  44  43 218  41 217 [42] 219 220 165 166 167 168
 72  71  70  69  60  59 202  57 201 [58] 203 204 181 182 183 184
177 178 179 180 205 206  63 208  64 207  62  61  68  67  66  65
161 162 163 164 221 222  47 224  48 223  46  45  84  83  82  81
160 159 158 157 228 227  18 225  17 226  19  20 [109][110][111][112]
129 130 131 132 253 254  15 256  16 255  14  13 116 115 114 113
144 143 142 141 244 243   2 241   1 242   3   4 125 126 127 128
[145][146][147][148] 237 238  31 240  32 239  30  29 100  99  98  97
176 175 174 173 212 211  34 209  33 210  35  36  93  94  95  96
192 191 190 189 196 195  50 193  49 194  51  52  77  78  79  80
 73  74  75  76  53  54 [199]  56 200  55 198 197 188 187 186 185
 89  90  91  92  37  38 [215]  40 216  39 214 213 172 171 170 169
105 106 107 108  21  22 [231]  24 232  23 230 229 156 155 154 153
121 122 123 124   5   6 [247]   8 248   7 246 245 140 139 138 137
```

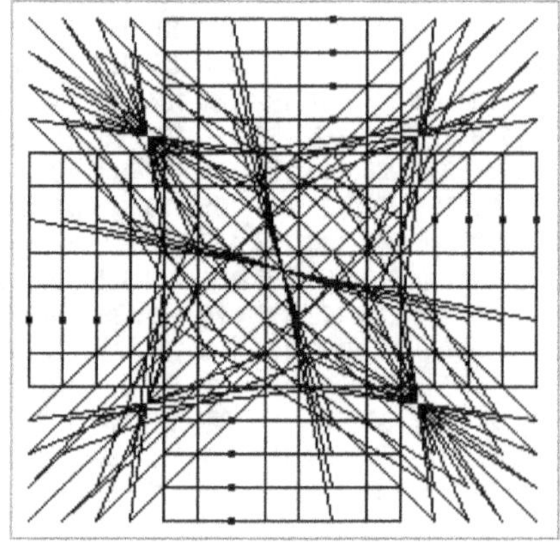

```
120 119 134 133 252 251  10   9 249 250 [11][12] 117 118 135 136
104 103 150 149 236 235  26  25 233 234 [27][28] 101 102 151 152
 89  90 171 172 213 214 [39][40] 216 215  38  37  92  91 170 169
 73  74 187 188 197 198 [55][56] 200 199  54  53  76  75 186 185
192 191  78  77  52  51 194 193  49  50 195 196 189 190 [79][80]
176 175  94  93  36  35 210 209  33  34 211 212 173 174 [95][96]
145 146 [99][100]  29  30 239 240  32  31 238 237 148 147  98  97
129 130 [115][116]  13  14 255 256  16  15 254 253 132 131 114 113
144 143 126 125   4   3 242 241   1   2 243 244 [141][142] 127 128
160 159 110 109  20  19 226 225  17  18 227 228 [157][158] 111 112
[161][162]  83  84  45  46 223 224  48  47 222 221 164 163  82  81
[177][178]  67  68  61  62 207 208  64  63 206 205 180 179  66  65
 72  71 182 181 204 203  58  57 [201][202]  59  60  69  70 183 184
 88  87 166 165 220 219  42  41 [217][218]  43  44  85  86 167 168
105 106 155 156 [229][230]  23  24 232 231  22  21 108 107 154 153
121 122 139 140 [245][246]   7   8 248 247   6   5 124 123 138 137
```

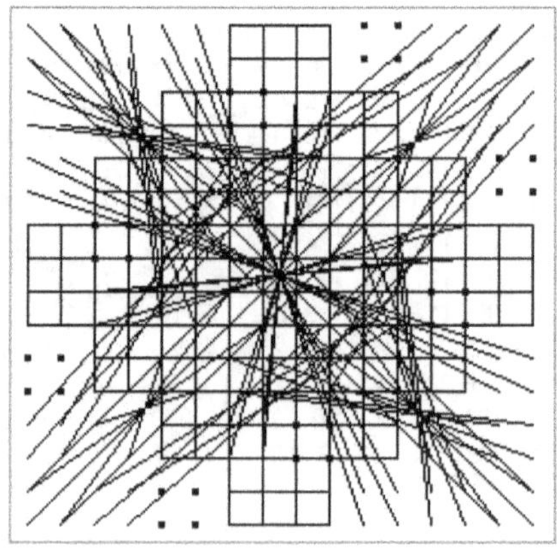

239	17	237	19	235	21	233	23	26	232	28	230	30	228	32	226
2	256	4	254	6	252	8	250	247	9	245	11	243	13	241	15
207	49	205	51	203	53	201	55	58	200	60	198	62	196	64	194
34	224	36	222	38	220	40	218	215	41	213	43	211	45	209	47
175	81	173	83	171	85	169	87	90	168	92	166	94	164	96	162
66	192	68	190	70	188	72	186	183	73	181	75	179	77	177	79
143	113	141	115	139	117	137	119	122	136	124	134	126	132	128	130
98	160	100	158	102	156	104	154	151	105	149	107	147	109	145	111
146	112	148	110	150	108	152	106	103	153	101	155	99	157	97	159
127	129	125	131	123	133	121	135	138	120	140	118	142	116	144	114
178	80	180	78	182	76	184	74	71	185	69	187	67	189	65	191
95	161	93	163	91	165	89	167	170	88	172	86	174	84	176	82
210	48	212	46	214	44	216	42	39	217	37	219	35	221	33	223
63	193	61	195	59	197	57	199	202	56	204	54	206	52	208	50
242	16	244	14	246	12	248	10	7	249	5	251	3	253	1	255
31	225	29	227	27	229	25	231	234	24	236	22	238	20	240	18

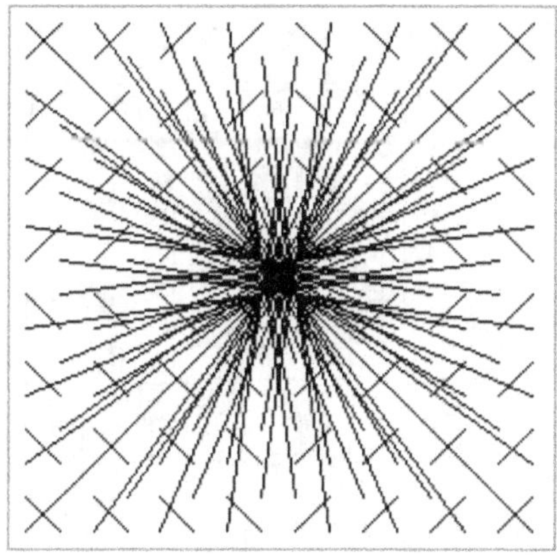

256	240	224	208	56	40	24	8	9	25	41	57	193	209	225	241
255	239	223	207	55	39	23	7	10	26	42	58	194	210	226	242
254	238	222	206	54	38	22	6	11	27	43	59	195	211	227	243
253	237	221	205	53	37	21	5	12	28	44	60	196	212	228	244
116	100	84	68	188	172	156	140	133	149	165	181	77	93	109	125
115	99	83	67	187	171	155	139	134	150	166	182	78	94	110	126
114	98	82	66	186	170	154	138	135	151	167	183	79	95	111	127
113	97	81	65	185	169	153	137	136	152	168	184	80	96	112	128
129	145	161	177	73	89	105	121	120	104	88	72	192	176	160	144
130	146	162	178	74	90	106	122	119	103	87	71	191	175	159	143
131	147	163	179	75	91	107	123	118	102	86	70	190	174	158	142
132	148	164	180	76	92	108	124	117	101	85	69	189	173	157	141
13	29	45	61	197	213	229	245	252	236	220	204	52	36	20	4
14	30	46	62	198	214	230	246	251	235	219	203	51	35	19	3
15	31	47	63	199	215	231	247	250	234	218	202	50	34	18	2
16	32	48	64	200	216	232	248	249	233	217	201	49	33	17	1

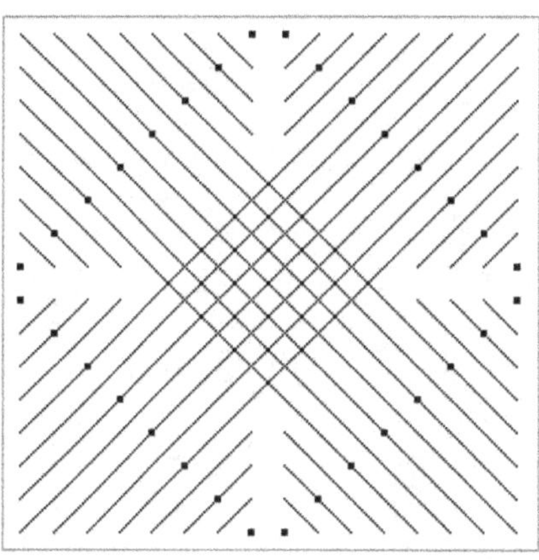

The following are regular 16th order magic squares. Odd numbers are shaded on the left, and lines are drawn in sequence on the right (from 1 to 2 to 3, etc.)

1	2	3	4	247	251	245	249	128	127	126	125	138	134	140	136
17	18	19	20	231	235	229	233	192	191	190	189	74	70	76	72
33	34	35	36	215	219	213	217	96	95	94	93	170	166	172	168
49	50	51	52	199	203	197	201	160	159	158	157	106	102	108	104
112	111	110	109	154	150	156	152	193	194	195	196	55	59	53	57
176	175	174	173	90	86	92	88	209	210	211	212	39	43	37	41
80	79	78	77	186	182	188	184	225	226	227	228	23	27	21	25
144	143	142	141	122	118	124	120	241	242	243	244	7	11	5	9
248	252	246	250	13	14	15	16	137	133	139	135	116	115	114	113
232	236	230	234	29	30	31	32	73	69	75	71	180	179	178	177
216	220	214	218	45	46	47	48	169	165	171	167	84	83	82	81
200	204	198	202	61	62	63	64	105	101	107	103	148	147	146	145
153	149	155	151	100	99	98	97	56	60	54	58	205	206	207	208
89	85	91	87	164	163	162	161	40	44	38	42	221	222	223	224
185	181	187	183	68	67	66	65	24	28	22	26	237	238	239	240
121	117	123	119	132	131	130	129	8	12	6	10	253	254	255	256

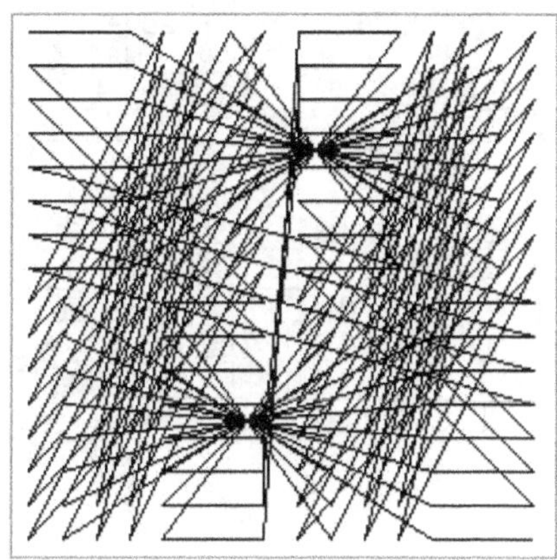

1	2	3	253	252	251	250	8	9	247	246	245	244	14	15	16
17	18	19	20	236	235	234	233	232	231	230	229	29	30	31	32
33	34	222	36	37	219	218	217	216	215	214	44	45	211	47	48
208	50	51	205	53	54	202	201	200	199	59	60	196	62	63	193
192	191	67	68	188	70	71	185	184	74	75	181	77	78	178	177
176	175	174	84	85	171	87	88	89	90	166	92	93	163	162	161
160	159	158	157	101	102	103	104	105	106	107	108	148	147	146	145
113	143	142	141	140	118	119	120	121	122	123	133	132	131	130	128
129	127	126	125	124	134	135	136	137	138	139	117	116	115	114	144
112	111	110	109	149	150	151	152	153	154	155	156	100	99	98	97
96	95	94	164	165	91	167	168	169	170	86	172	173	83	82	81
80	79	179	180	76	182	183	73	72	186	187	69	189	190	66	65
64	194	195	61	197	198	58	57	56	55	203	204	52	206	207	49
209	210	46	212	213	43	42	41	40	39	38	220	221	35	223	224
225	226	227	228	28	27	26	25	24	23	22	21	237	238	239	240
241	242	243	13	12	11	10	248	249	7	6	5	4	254	255	256

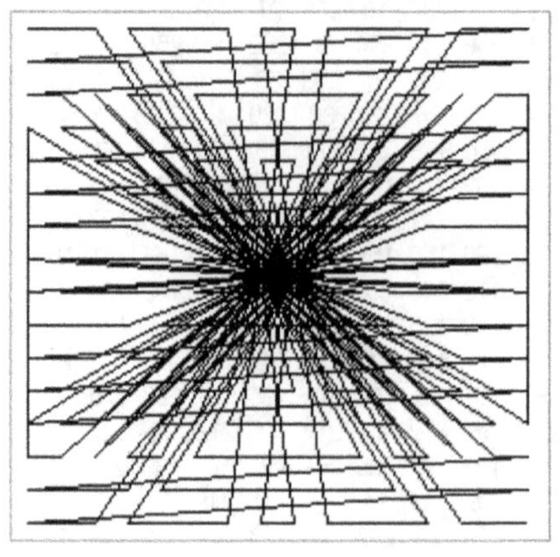

1	15	14	4	209	223	222	212	225	239	238	228	49	63	62	52
12	6	7	9	220	214	215	217	236	230	231	233	60	54	55	57
8	10	11	5	216	218	219	213	232	234	235	229	56	58	59	53
13	3	2	16	221	211	210	224	237	227	226	240	61	51	50	64
113	127	126	116	161	175	174	164	145	159	158	148	65	79	78	68
124	118	119	121	172	166	167	169	156	150	151	153	76	70	71	73
120	122	123	117	168	170	171	165	152	154	155	149	72	74	75	69
125	115	114	128	173	163	162	176	157	147	146	160	77	67	66	80
177	191	190	180	97	111	110	100	81	95	94	84	129	143	142	132
188	182	183	185	108	102	103	105	92	86	87	89	140	134	135	137
184	186	187	181	104	106	107	101	88	90	91	85	136	138	139	133
189	179	178	192	109	99	98	112	93	83	82	96	141	131	130	144
193	207	206	196	17	31	30	20	33	47	46	36	241	255	254	244
204	198	199	201	28	22	23	25	44	38	39	41	252	246	247	249
200	202	203	197	24	26	27	21	40	42	43	37	248	250	251	245
205	195	194	208	29	19	18	32	45	35	34	48	253	243	242	256

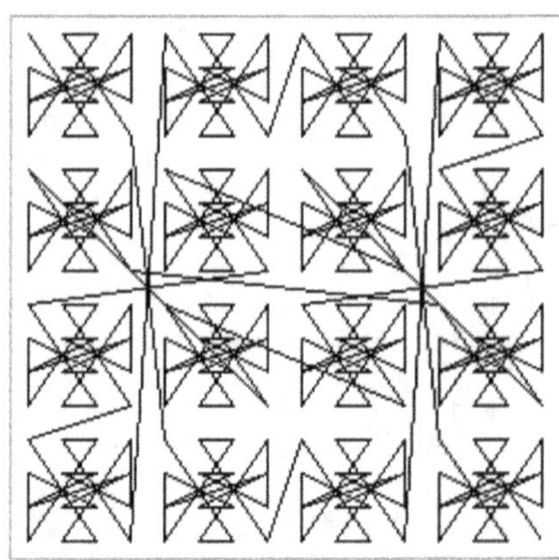

1	15	228	238	225	239	4	14	63	49	222	212	223	209	62	52
12	6	233	231	236	230	9	7	54	60	215	217	214	220	55	57
189	179	96	82	93	83	192	178	131	141	98	112	99	109	130	144
184	186	85	91	88	90	181	187	138	136	107	101	106	104	139	133
177	191	84	94	81	95	180	190	143	129	110	100	111	97	142	132
188	182	89	87	92	86	185	183	134	140	103	105	102	108	135	137
13	3	240	226	237	227	16	2	51	61	210	224	211	221	50	64
8	10	229	235	232	234	5	11	58	56	219	213	218	216	59	53
204	198	41	39	44	38	201	199	246	252	23	25	22	28	247	249
193	207	36	46	33	47	196	206	255	241	30	20	31	17	254	244
120	122	149	155	152	154	117	123	74	72	171	165	170	168	75	69
125	115	160	146	157	147	128	114	67	77	162	176	163	173	66	80
124	118	153	151	156	150	121	119	70	76	167	169	166	172	71	73
113	127	148	158	145	159	116	126	79	65	174	164	175	161	78	68
200	202	37	43	40	42	197	203	250	248	27	21	26	24	251	245
205	195	48	34	45	35	208	194	243	253	18	32	19	29	242	256

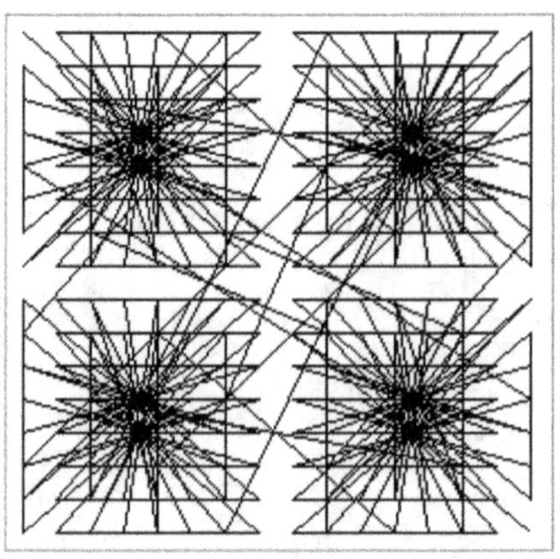

1	33	210	242	65	97	146	178	191	159	112	80	255	223	48	16
3	35	212	244	67	99	148	180	189	157	110	78	253	221	46	14
5	37	214	246	69	101	150	182	187	155	108	76	251	219	44	12
7	39	216	248	71	103	152	184	185	153	106	74	249	217	42	10
26	58	201	233	90	122	137	169	168	136	119	87	232	200	55	23
28	60	203	235	92	124	139	171	166	134	117	85	230	198	53	21
30	62	205	237	94	126	141	173	164	132	115	83	228	196	51	19
32	64	207	239	96	128	143	175	162	130	113	81	226	194	49	17
240	208	63	31	176	144	127	95	82	114	129	161	18	50	193	225
238	206	61	29	174	142	125	93	84	116	131	163	20	52	195	227
236	204	59	27	172	140	123	91	86	118	133	165	22	54	197	229
234	202	57	25	170	138	121	89	88	120	135	167	24	56	199	231
247	215	40	8	183	151	104	72	73	105	154	186	9	41	218	250
245	213	38	6	181	149	102	70	75	107	156	188	11	43	220	252
243	211	36	4	179	147	100	68	77	109	158	190	13	45	222	254
241	209	34	2	177	145	98	66	79	111	160	192	15	47	224	256

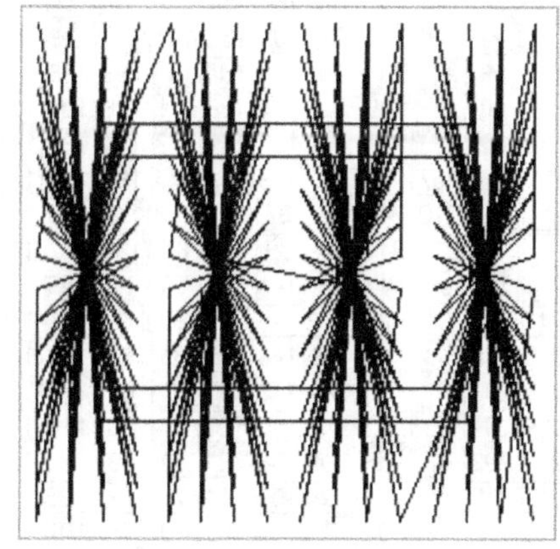

1	255	243	253	245	11	7	8	9	10	6	252	244	254	242	16
240	18	227	237	229	27	23	24	25	26	22	236	228	238	31	225
48	47	222	36	37	219	215	216	217	218	214	44	45	211	34	33
208	207	51	205	53	54	58	200	201	55	59	60	196	62	194	193
80	79	67	68	188	187	183	185	184	186	182	181	77	78	66	65
161	162	174	84	172	86	90	89	88	87	91	165	93	163	175	176
97	98	110	148	108	150	154	153	152	151	155	101	157	99	111	112
113	114	126	125	140	134	138	137	136	135	139	133	116	115	127	128
129	130	142	141	124	118	122	121	120	119	123	117	132	131	143	144
145	146	158	100	156	102	106	105	104	103	107	149	109	147	159	160
81	82	94	164	92	166	170	169	168	167	171	85	173	83	95	96
192	191	179	180	76	75	71	73	72	74	70	69	189	190	178	177
64	63	195	61	197	198	202	56	57	199	203	204	52	206	50	49
224	223	46	212	213	43	39	40	41	42	38	220	221	35	210	209
32	226	19	29	21	235	231	232	233	234	230	28	20	30	239	17
241	15	3	13	5	251	247	248	249	250	246	12	4	14	2	256

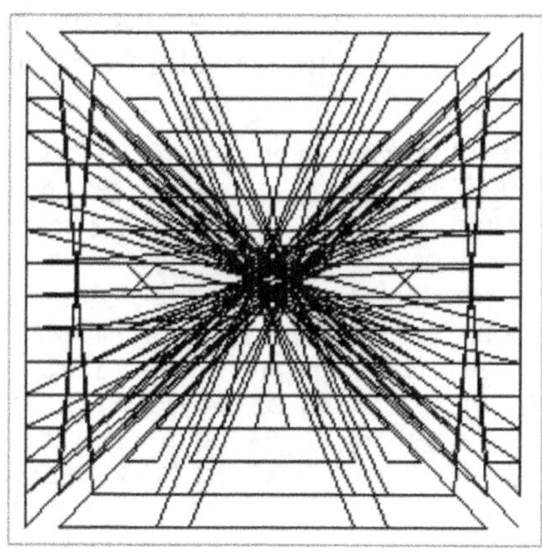

2	4	6	8	247	245	243	241	239	237	235	233	26	28	30	32
34	36	38	40	215	213	211	209	207	205	203	201	58	60	62	64
66	68	70	72	183	181	179	177	175	173	171	169	90	92	94	96
98	100	102	104	151	149	147	145	143	141	139	137	122	124	126	128
127	125	123	121	138	140	142	144	146	148	150	152	103	101	99	97
95	93	91	89	170	172	174	176	178	180	182	184	71	69	67	65
63	61	59	57	202	204	206	208	210	212	214	216	39	37	35	33
31	29	27	25	234	236	238	240	242	244	246	248	7	5	3	1
256	254	252	250	9	11	13	15	17	19	21	23	232	230	228	226
224	222	220	218	41	43	45	47	49	51	53	55	200	198	196	194
192	190	188	186	73	75	77	79	81	83	85	87	168	166	164	162
160	158	156	154	105	107	109	111	113	115	117	119	136	134	132	130
129	131	133	135	120	118	116	114	112	110	108	106	153	155	157	159
161	163	165	167	88	86	84	82	80	78	76	74	185	187	189	191
193	195	197	199	56	54	52	50	48	46	44	42	217	219	221	223
225	227	229	231	24	22	20	18	16	14	12	10	249	251	253	255

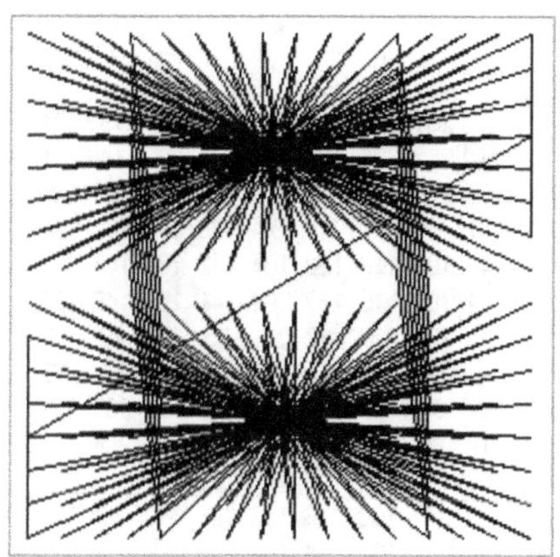

5	37	214	246	69	101	150	182	187	155	108	76	251	219	44	12
7	39	216	248	71	103	152	184	185	153	106	74	249	217	42	10
26	58	201	233	90	122	137	169	168	136	119	87	232	200	55	23
28	60	203	235	92	124	139	171	166	134	117	85	230	198	53	21
30	62	205	237	94	126	141	173	164	132	115	83	228	196	51	19
32	64	207	239	96	128	143	175	162	130	113	81	226	194	49	17
1	33	210	242	65	97	146	178	191	159	112	80	255	223	48	16
3	35	212	244	67	99	148	180	189	157	110	78	253	221	46	14
243	211	36	4	179	147	100	68	77	109	158	190	13	45	222	254
241	209	34	2	177	145	98	66	79	111	160	192	15	47	224	256
240	208	63	31	176	144	127	95	82	114	129	161	18	50	193	225
238	206	61	29	174	142	125	93	84	116	131	163	20	52	195	227
236	204	59	27	172	140	123	91	86	118	133	165	22	54	197	229
234	202	57	25	170	138	121	89	88	120	135	167	24	56	199	231
247	215	40	8	183	151	104	72	73	105	154	186	9	41	218	250
245	213	38	6	181	149	102	70	75	107	156	188	11	43	220	252

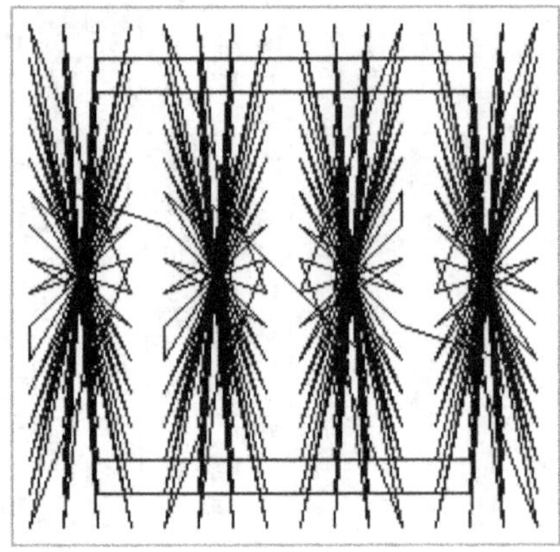

The following are 16th order pan-diagonal magic squares. The shaded numbers on the left are in their original positions. The right squares show the movement of the numbers from their original to their current position.

1	2	251	252	249	250	3	4	13	14	247	248	245	246	15	16
17	18	235	236	233	234	19	20	29	30	231	232	229	230	31	32
176	175	86	85	88	87	174	173	164	163	90	89	92	91	162	161
192	191	70	69	72	71	190	189	180	179	74	73	76	75	178	177
144	143	118	117	120	119	142	141	132	131	122	121	124	123	130	129
160	159	102	101	104	103	158	157	148	147	106	105	108	107	146	145
33	34	219	220	217	218	35	36	45	46	215	216	213	214	47	48
49	50	203	204	201	202	51	52	61	62	199	200	197	198	63	64
193	194	59	60	57	58	195	196	205	206	55	56	53	54	207	208
209	210	43	44	41	42	211	212	221	222	39	40	37	38	223	224
112	111	150	149	152	151	110	109	100	99	154	153	156	155	98	97
128	127	134	133	136	135	126	125	116	115	138	137	140	139	114	113
80	79	182	181	184	183	78	77	68	67	186	185	188	187	66	65
96	95	166	165	168	167	94	93	84	83	170	169	172	171	82	81
225	226	27	28	25	26	227	228	237	238	23	24	21	22	239	240
241	242	11	12	9	10	243	244	253	254	7	8	5	6	255	256

1	17	251	235	234	250	20	4	13	29	247	231	230	246	32	16
2	18	252	236	233	249	19	3	14	30	248	232	229	245	31	15
176	192	86	70	71	87	189	173	164	180	90	74	75	91	177	161
175	191	85	69	72	88	190	174	163	179	89	73	76	92	178	162
159	143	101	117	120	104	142	158	147	131	105	121	124	108	130	146
160	144	102	118	119	103	141	157	148	132	106	122	123	107	129	145
50	34	204	220	217	201	35	51	62	46	200	216	213	197	47	63
49	33	203	219	218	202	36	52	61	45	199	215	214	198	48	64
193	209	59	43	42	58	212	196	205	221	55	39	38	54	224	208
194	210	60	44	41	57	211	195	206	222	56	40	37	53	223	207
112	128	150	134	135	151	125	109	100	116	154	138	139	155	113	97
111	127	149	133	136	152	126	110	99	115	153	137	140	156	114	98
95	79	165	181	184	168	78	94	83	67	169	185	188	172	66	82
96	80	166	182	183	167	77	93	84	68	170	186	187	171	65	81
242	226	12	28	25	9	227	243	254	238	8	24	21	5	239	255
241	225	11	27	26	10	228	244	253	237	7	23	22	6	240	256

18	255	159	114	82	191	223	50	194	47	79	162	130	111	15	226
240	1	97	144	176	65	33	208	64	209	177	96	128	145	241	32
234	7	103	138	170	71	39	202	58	215	183	90	122	151	247	26
24	249	153	120	88	185	217	56	200	41	73	168	136	105	9	232
22	251	155	118	86	187	219	54	198	43	75	166	134	107	11	230
236	5	101	140	172	69	37	204	60	213	181	92	124	149	245	28
238	3	99	142	174	67	35	206	62	211	179	94	126	147	243	30
20	253	157	116	84	189	221	52	196	45	77	164	132	109	13	228
29	244	148	125	93	180	212	61	205	36	68	173	141	100	4	237
227	14	110	131	163	78	46	195	51	222	190	83	115	158	254	19
229	12	108	133	165	76	44	197	53	220	188	85	117	156	252	21
27	246	150	123	91	182	214	59	203	38	70	171	139	102	6	235
25	248	152	121	89	184	216	57	201	40	72	169	137	104	8	233
231	10	106	135	167	74	42	199	55	218	186	87	119	154	250	23
225	16	112	129	161	80	48	193	49	224	192	81	113	160	256	17
31	242	146	127	95	178	210	63	207	34	66	175	143	98	2	239

46	45	216	215	213	214	47	48	33	34	219	220	218	217	36	35
62	61	200	199	197	198	63	64	49	50	203	204	202	201	52	51
131	132	121	122	124	123	130	129	144	143	118	117	119	120	141	142
147	148	105	106	108	107	146	145	160	159	102	101	103	104	157	158
179	180	73	74	76	75	178	177	192	191	70	69	71	72	189	190
163	164	89	90	92	91	162	161	176	175	86	85	87	88	173	174
30	29	232	231	229	230	31	32	17	18	235	236	234	233	20	19
14	13	248	247	245	246	15	16	1	2	251	252	250	249	4	3
254	253	8	7	5	6	255	256	241	242	11	12	10	9	244	243
238	237	24	23	21	22	239	240	225	226	27	28	26	25	228	227
83	84	169	170	172	171	82	81	96	95	166	165	167	168	93	94
67	68	185	186	188	187	66	65	80	79	182	181	183	184	77	78
99	100	153	154	156	155	98	97	112	111	150	149	151	152	109	110
115	116	137	138	140	139	114	113	128	127	134	133	135	136	125	126
206	205	56	55	53	54	207	208	193	194	59	60	58	57	196	195
222	221	40	39	37	38	223	224	209	210	43	44	42	41	212	211

52	51	202	201	204	203	50	49	64	63	198	197	200	199	62	61
36	35	218	217	220	219	34	33	48	47	214	213	216	215	46	45
157	158	103	104	101	102	159	160	145	146	107	108	105	106	147	148
141	142	119	120	117	118	143	144	129	130	123	124	121	122	131	132
189	190	71	72	69	70	191	192	177	178	75	76	73	74	179	180
173	174	87	88	85	86	175	176	161	162	91	92	89	90	163	164
20	19	234	233	236	235	18	17	32	31	230	229	232	231	30	29
4	3	250	249	252	251	2	1	16	15	246	245	248	247	14	13
244	243	10	9	12	11	242	241	256	255	6	5	8	7	254	253
228	227	26	25	28	27	226	225	240	239	22	21	24	23	238	237
93	94	167	168	165	166	95	96	81	82	171	172	169	170	83	84
77	78	183	184	181	182	79	80	65	66	187	188	185	186	67	68
125	126	135	136	133	134	127	128	113	114	139	140	137	138	115	116
109	110	151	152	149	150	111	112	97	98	155	156	153	154	99	100
212	211	42	41	44	43	210	209	224	223	38	37	40	39	222	221
196	195	58	57	60	59	194	193	208	207	54	53	56	55	206	205

The following are regular magic squares with circular plots.

1	2	126	125	133	134	250	249	248	247	139	140	116	115	15	16
17	18	110	109	149	150	234	233	232	231	155	156	100	99	31	32
176	175	35	36	220	219	87	88	89	90	214	213	45	46	162	161
192	191	51	52	204	203	71	72	73	74	198	197	61	62	178	177
193	194	190	189	69	70	58	57	56	55	75	76	180	179	207	208
209	210	174	173	85	86	42	41	40	39	91	92	164	163	223	224
160	159	19	20	236	235	103	104	105	106	230	229	29	30	146	145
144	143	3	4	252	251	119	120	121	122	246	245	13	14	130	129
128	127	243	244	12	11	135	136	137	138	6	5	253	254	114	113
112	111	227	228	28	27	151	152	153	154	22	21	237	238	98	97
33	34	94	93	165	166	218	217	216	215	171	172	84	83	47	48
49	50	78	77	181	182	202	201	200	199	187	188	68	67	63	64
80	79	195	196	60	59	183	184	185	186	54	53	205	206	66	65
96	95	211	212	44	43	167	168	169	170	38	37	221	222	82	81
225	226	158	157	101	102	26	25	24	23	107	108	148	147	239	240
241	242	142	141	117	118	10	9	8	7	123	124	132	131	255	256

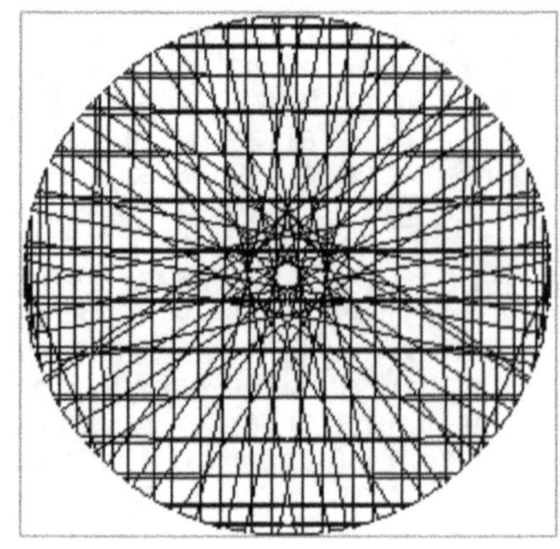

Circular Std Sequence

1	17	33	49	192	176	160	144	128	112	96	80	193	209	225	241
2	18	34	50	191	175	159	143	127	111	95	79	194	210	226	242
3	19	35	51	190	174	158	142	126	110	94	78	195	211	227	243
4	20	36	52	189	173	157	141	125	109	93	77	196	212	228	244
252	236	220	204	69	85	101	117	133	149	165	181	60	44	28	12
251	235	219	203	70	86	102	118	134	150	166	182	59	43	27	11
250	234	218	202	71	87	103	119	135	151	167	183	58	42	26	10
249	233	217	201	72	88	104	120	136	152	168	184	57	41	25	9
248	232	216	200	73	89	105	121	137	153	169	185	56	40	24	8
247	231	215	199	74	90	106	122	138	154	170	186	55	39	23	7
246	230	214	198	75	91	107	123	139	155	171	187	54	38	22	6
245	229	213	197	76	92	108	124	140	156	172	188	53	37	21	5
13	29	45	61	180	164	148	132	116	100	84	68	205	221	237	253
14	30	46	62	179	163	147	131	115	99	83	67	206	222	238	254
15	31	47	63	178	162	146	130	114	98	82	66	207	223	239	255
16	32	48	64	177	161	145	129	113	97	81	65	208	224	240	256

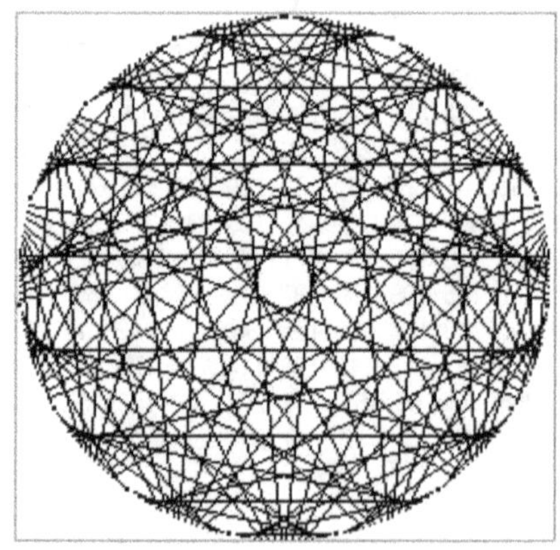

Circular Orig to Curr

1	17	33	49	192	176	160	144	248	232	216	200	73	89	105	121
2	18	34	50	191	175	159	143	247	231	215	199	74	90	106	122
3	19	35	51	190	174	158	142	246	230	214	198	75	91	107	123
4	20	36	52	189	173	157	141	245	229	213	197	76	92	108	124
252	236	220	204	69	85	101	117	13	29	45	61	180	164	148	132
251	235	219	203	70	86	102	118	14	30	46	62	179	163	147	131
250	234	218	202	71	87	103	119	15	31	47	63	178	162	146	130
249	233	217	201	72	88	104	120	16	32	48	64	177	161	145	129
128	112	96	80	193	209	225	241	137	153	169	185	56	40	24	8
127	111	95	79	194	210	226	242	138	154	170	186	55	39	23	7
126	110	94	78	195	211	227	243	139	155	171	187	54	38	22	6
125	109	93	77	196	212	228	244	140	156	172	188	53	37	21	5
133	149	165	181	60	44	28	12	116	100	84	68	205	221	237	253
134	150	166	182	59	43	27	11	115	99	83	67	206	222	238	254
135	151	167	183	58	42	26	10	114	98	82	66	207	223	239	255
136	152	168	184	57	41	25	9	113	97	81	65	208	224	240	256

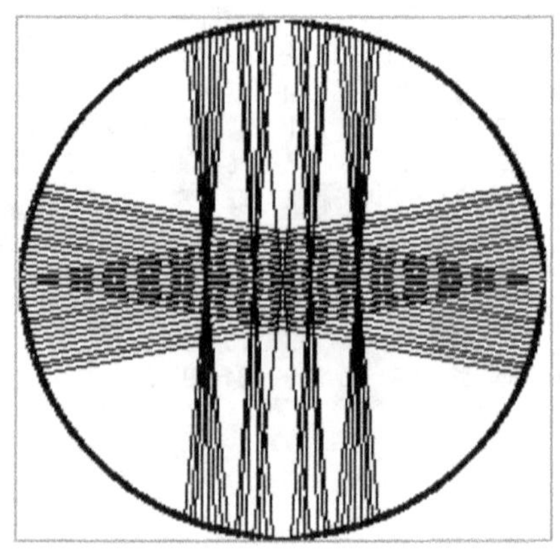

Circular Std Sequence

1	17	209	193	192	176	112	128	144	160	96	80	49	33	225	241
2	18	210	194	191	175	111	127	143	159	95	79	50	34	226	242
14	30	222	206	179	163	99	115	131	147	83	67	62	46	238	254
13	29	221	205	180	164	100	116	132	148	84	68	61	45	237	253
246	230	38	54	75	91	155	139	123	107	171	187	198	214	22	6
245	229	37	53	76	92	156	140	124	108	172	188	197	213	21	5
247	231	39	55	74	90	154	138	122	106	170	186	199	215	23	7
248	232	40	56	73	89	153	137	121	105	169	185	200	216	24	8
249	233	41	57	72	88	152	136	120	104	168	184	201	217	25	9
250	234	42	58	71	87	151	135	119	103	167	183	202	218	26	10
252	236	44	60	69	85	149	133	117	101	165	181	204	220	28	12
251	235	43	59	70	86	150	134	118	102	166	182	203	219	27	11
4	20	212	196	189	173	109	125	141	157	93	77	52	36	228	244
3	19	211	195	190	174	110	126	142	158	94	78	51	35	227	243
15	31	223	207	178	162	98	114	130	146	82	66	63	47	239	255
16	32	224	208	177	161	97	113	129	145	81	65	64	48	240	256

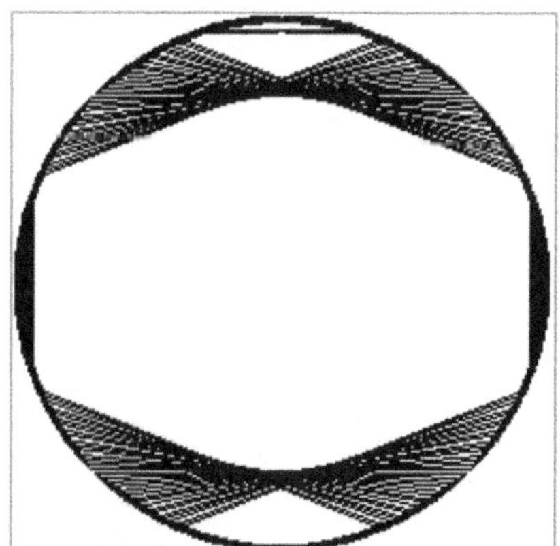

Circular Std Sequence

1	33	32	200	64	232	217	248	249	216	233	49	201	17	48	16
3	35	30	198	62	230	219	246	251	214	235	51	203	19	46	14
242	210	239	55	207	23	42	7	10	39	26	194	58	226	223	255
125	93	100	188	68	156	165	140	133	172	149	77	181	109	84	116
244	212	237	53	205	21	44	5	12	37	28	196	60	228	221	253
127	95	98	186	66	154	167	138	135	170	151	79	183	111	82	114
142	174	147	75	179	107	86	123	118	91	102	190	70	158	163	131
128	96	97	185	65	153	168	137	136	169	152	80	184	112	81	113
144	176	145	73	177	105	88	121	120	89	104	192	72	160	161	129
126	94	99	187	67	155	166	139	134	171	150	78	182	110	83	115
143	175	146	74	178	106	87	122	119	90	103	191	71	159	162	130
4	36	29	197	61	229	220	245	252	213	236	52	204	20	45	13
141	173	148	76	180	108	85	124	117	92	101	189	69	157	164	132
2	34	31	199	63	231	218	247	250	215	234	50	202	18	47	15
243	211	238	54	206	22	43	6	11	38	27	195	59	227	222	254
241	209	240	56	208	24	41	8	9	40	25	193	57	225	224	256

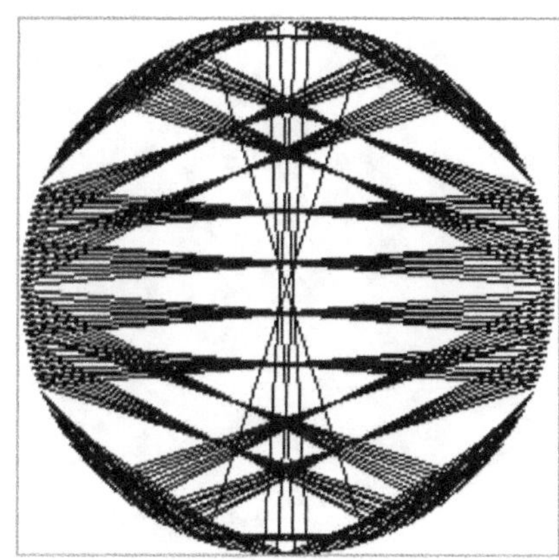

Circular Std Sequence

2	4	6	8	247	245	243	241	239	237	235	233	26	28	30	32
34	36	38	40	215	213	211	209	207	205	203	201	58	60	62	64
66	68	70	72	183	181	179	177	175	173	171	169	90	92	94	96
98	100	102	104	151	149	147	145	143	141	139	137	122	124	126	128
127	125	123	121	138	140	142	144	146	148	150	152	103	101	99	97
95	93	91	89	170	172	174	176	178	180	182	184	71	69	67	65
63	61	59	57	202	204	206	208	210	212	214	216	39	37	35	33
31	29	27	25	234	236	238	240	242	244	246	248	7	5	3	1
256	254	252	250	9	11	13	15	17	19	21	23	232	230	228	226
224	222	220	218	41	43	45	47	49	51	53	55	200	198	196	194
192	190	188	186	73	75	77	79	81	83	85	87	168	166	164	162
160	158	156	154	105	107	109	111	113	115	117	119	136	134	132	130
129	131	133	135	120	118	116	114	112	110	108	106	153	155	157	159
161	163	165	167	88	86	84	82	80	78	76	74	185	187	189	191
193	195	197	199	56	54	52	50	48	46	44	42	217	219	221	223
225	227	229	231	24	22	20	18	16	14	12	10	249	251	253	255

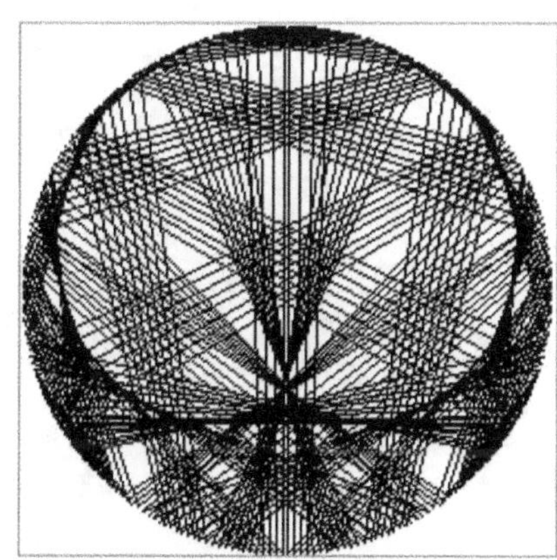

Circular Orig to Curr

285